# Philosophy and Medicine

Volume 130

More information about this series at http://www.springer.com/series/6414

Peter J. Cataldo • Dan O'Brien
Editors

# Palliative Care and Catholic Health Care

## Two Millennia of Caring for the Whole Person

Foreword by Ira Byock

*Editors*
Peter J. Cataldo
Providence St. Joseph Health
Renton, WA, USA

Dan O'Brien
Ethics, Discernment and Church Relations
Ascension
St. Louis, MO, USA

ISSN 0376-7418 ISSN 2215-0080 (electronic)
Philosophy and Medicine
ISBN 978-3-030-05004-7 ISBN 978-3-030-05005-4 (eBook)
https://doi.org/10.1007/978-3-030-05005-4

Library of Congress Control Number: 2019931528



This Springer imprint is published by the registered company Springer Nature Switzerland AG.
The registered company address is: Gewerbestrasse 11, 6330 Cham, Switzerland

***In Memoriam:***

*Pope St. John Paul II and Joseph Cardinal Bernardin, beloved and appreciated for teaching us all how to die well.*

# Foreword

## The Catholicity of Caring

This anthology is a worthy contribution to the literatures of both palliative care and Catholic health care. Editors Peter Cataldo and Dan O'Brien, scholars of Catholic theology and clinical ethics, have recruited recognized experts to address key issues of caring for people through the end of life. While the editors' and contributors' perspectives are anchored in the history and principles of ethics and therapeutics, their reasoned insights illuminate today's headlines. As a result, this collection will advance the ongoing debates related to health-care ethics, policy, and systems reform.

I come to this book as a palliative care physician who has long been interested in the origins and evolution of communal responses to illness and dying. It was, therefore, no surprise that I was enthralled by its content and agree, as this volume effectively shows, that palliative care and Catholic health care are highly aligned. They share values of service, clinical excellence, compassion, and respect for vulnerable people's dignity and worth. Admirable, but hardly unique. Are not these same values held by countless health-care professionals across every specialty and organization? I firmly believe they are.

The relation of palliative care and Catholic health care is well worth examining; yet, time and again, while reading the analyses and arguments within this book, I found myself thinking that what was being discussed was just good care.

I've had similar reveries before. Sometimes, while poring through a published report of a health-care innovation or quality improvement initiative, the categorical distinctions between the professions, delivery systems, and specialties of health care dissolve before my mind's eye like smoke, and what remains, starkly apparent, are the basic elements of human caring. In those moments, I reflect on what we know of our early forbearers. Millennia before there were religions and professions, priests or doctors, human beings were caring for one another.

Of course, caring is an impulse shared by many species. Most animals, and all mammals, care for their young. However, human babies and toddlers are utterly dependent and vulnerable for far longer than youth of other species. Other mammals also tend to their injured and ill. But here again, *Homo sapiens* are unique in being able to devote so much effort, over prolonged periods of time, tending to their most frail and dying members. It is part of what makes us human.

Earliest humans lived in tribes and rudimentary communities. Communal bonds were most evident in times of danger. When hurricanes or blizzards threatened people's lives, they turned to one another to weather the storm. Primitive people instinctively knew that the safest way to get through the most difficult and dangerous situations was *together*. It felt natural and right to do so; it would have felt unnatural, unwise, and wrong not to look after each other.

The biologist in me would say that primates evolved to look after one another; it was a survival trait. Caring was written in our genetic code before being etched in Biblical commandments and covenants. As a Jew, I came to understand the Abrahamic covenant to mean that we belong not only to God but also to one another. Whether one considers it the result of evolutionary pressures or God's will, it is indisputable that we humans matter to one another.

Two corollaries to this basic fact of life are foundational to the ethics and practice of caring. First, it is natural for people to live *in community* with one another, rather than merely in proximity to one another. Second, within human communities, social responsibilities and self-interests are intertwined. In a healthy human society, your well-being matters to me – and my well-being matters to you. I cannot be entirely well if you are suffering. Humans come to each other's aid in times of need, not merely because they were taught that it is the right thing to do but also because it would feel unnatural and unhealthy to do otherwise.

## *Love Has a Lot to Do with It*

To this point, I've described human behavior in objective terms, as observed reality. In understanding the origins of caring, it is also important to understand the subjective reality of what people *feel* at times of their own or another's need. Emotionally, the primal human drive to come to the aid of others is love. This statement is not an assertion, nor a romantic, philosophical notion. It is, instead, a description of the synthesis of findings from the sciences of anthropology, ethology (animal behavior), neuroanatomy, functional brain physiology, and brain chemistry. Within the physiological substrate of *Homo sapiens*, the instinct of "mattering to one another" is subjectively experienced as a deep filial affinity. In plain speak, love.

In contemporary ethics and health care, love is infrequently discussed as a motivation for human caring. I suspect this reflects an unspoken assumption among clinicians and ethicists that basing academic discussions on emotional drivers of behavior, particularly love, risks appearing nonscientific. It is, however, an inarguable fact of human life that people have an intrinsic need to love and feel loved (Lewis et al. 2001).

## *Love May Not Be All We Need; But It Is Essential to Human Well-Being*

Experienced clinicians recognize that even in the context of significant physical discomfort or social deprivation, subjective fulfillment of love can engender a profound sense of well-being. Conversely, even in the context of physical health and social privilege, feeling unloved can engender profound suffering – as can being frustrated by one's inability to express love. In early life, a paucity of love gives rise to failure to thrive, which can be lethal in spite of fully adequate nutrition and hygiene (Spitz 1945; Harlow and Mears 1979). So too, during the normal dependency that often accompanies late life, an absence of love can make life not worth living (Byock 2004; Thomas 1996).

Recognizing love as essential to human well-being and as the motivation of service integrates biological, philosophical, and theological lines of thought and underlies the catholicity, or universality, of caring ethics and practice. The story of Abraham in the Old Testament emphasizes the covenant with God that binds people in community. In the New Testament, Jesus went further in emphasizing love as the most authentic emotion and response to human suffering and service as the healthiest manner of expressing love to one another within community.

In antiquity, love and service gave rise to the caring professions of medicine and ministry. Thankfully, these highest of impulses remain driving forces within health care in contemporary times.

## The Basics of Human Caring

The basics of human caring are not confined to any profession. The rudimentary elements of service to a person who is seriously ill, injured, or aged are held collectively. Providing shelter from the elements, offering of food and drink, helping with elimination, and keeping company need not be delegated. Instead, in a healthy community, people say to an ill person in fellowship, "We will keep you warm and dry. We will help you to eat and drink. We will help you with your bowels and bladder. We will keep you clean. And we will accompany you in your time of need."

Contemporary health care complements these basics by providing treatments to counter the effects of trauma and disease, curing and preserving life and function when possible, and comforting always. Although excellent health care is obviously important for providing the benefits of medical science and technology to a person who is seriously ill, it is not sufficient. Essential components of human caring are more important still. In fact, they define fundamental social responsibilities, responding to our instinctual obligations, as well as the core teachings of our faith traditions. Yet, these elemental components represent the minimum obligations rather than the full potential of human caring.

## Beyond the Basics

While by no means exclusive to Judaism or Christianity, nor any health-care discipline, to my mind, palliative care and Catholic health care are distinguished in contemporary Western society by the emphases they place on inherent human dignity and worth and by the extent to which they consciously embrace values of love, service, and community. While mainstream health care is largely defined by the transactional diagnostic and therapeutic services it brings to patients' medical problems, Catholic health care and palliative care advance relation-based models of service. They stand out in contemporary health care by seeing each patient as a whole person, who deserves not only competent diagnostic and therapeutic care but also comprehensive attention to his or her bodily comfort and functional capacities as well as his or her emotional, interpersonal, social, and spiritual well-being.

Ethicist Laurie Dorfman observed the remarkable potential of this relational mode of caring, "The blessings of friendship in the human endeavor, the responsibility for the bearing of collective burden of the ill and the vulnerable, must be borne by the theology and purpose of the faith communities. That each of us will die is inevitable; what must come to be understood as miraculous is our ability to love and bear the weight of the dying in fellowship" (Zoloth 1993).

In meeting each seriously ill patient as a whole person, our basic responsibility to keep company expands to encompass the *response-ability* to bear witness. Beyond providing for the biological necessities of life and competent medical care, in fellowship we say to each person whose life is threatened, "We will bear witness to your pain and sorrows, your disappointments, as well as your accomplishments. We will listen to the stories of your life and will remember the story of your passing."

This perspective enables clinicians to recognize the potential for human development persists during illness, debility, dependence, and dying. Beyond the basic obligations we have to our fellow beings, our *response-abilities* extend to preserving opportunities for people to grow – individually and together – through the end of life. Death is inevitable; however, forgiveness, reconciliation, and healing all continue to be possible. Years ago in my palliative care practice, before concluding a review of a patient's symptoms and concerns, I learned to ask, "And how are you within yourself?" In listening to the responses to my questions, I discovered that people can remain or become well in their dying. This fuller understanding of the human condition engenders a more complete concept of beneficence which aspires to preserve opportunities for people to flourish within their families and communities through the very end of life (Byock 1997).

## Non-killing

It is sadly ironic that within contemporary debates of health-care ethics and policies, Catholic health care and palliative care are most commonly recognized for honoring the core principle of non-killing. To my mind, here again, these ethics seem catholic in the sense of being universal. Non-maleficence and non-killing were among the first principles of medicine and, even where it is legal, the proscription against killing applies to all of health care today (Miles 2005). It is worth recalling that Dame Cicely Saunders and the other founders of hospice and palliative care specifically reaffirmed non-killing was a core principle of the specialty (Saunders 1984; Twycross 1996; Byock 2009). The World Health Organization explicitly stated within its first definition of palliative care that the specialty neither hastens nor prolongs dying (WHO 1990).

It is true that articulate voices within health care, including even some within Catholic and palliative care, advocate for "medical aid in dying"; however, the intentional ending of human life remains beyond the scope of legitimate health care practice (Byock 2016). Even in jurisdictions in which physician-assisted suicide or euthanasia have become legal, the vast majority of professionals within Catholic health-care and palliative care programs respectfully and lawfully decline to participate.

## Conclusion

Although it may be a shared opposition to physician-hastened death that makes us exceptional in the eyes of proponents of such practices, it is *what we are for* that truly sets both Catholic health care and palliative care apart. What makes us exemplary within contemporary health care is our intention to go beyond the basics, to meet each patient as a whole person, within their family and community, accepting each as inherently worthy and dignified, deserving of our knowledge and technical skill, but also our love.

These attributes of Catholic health care and palliative care affirm the healthiest of human drives and core tenets of our faith traditions. In our commitment to care for others expertly, as well as tenderly and lovingly, we evince the fullness of human life, the highest therapeutic principles, and, in so doing, we complete the covenant inscribed within our genetic code and Biblical commandments.

Institute for Human Caring
Providence St. Joseph Health Gardena
CA, USA

Ira Byock

## References

Byock, I. 2009. Principles of palliative medicine. In *Palliative medicine*, ed. T.D. Walsh, 33–41. Philadelphia: Saunders Elsevier.

Byock, I. 2016. The case against physician-assisted suicide and euthanasia. In *The Oxford handbook of ethics at the end of life*, eds. S.J. Younger and R.M. Arnold, 366–382. New York: Oxford University Press.

Byock, I. 2004. The ethics of loving care. *Health Progress* 85 (4):12–19, 57.
Byock, I. R. 1997. *Dying well: The prospect for growth at the end of life*. New York: Putnam.
Harlow, H.F. and C. Mears 1979. *The human model: Primate perspectives*. Washington, DC: V. H. Winston & Sons.
Lewis T., F. Amini, and R.A. Lannon. 2001. *General theory of love*. New York: Vintage.
Miles, Steven H. 2005. *The Hippocratic oath and the ethics of medicine*. New York: Oxford University Press.
Saunders, Cicely. 1984. Facing death. *The Way* 24 (4): 296–304.
Spitz R. 1945. Hospitalism: An inquiry into the genesis of psychiatric conditions of early childhood. *The Psychoanalytic Study of the Child* 1: 53–74.
Thomas W.H. 1996. *Life worth living: How someone you love can still enjoy life in a nursing home.* Acton: Vanderwyk & Burnham.
Twycross R.G. 1996. *Euthanasia: going* Dutch? *Journal of the Royal Society of Medicine*. 89 (2):61–63.
World Health Organization. 1990. *Cancer pain relief and palliative care*. Geneva: World Health Organization.
Zoloth L. 1993. First, make meaning: An ethics of encounter for health care reform. *Tikkun* 8 (4):133–135.

# Acknowledgments

We dedicate this book to the memory of Pope St. John Paul II (*d. April 2, 2005*) and Joseph Cardinal Bernardin (*d. November 14, 1996*) who, each in his own way, taught us much about how to live and how to die. Pope John Paul died at home of sepsis, after several years of struggling with Parkinson's disease and after a long and profound pontificate. Cardinal Bernardin died at home as well, of pancreatic cancer, spending his last days writing to the United States Supreme Court against assisted suicide. Each man died consistent with how he lived: confident in God's grace and mercy. Both, in their own unique ways, died exactly how they lived and preached: deeply dedicated to the truth that life was worth living, and worthy of our deepest respect. One can also make the case that these leaders profoundly influenced each other's teachings and writings. Both were deeply committed to the sacredness and inherent dignity of every human life. They taught that human dignity is the common thread of our existence, which connects all life issues and places a moral burden on all society to promote and protect its members – especially its most vulnerable – and to address the structures of injustice that stand in the way of peace, reconciliation, and human flourishing.

Pope John Paul is particularly known for his encyclical *The Gospel of Life* (published on March 25, 1995) and Cardinal Bernardin for his *Consistent Ethic of Life* thesis (first introduced at Fordham University in the Gannon Lecture on December 6, 1983). Some people held these two men and their teaching in opposition, as though they represented competing ideologies. Nothing can be further from the truth. In fact, both pointed to the Second Vatican Council, in its *Pastoral Constitution on the Church in the Modern World*, as providing the foundation and the key hermeneutic for understanding the common source of their teaching:

> Whatever is opposed to life itself, such as any type of murder, genocide, abortion, euthanasia, or willful self-destruction, whatever violates the integrity of the human person, such as mutilation, torments inflicted on body or mind, attempts to coerce the will itself; whatever insults human dignity, such as subhuman living conditions, arbitrary imprisonment, deportation, slavery, prostitution, the selling of women and children; as well as disgraceful working conditions, where people are treated as mere instruments of gain rather than as free and responsible persons; all these things and others like them are infamies indeed. They poison

> human society, and they do more harm to those who practice them than to those who suffer from the injury. Moreover, they are a supreme dishonor to the Creator. (*Pastoral Constitution on the Church in the Modern World*, *Gaudium et Spes*, 27, December 7, 1965)

These various offenses against human dignity identified by the Second Vatican Council demonstrate that the dignity of the person is a seamless garment, and, therefore, so is our respect for human life.

How we treat our most vulnerable members of society, our brothers and sisters, is a testament to or an indictment against our humanity. And so each of us here, in this volume, has, in our own words, shown why the work of palliative care is not only consistent with the Church's long history of care for the most vulnerable and poor among us, but is an essential characteristic of the care we ought to provide in response to suffering persons if we are truly committed to a consistent ethic of life.

This volume had its origins a number of years ago from two distinct but interrelated lectures that we delivered. We were convinced that the eight domains of palliative care, so widely recognized, were also consistent with the Church's understanding of compassionate care for the seriously ill and dying and with the inherent dignity of the human person. We began sharing an idea for a book with this thesis, with our friends and colleagues in Catholic health care, and in the larger medical community. We are deeply grateful for the contributions to this book made by our colleagues. The pages of this volume are graced with their wisdom which will be of great benefit to its readers.

This volume tells a story. It is the story of love, care, and compassion that connects us to each other, that lifts up our spirits, gives hope to the hopeless, and restores and reconciles people with each other's humanity. It is the story of palliative care and Catholic health care.

Renton, WA, USA Peter J. Cataldo
St. Louis, MO, USA Dan O'Brien
February 2, 2019

# Contents

# Chapter 1
# Introduction

**Peter J. Cataldo and Dan O'Brien**

Christianity is aptly described as a religion of healing. Healing the human person permeates every aspect of the Christian faith: in its sacramental life and in its moral and social vision of the human person and society. The healing essence of Christianity is grounded in its understanding of and faith in Jesus Christ as the one who brings total healing– in an integrated way – to individuals and societies, and to the world. Throughout its history, Christianity has adopted the model of Jesus' healing ministry by extending that ministry to all dimensions of the human person: physical, spiritual, emotional, moral, and social. Even with a cursory reading of the healing accounts in the Gospels, one can readily see that Jesus healed *whole* persons. Whether touching the body or the spirit of a person, he ministered to every aspect of their brokenness. The holistic character of the healing ministry of the Christian faith is paralleled in modern concepts of palliative care. For example, the Institute of Medicine defines palliative care as

> Care that provides relief from pain and other symptoms, supports quality of life, and is focused on patients with serious advanced illness and their families. Palliative care may begin early in the course of treatment for a serious illness and may be delivered in a number of ways across the continuum of health care settings, including in the home, nursing homes, long-term acute care facilities, acute care hospitals, and outpatient clinics. Palliative care encompasses hospice and specialty palliative care, as well as basic palliative care. (2015, p. 27)

The Institute of Medicine explains that this definition of palliative care incorporates the essential elements of other well-known definitions, all of which explicitly or

P. J. Cataldo (✉)
Providence St. Joseph Health, Renton, WA, USA
e-mail: peter.cataldo@providence.org

D. O'Brien
Ethics, Discernment and Church Relations, Ascension, St. Louis, MO, USA
e-mail: dobrien@ascension.org



P. J. Cataldo, D. O'Brien (eds.), *Palliative Care and Catholic Health Care*, Philosophy and Medicine 130, https://doi.org/10.1007/978-3-030-05005-4_1

implicitly recognize the multiple needs of persons and their families that are addressed by palliative care, including physical, spiritual, psychological, emotional, and social needs (2015, pp. 58–59).

The healing mission of the Christian faith may also be described in terms of serving human dignity. For the Christian faith, human dignity is derived from the fact that the human person is created in the image and likeness of God, which, as the *Catechism of the Catholic Church* explains, makes each and every human life sacred, "because from its beginning it involves the creative action of God and it remains forever in a special relationship with the Creator, who is its sole end" (n. 2258). According to the Christian tradition, the locus of human dignity in a person is the spiritual and immortal soul from which are derived the powers of intellect and free will. With these powers, the human person, in and through the body-soul unity (*corpore et anima unus*), is able to know truth, choose wisely, and direct her or his physical, spiritual, emotional and social well-being. In this understanding, human dignity is not something that can be bestowed on a person by others, nor is it simply an instrumental value that can be manipulated by the individual or by society. Rather, human dignity is an intrinsic reality that is fully present in every human being by the simple fact of being human. Human dignity therefore is not reducible to the particular circumstances of a person's life. Even though a person can be neglected and harmed or esteemed and valued by others, his or her inherent dignity—in the sense understood here—cannot be lost, diminished, or eliminated as a result of such circumstances. Recall John Merrick's famous line from the 1980 film, *The Elephant Man*, "I am not an elephant! I am not an animal! I am a *human being*! I am a man!" From the perspective of Christian faith, the physical, spiritual, emotional, and social aspects of human dignity are integrally bound together within the human person. This fundamental unity of the body and the spiritual, immortal soul is what makes the individual person a human being (*Gaudium et Spes*, n. 14).

It is this view of human dignity and the unity of the human person that has informed the Catholic tradition in healing—and so palliative care—for two millennia. The unity and intrinsic dignity of the patient as a human person is at the core of both the Catholic approach to palliative care and of palliative care in general. Recent papal teaching has reaffirmed this truth about palliative care. Reflecting on the meaning of the word, "health," St. John Paul II captured its integral nature when he wrote that: "the word is intended to refer to all the dimensions of the person, in their harmony and reciprocal unity: the *physical, the psychological,* and *the spiritual and moral* dimensions" (St. John Paul II 2005). Pope Benedict XVI often cited the social obligation to provide palliative care and recognized its integral nature: "It is necessary to maintain the development of palliative care that offers an integral assistance and furnishes the incurably ill with that human support and spiritual guide they greatly need" (Pope Benedict XVI 2007). Pope Francis, likewise, speaking in the context of palliative care for the elderly, describes palliative care as "truly human assistance," as meeting the needs of patients, as "appropriate human accompaniment," and as valuing the person (Pope Francis 2015). In all these accounts of palliative care, the popes have recognized that the holistic, unified approach to care in palliative care corresponds to the unity and dignity of the human person.

This holistic view of human dignity and whole-person care in the Catholic tradition is also paralleled in the historical development of modern hospice care. Dame Cicely Saunders, the founder of the modern hospice movement, developed the notion of "total pain experience." In 1963, Saunders asked a patient to describe her pain. Through her answer, Saunders recognized that a sufficient understanding of the experience of pain must be inclusive of multiple elements. She recounted that within one answer the patient had identified, "physical, emotional and social pain and the spiritual need for security, meaning and self-worth" (Saunders 2000). The multidimensional nature of total pain reflects the integral unity of the human person, which is at the center of the Catholic understanding of palliative care.

Since the opening of St. Christopher's Hospice by Cicely Saunders in 1967, and later the opening of the first U.S. hospice in Connecticut in 1974 by Florence Wald, there has been a steady growth and acceptance of palliative care in the United States. The Institute of Medicine reports in its *Dying in America* that by 2011, 67% of hospitals with at least 50 beds had palliative care programs. Among hospitals with more than 300 beds, 85% had palliative care programs (2015, pp. 63–64). As of 2011, 45% of individuals who died in the United States were under the care of hospice at the time of death, and in 2012, there were 5500 hospice programs (some with multiple programs per entity) in the U.S. with a (mean) average daily census of 149 patients (2015, pp. 60–61). As of 2013, 90% of Catholic hospitals with 50 or more beds reported that they had palliative care programs, and among Catholic hospitals with 300 or more beds, 95% reported having palliative care programs (Dumanovsky et al. 2016, pp. 6–7). Such statistics indicate in a concrete way Catholic health care's commitment to palliative care and the fundamental compatibility between the Catholic vision of the human person and palliative care.

The aim of this volume is to show how the integral and holistic character of palliative care parallels Catholic moral teaching and tradition on health care and the Catholic vision of the dignity and unity of the human person. To this end, the book brings together experts in medicine, (including palliative medicine), moral theology, bioethics, spiritual care, palliative care services, and church history to explore under three central themes the compatibility of palliative care with Catholic teaching on human dignity and health care. These themes are: Catholic theological and moral tradition and teaching on palliative care; the treatment of body, mind, and spirit in palliative care; and institutional and societal issues relating to palliative care as considered within the context of Catholic social teaching. As a supplement to these topics, the book provides magisterial and pastoral episcopal texts on palliative care and advance care planning.

This collection of essays begins by explaining the theological and ethical foundation for the Catholic tradition in palliative care. In Chap. 2, Dan O'Brien explores the bases for the Catholic approach to palliative care in the Gospels and in Catholic theology. He accomplishes this by focusing on some key scriptural passages from the New Testament regarding Jesus' miracles and teaching, on the history of the response of the early Christians to victims of plagues, and by demonstrating how the constitutive features of the Catholic healing ministry can be discovered in this history and in these Scriptures. Peter Cataldo provides an overview in Chap. 3 of the central concepts and principles pertaining to the Catholic moral tradition in health care as they correspond to the standard eight major domains of palliative care. Among the moral principles explained is the principle of proportionate and disproportionate means of sustaining life and its history.

In Chap. 4, Charles Bouchard shows parallels between the essential concepts in the Christian *ars moriendi* (art of dying) tradition regarding the experience of a good death and modern Catholic palliative care. His analysis includes an adaptation of various virtues espoused in the *ars moriendi* tradition to the contemporary approach to suffering in palliative care. As a result, Bouchard makes a helpful contribution to the spirituality of palliative care. One obstacle to the acceptance of palliative care for some individuals is the perception that palliative care is a form of "stealth euthanasia." Ron Hamel examines the arguments for this claim in Chap. 5 and shows how they are both incorrect on their merits and are inconsistent with Catholic moral teaching and tradition. Hamel rightly points out that palliative care is the only practical alternative to and defense against physician assisted suicide, and as such, even accounting for the potential for abuse, it is important that palliative care as a whole not be mischaracterized as stealth euthanasia.

The particular issue of whether palliative care is really stealth euthanasia points to the reality of a spectrum of understanding, interpretation, and application of Catholic teaching and tradition on end of life ethical issues that has developed since the 1960s across all the populated continents of the globe.[1] This field of theological and ethical opinion has both influenced and responded to a growing body of Catholic teaching and to advances in palliative care. A range of interpretation and application is evident, for example, with issues such as advance care planning and medically administered nutrition and hydration. An exploration of the convergence and divergence of views along a spectrum regarding these kinds of issues was recently undertaken in the Pathways to Convergence project (The Center for Practical Bioethics 2017). This endeavor brought together experts in palliative medicine, theology, bioethics, and administration within Catholic health care to identify points of agreement and disagreement and to indicate a pathway toward greater clarity and a common understanding of Catholic teaching and tradition on palliative care and end of life issues. One of the outcomes of this project was the realization that while there exists a difference of interpretation and application of teaching and tradition in some aspects of the issues examined, there is concurrence with respect to key principles and many areas of application.

One way in which this volume demonstrates how the essence of palliative care and the tenets of Catholic health care are coherent despite some differences in interpretation and application of Catholic teaching is with respect to specific issues pertaining to the triad of body, mind and spirit in palliative care, which several chapters

[1] For example, on the topic of medically administered nutrition and hydration, see the work of Germain Grisez, John Finnis, Joseph Boyle, William E. May, John Keown, Luke Gormally, Nicholas Tonti-Filippini, Kevin O'Rourke, O.P., Benedict Ashley, O.P., Albert S. Moraczewski, O.P., Richard A. McCormick, S.J., Thomas A. Shannon, and James J. Walter. For a diversity of European opinion on the role of palliative care in euthanasia see Jones et al. (2017).

address. In Chap. 6, Kevin Donovan applies Catholic moral teaching and tradition to the medical data regarding the use of analgesics and sedatives in palliative care. He also analyzes the issue of palliative sedation and generally considers the compatibility of the goal of pain and symptom management with human dignity. David Lichter, in Chap. 7, presents the role of spiritual care in palliative care. He provides an overview of the work of Catholic pastoral care in palliative care and explains how spiritual needs are assessed and what sorts of spiritual care services are provided within palliative care. Lichter also explains national standards in palliative care spirituality and weaves helpful patient narratives into his presentation.

In Chap. 8, Daniel Dwyer explores the psychological dimension of palliative care and its compatibility with the Catholic view of the human person. In particular, he examines the reality of depression and hopelessness in dying patients, how these can often be correlated with the request for hastened death, and how hopelessness and depression may be effectively treated within the context of palliative care. Sarah Hetue Hill examines the role of the family in palliative care in Chap. 9. Drawing especially on data and experiences in perinatal palliative care, she presents the assessment of family and social needs in palliative care and explores how to respond to those needs in the context of Catholic teaching on the family. In Chap. 10, Diane Meier and Christopher Lawton provide an explanation of the core elements of palliative care and its significant benefits. They also examine the various challenges faced by the attempt to integrate palliative care into how we approach care of the seriously ill. These barriers include an inadequate understanding of mortality within our health care system, insufficient education, emotional barriers and lack of access. Meier and Lawton emphasize how Catholic health care can play a critical role in overcoming these barriers and in the advancement of palliative care.

The remaining set of chapters cover Catholic social teaching and institutional and societal issues relating to palliative care. In Chap. 11, James Bailey explores issues affecting the relationship between the common good and palliative care. To this end, he applies Catholic social teaching to examine issues such as the provision of palliative care as part of the common good, social responsibilities and palliative care, and ensuring access to palliative care within society. Tina Picchi, in Chap. 12, provides an overview of palliative care excellence across Catholic health care, which includes a consideration of models of Catholic palliative care programs employed by several institutional members of the Supportive Care Coalition. She shows that the palliative care movement has much to learn from Catholic health care models of palliative care delivery because these are some of the best palliative care programs in the United States.

In Chap. 13, MC Sullivan addresses the need to educate the public and Catholics in particular about palliative care. She shows how this education effort is critically important for patients, families, and for the work of health professionals. In particular, she examines the issue of palliative care education for laypersons within a Catholic setting, and the need for collaboration with others in this effort. In Chap. 14, Mark Repenshek and Leslie Schmidt consider the compatibility between Catholic moral teaching and tradition and advance care planning (ACP), both from a general perspective and with respect to planning for palliative care. This is another area full of misconceptions both inside and outside the Catholic Church. Repenshek and Schmidt provide a historical overview of advance care planning, an analysis of Catholic teaching on this subject, and the influence that palliative care can have on a proper utilization of advance care planning. Elliott Bedford, in Chap. 15, closes the book's examination of palliative care within the Catholic tradition by exploring and contrasting the goals and actions of palliative care and physician-assisted suicide. Despite attempts by many medical and advocacy groups to depict palliative care and physician-assisted suicide as being within the same ethical and medical spectrum of care for persons with serious progressive or terminal illnesses, Bedford shows how these two approaches to illness are radically different. He also demonstrates how palliative care—not assisted suicide—is the quintessential example of medical mercy.

The volume ends with an appendix that provides selected magisterial and pastoral episcopal texts on palliative care, hospice, and advance care planning. These texts serve to strengthen the chapters of the volume and reinforce the message that Catholic teaching fully supports both advance care planning and palliative care for the seriously ill.

## References

Benedict XVI, Pope. 2007. Angelus. February 11, 2007. http://w2.vatican.va/content/benedict-xvi/en/angelus/2007/documents/hf_ben-xvi_ang_20070211.html. Accessed 15 Feb 2017.

Catechism of the Catholic Church. 1997. 2nd ed. Vatican. Libreria Editrice Vaticana.

Dumanovsky, T.R., M. Augustin, K. Rogers, D.E. Meier Lettang, and R.S. Morrison. 2016. The growth of palliative care in U.S. hospitals: A status report. *Journal of Palliative Medicine* 19 (1): 8–15.

Francis, Pope. 2015. Address to participants in the plenary of the pontifical academy for life. http://w2.vatican.va/content/francesco/en/speeches/2015/march/documents/papa-francesco_20150305_pontificia-accademia-vita.html. Accessed 15 Feb 2017.

Institute of Medicine of the National Academies (IOM). 2015. *Dying in America: improving quality and honoring individual preferences near the end of life*. Washington, DC: The National Academies Press.

John Paul II, St. 2005. Letter to the president of the Pontifical Academy for Life on the occasion of a study congress on auality of life and ethics of health. 2005. http://w2.vatican.va/content/john-paul-ii/en/letters/2005/documents/hf_jp-ii_let_20050219_pont-acad-life.html. Accessed 15 Feb 2017.

Jones, David Albert, Chris Gastmans, and Calum Mackellar. 2017. *Euthanasia and assisted suicide: Lessons from Belgium*. Cambridge: Cambridge University Press.

Saunders, Cicely. 2000. The evolution of palliative care. *Patient Education and Counseling* 41: 7–13.

The Center for Practical Bioethics. 2017. *Pathways to convergence*. https://www.practicalbioethics.org/resources/spirituality-and-religion.html. Accessed 14 Oct 2018.

The pastoral constitution on the church in the modern world (Gaudium et spes). 1965. http://www.vatican.va/archive/hist_councils/ii_vatican_council/documents/vat-ii_const_19651207_gaudium-et-spes_en.html. Accessed 15 Feb 2017.

# Part I
# Catholic Theological and Moral Tradition and Teaching on Palliative Care

# Chapter 2
# Palliative Care and the Catholic Healing Ministry: Biblical and Historical Roots

**Dan O'Brien**

You are likely already familiar with the iceberg metaphor. It's a great way of initially framing the question of "what is that something else" that the Catholic healing ministry brings to care for those who are most vulnerable, especially at the end of life. The medical and nursing skills and knowledge that it takes to do effective general care, end of life care – including *hospice* and *palliative care* – are all above the surface; they can be learned and replicated. However, the *competencies* needed to apply that knowledge and those skills *well* – in an interdisciplinary, coordinated and collaborative manner – are below the surface: they require experience, ongoing training and continuous professional development. We have superb examples of people and programs both inside and outside of Catholic health care that demonstrate those competencies. But there is nothing necessarily "Catholic" about those competencies.

We need to go deeper still into the *culture* of the Catholic healing ministry – into the depths of Scripture and the Church's history and teaching, to uncover the vision, meaning and purpose of what Catholic health care does, and how that distinguishes it from other forms of care. There we will discover what created and sustained the rise of the healing ministry within the Catholic Christian tradition, history and culture. Seen in this context, the Catholic healing ministry is not just another set of technical skills to which Catholic health care aspires; nor is it simply a great way to improve quality and patient satisfaction or to lower out-of-control spending. All these things are good, and there are many other goods that we could name; but these would not get us to the heart of the matter for the Catholic healing ministry. That is what I want to explore in this chapter.

In the *Introduction* to Part Five of the *Ethical and Religious Directives for Catholic Health Care Services, Sixth Edition* ("Ethical and Religious Directives")

D. O'Brien (✉)
Ethics, Discernment and Church Relations, Ascension, St. Louis, MO, USA
e-mail: dobrien@ascension.org



P. J. Cataldo, D. O'Brien (eds.), *Palliative Care and Catholic Health Care*, Philosophy and Medicine 130, https://doi.org/10.1007/978-3-030-05005-4_2

produced by the United States Conference of Catholic Bishops (USCCB 2018), we find three critically important statements. First, it states that:

> Above all, as a witness to its faith, a Catholic health care institution will be a community of respect, love, and support to patients or residents and their families as they face the reality of death. What is hardest to face is the process of dying itself, especially the dependency, the helplessness, and the pain that so often accompany terminal illness. *One of the primary purposes of medicine* in caring for the dying is the **relief of pain and the suffering** *caused by it*. Effective management of pain in all its forms is critical in the appropriate care of the dying.

Secondly,

> The truth that life is a precious gift from God has profound implications for the question of **stewardship over human life**. We are not the owners of our lives and, hence, do not have absolute power over life. We have a duty to preserve our life and to use it for the glory of God, but the duty to preserve life is not absolute, for we may reject life-prolonging procedures that are insufficiently beneficial or excessively burdensome. Suicide and euthanasia are never morally acceptable options.

And Thirdly,

> The task of medicine is **to care even when it cannot cure**. Physicians must evaluate the use of the technology at their disposal. Reflection on the innate dignity of human life in all its dimensions and on the purpose of medical care is indispensable for formulating a true moral judgment about the use of technology to maintain life. The use of life-sustaining technology is judged in light of the Christian meaning of life, suffering, and death. In this way two extremes are avoided: on the one hand, an insistence on useless or burdensome technology even when a patient may legitimately wish to forgo it and, on the other hand, the withdrawal of technology with the intention of *causing* death. (USCCB 2018; emphasis added)

These are three distinguishing characteristics of Catholic health care in that *what* we emphasize in this approach to care represents the best of what the Catholic faith has consistently believed and emphasized throughout the centuries: about who we are as human beings; about our relationship to God and to each other; about our destiny; and about the meaning and purpose of carrying on this healing ministry.

In this chapter, I will explore how these three key characteristics of care – especially end of life care – represent what Catholic Christians have been *inspired* to do for 2000 years. They reached out and cared for the sick, the poor, the marginalized, the vulnerable and the dying, because they recognized in them their sisters and brothers in Christ. So too, Catholics believe today, as a matter of that same *faith* that has been handed down to us, that when we heal and care for the sick, the marginalized, the vulnerable and the dying, we are touching and caring for not only our sisters and brothers, but we are touching the face of *God*.

## 2.1 The Faith That Has Been Handed Down to Us

Where did this belief come from? I would propose that the Church's fundamental commitment to the healing ministry – characterized especially by its commitment to care for poor and vulnerable persons – springs first from its belief in the *Incarnation*. The Incarnation is the belief that God took on our human nature in the historical man Jesus of Nazareth, and in so doing, has forever changed not only the relationship between humanity and God, but has forever changed our relationship with each other. This change entails a new order of creation – a new union between God and creation in the person of Jesus Christ. This new union inserts itself into the old order of creation, and continually *calls the old order* into a new way of being with God and of being with each other.

Sin, suffering, injustice, war, sickness and death are not done away with by the Incarnation – we are not making such a fantastic claim. Rather, the claim is that through the Incarnation, God takes on our sin and suffering, our sickness and death – and transforms them in and through the man Jesus, who is crucified and risen. Because of Christ, we believe that God suffers in and through our humanity, in and through our injustices and neglect, in and through his own divine-human unity – *which is our transformed humanity*, destined for eternal union with God.

There is another way to express this. The Catholic faith teaches that not only does God suffer in and through our suffering humanity and through our injustices, neglect and foul treatment of each other; but we also believe that God loves through our love; forgives through our mercy; heals through our touch; and comforts through our compassion.[1] The account of St. Paul's conversion drives the point home. In Acts 22, Paul recalls,

> On that journey as I drew near to Damascus, about noon a great light from the sky suddenly shone around me. I fell to the ground and heard a voice saying to me, 'Saul, Saul, why are you persecuting me?' I replied, 'Who are you, sir?' And he said to me, 'I am Jesus the Nazorean whom you are persecuting.' (NAB 1991)

Paul had never met the earthly Jesus, and yet there Jesus was, telling him how he was persecuting him. Christians believe that Jesus was not merely speaking figuratively in this liminal encounter. What Paul was doing to the followers of Jesus, he was doing to Jesus himself – on account of the Incarnation.

It is also driven home by Jesus' account of the Last Judgment as described in Matthew's Gospel, chapter 25, where he depicts the Son of Man saying,

> Come, you who are blessed by my Father, inherit the kingdom prepared for you from the foundation of the world. For I was hungry and you gave me food, I was thirsty and you gave me drink, I was a stranger and you welcomed me, I was naked and you clothed me, I was

[1] I do not mean to imply that Christ suffers in his divine nature, but only in and through the hypostatic union of his two natures. Christ, according to ancient Christian teaching, is one divine person with two natures – one human and one divine – in one hypostasis, or one individual concrete existence. In and through this hypostatic union, Christ – on account of his human nature – has truly suffered and died. In a certain analogical sense, we can also say that through his union with humanity, Christ continues to suffer here on earth.

> sick and you visited me, I was in prison and you came to me.' Then the righteous will answer him, saying, 'Lord, when did we see you hungry and feed you, or thirsty and give you drink? And when did we see you a stranger and welcome you, or naked and clothe you? When did we see you ill or in prison, and visit you?' And the king will say to them in reply, 'Amen, I say to you, whatever you did for one of these least brothers [and sisters] of mine, you did it for me.' (NAB 1991)

The words of St. Paul and of the account of Christ's parable of the Last Judgment are more than a poetic metaphor for the body of Christ. They refer to the *divinization of humanity* – the drawing of our humanity into the inner life of God[2] (Schaff 2017). If we truly believe that God so loved the world – including you and me, here, now, in our time, in our place – then the healing ministry that Jesus set out to do 2000 years ago is *our* work today. This belief has inspired Christians, individually and collectively, for *centuries*, to reach out to the sick, the hungry, the lame, the imprisoned, in order to heal, to comfort, to care, to console – and so touch the face of God.

I am reminded of the founders of our Catholic health ministries. Those women and men who began our healing ministries centuries ago were quite conscious of the fact that the work they were doing, that they were carrying on, was the healing ministry of Jesus and his Apostles. Whether the Alexian Brothers, founded in the twelfth century; the Sisters of St. Joseph or the Daughters of Charity, founded in the seventeenth century; or the Sisters of the Sorrowful Mother, founded in the nineteenth century – they were all driven by the same motto: "The love of Christ compels us" (2 Corinthians 5:14). They took seriously the command of Christ "to proclaim the reign of God and heal the afflicted" (Luke 9:2), "to expel unclean spirits and to cure sickness and disease of every kind" (Matthew 10:1), "to raise the dead and heal the leprous…" (Matthew 10:8). They took to heart Jesus' teaching that whatsoever we do to the least, we do to him (Matthew 25: 40). And so they believed and taught that if we truly love *him*, then we *must* feed the hungry, welcome the stranger, clothe the naked, comfort the sick, and visit the imprisoned.

This is the primacy of love of neighbor – of a faith that leads into action, just as they and we are taught in the letter of James (2: 14–17):

[2] "Divinization" is an ancient concept. It refers to the mystery of humans becoming divine through the action of Divine grace. During every Liturgy of Eucharist celebrated today, Catholics pray with the priest during the preparation of the gifts: "May we come to share in the divinity of Christ who humbled himself to share in our humanity." This same concept is found in the earliest writings of the Church Fathers, for example, in Ireneus of Lyon (c. 130–200), who wrote, "The Word of God, our Lord Jesus Christ, through His transcendent love, became what we are, that He might bring us to be even what He is Himself." (Book Five, Preface, in *Against the Heresies*, in Philip Schaff, Ante-Nicene Fathers, Vol. 1). Also: "For it was necessary … that what was mortal should be conquered and swallowed up by immortality, and the corruptible by incorruptibility, and that man should be made after the image and likeness of God" (Book 4, Chapter 38, in *Against the Heresies*). Likewise, Clement of Alexandria (c. 150–215) wrote, "The Word of God became man, that you may learn from man how man may become God" (Chapter I, *Exhortation to the Heathen*), and, "For if one knows himself, he will know God; and knowing God, he will be made like God" (Book III, Chapter I, *The Instructor*).

> My brothers and sisters, what good is it to profess faith without practicing it? Such faith has no power to save one, has it? If a brother or sister has nothing to wear and no food for the day, and you say to them, "Good-bye and good luck! Keep warm and well fed," but do not meet their bodily needs, what good is that? So it is with the faith that does nothing in practice. It is thoroughly lifeless. (NAB 1991)

So, let's turn now to the healing ministry of Jesus, to examine the ways that he responded to the sick and the vulnerable. If we are to make a claim that we are the *body of Christ*, and that the work we do in caring for the sick, the dying, the poor and vulnerable is the heart and soul of Catholic health care, then we need to know and understand how Jesus healed and what Jesus teaches us about the primacy of love of neighbor.

## 2.2 The Healing Ministry of Jesus

In Jesus' own day and culture, illness was generally understood to be related to and even caused in some way by sin – a manifestation of spiritual ailment. This is why in John's Gospel (9:2), upon seeing a blind man, the disciples of Jesus asked him, "Rabbi, was it his sin or that of his parents that caused him to be born blind?" The spiritual, the physical and the social are intertwined, inseparable. And so it is with Jesus' healing ministry. The seventeenth century Cartesian dualism and dichotomy between body and spirit, which still influences the way we tend to think of our bodies today, especially in modern medicine, doesn't exist in the biblical stories we are going to examine. The healing stories of Jesus in the Gospels still form the foundation for our understanding of care in the Catholic healing tradition. A closer look at just one of those stories will suffice for illustrating the point.

### *2.2.1 The Woman with the Hemorrhage: How Did Jesus Heal?*

The story of the woman with the hemorrhage is recounted in the Gospel according to Mark (also in Matthew 9:18–26 and Luke 8:41–56). In Mark's account, Jesus meets this woman while he is on his way to see a synagogue official's daughter. It is actually two healing stories wrapped into one:

> One of the synagogue officials named Jairus came up, and on seeing Jesus, fell at his feet and implored him earnestly, saying, "My little daughter is at the point of death; please come and lay your hands on her, so that she will get well and live." And [Jesus] went off with him; and a large crowd was following him and pressing in on him.
>
> A woman who had had a hemorrhage for twelve years, and had endured much at the hands of many physicians, and had spent all that she had and was not helped at all, but rather had grown worse—after hearing about Jesus, she came up in the crowd behind him and touched his cloak. For she thought, "If I just touch his garments, I will get well." Immediately the flow of her blood was dried up; and she felt in her body that she was healed of her affliction. Immediately Jesus, perceiving in himself that power had gone out from

him, turned around in the crowd and said, "Who touched my garments?" His disciples said to him, "You see the crowd pressing in on you, and yet you say, 'Who touched me?'" And he looked around to see the woman who had done this. But the woman fearing and trembling, aware of what had happened to her, came and fell down before him and told him the whole truth. And he said to her, "Daughter, your faith has made you well; go in peace and be healed of your affliction."

While he was still speaking, they came from the house of the synagogue official, saying, "Your daughter has died; why trouble the Teacher anymore?" But Jesus, overhearing what was being spoken, said to the synagogue official, "Do not be afraid any longer, only believe." And he allowed no one to accompany him, except Peter and James, and John the brother of James.

They came to the house of the synagogue official; and he saw a commotion, and people loudly weeping and wailing. And entering in, he said to them, "Why make a commotion and weep? The child has not died, but is asleep." They began laughing at him. But putting them all out, Jesus took along the child's father and mother and his own companions, and entered the room where the child was. Taking the child by the hand, he said to her, "Talitha kum!" (which translated means, "Little girl, get up!"). Immediately the girl got up and began to walk, for she was twelve years old. And immediately they were completely astounded. And he gave them strict orders that no one should know about this, and he said that something should be given her to eat. (NAB 1991, Mark 5:22–43)

Scripture scholars tell us that the recounting of these two healing stories, as with other such healing stories, serves another purpose besides the simple recollection of miracles: the story demonstrates Christ's power over evil, over the world, over suffering, over life and death. It also demonstrates that he is greater than just another prophet. Notice, for example, that Jesus did not implore God to heal the woman (as the prophet Elijah did); and he did not implore God to raise the child from the dead (as the prophet Elisha did). Rather, *power goes out from him*, the story says, where there is *faith* placed *in him*; and the dead are brought back to life by the power of Jesus' *own* word, not in answer to a prayer of supplication. The difference is accentuated by Jairus' request to Jesus to *lay his hands on her*. Healing through the imposition of hands is not mentioned in the Old Testament or in rabbinical writings (Mally 1968).

Now, we need to be very careful here to avoid any claim that Jesus was simply healing people in order to *display* his power, his divine nature, or his authority. That would suggest that Jesus was *using* people for other purposes – something that does not suggest authentic love or compassion. We also need to be careful to distinguish between what Jesus did and taught, and the particular meaning that the author of the Gospel may have been trying to convey through the recounting of the story, as well as the deeper meaning that we discover and may bring to the story through our own study, culture, experience and inspired insight. The Scripture text must be approached as a *living Word* – one that brings new insights with every new encounter.

That being said, assuming this is an actual event which took place as described by Mark, Mark uses this passage – and all healing miracles – to establish Jesus' authoritative credentials among the members of the community for whom he writes. This would in turn give Mark's Jesus the power and authority to be creative concerning rules of ritual impurity, healing on the Sabbath, touching the dead, etc. Jesus is the Lord of the Sabbath; he is Lord over the Law itself – referring to his divine origins. Multiple interpretations should be expected from such healing stories, even beyond Mark's intention. Faith is what helps us to hold all of this together in tension.

Scholars often remind us that Jesus was an observant Jew – this is critical to remember in order to understand several points that are implied in the story. First, we must understand that the woman with the hemorrhage was ritually unclean, following the conventions of Jewish law. In the book of Leviticus (15:19), we read that anyone who touches a woman during her menstrual flow shall be unclean until evening. It goes on to say that "as long as she suffers this unclean flow she shall be unclean" (15:25).

In the book of Numbers, we hear God speaking to Moses, saying, "Speak to the children of Israel and tell them that they and their descendants must put tassels on the corners of their garments, fastening each corner tassel with a blue cord" (15:37–41). The passage goes on to say that these tassels were to remind the Israelites that they were bound to the Law of the Lord. So, touching the hem, a tassel, of Jesus' garment was like touching him – perhaps even *worse*; it was like a visible defiance of the Law of the Lord. For an unclean woman to touch anyone in the community, in this way especially, would have been highly offensive. Yet, she was desperate; she had been in that state for 12 years, cut off from society, from her family, from worship, from the market, from the water well.

Against this backdrop, she reaches out in great *hope* to touch the one whom she *believed* could *cleanse her* – unlike the many physicians for whom she spent everything she had. She had to have great courage and even a certain audacity, because in moving through the crowd with the hope of touching the hem, a tassel, of Jesus' garment, she was willing to risk bringing great offense – subjecting Jesus and all those around her, to ritual impurity, in a manner possibly suggesting open defiance of the Law. So it is no wonder that the text says she was afraid and trembling when Jesus asked, "Who touched me?" But she had great *faith* in him. She *trusted* that she wasn't going to be rejected; that she would receive what one author calls "a fresh well of mercy" (Scalia 2013).

The account of Jesus raising Jairus' daughter from the dead follows. At first glance, the two stories may appear to be unrelated. But notice that the theme of ritual impurity continues. In the Book of Numbers, we also read that anyone who touches a dead body or is in a room with a dead body was to be ritually unclean for 7 days (19:11,14). Notice also that at the end of the passage, Mark adds a kind of afterthought: he says the little girl was 12 years old. Remember that the woman with the hemorrhage had been sick for 12 years. The number 12 carries great significance for the first Jewish followers of Jesus reading or hearing this story. The number 12 represents completeness, especially in reference to the reign of God. The twelve tribes of Israel had already been dispersed long before Christ, mostly lost since the Assyrians conquered the northern Kingdom of Israel over seven centuries before. And so, Jesus' act of naming 12 Apostles, and this story of the woman with a hemorrhage for 12 years, and raising a 12 year old girl to life, are all powerful statements about restoring what was lost, and fulfilling his intent to re-establish God's reign.

The Gospel story of these two healings likewise tells us something about Jesus' desire to restore us. He addresses the woman as "daughter." One author suggests, "You might say that Jesus plunders the realm of the unclean and even the realm of the dead to restore these two women to abundant life" (Galloway 2012). They were both outcasts: unclean, outsiders, who have now been restored to their dignity, their inheritance, as Daughters of God. This theme of restoration is carried on throughout the Gospels – Christ bringing about the Reign of God. This is the work Catholics believe they are called to do in healing the sick, in end of life care, hospice care, palliative care, care for the elderly, and for all poor and vulnerable persons. When we reach out to heal, to comfort and to care—*even if there is no cure*—we are helping to restore people to their relationship with God, their loved ones and their communities.

### 2.2.2 The Parable of the Good Samaritan: What Did Jesus Teach?

There is another Gospel story that has had perhaps the most dramatic impact for centuries on how Catholics and other Christians have thought about their duty to reach out to and care for the poor, the sick and the suffering, in love and mercy. The encounter between Jesus and a spokesperson of the law discussing what is the most important commandment is recounted first by Mark (12: 28–34). The spokesman is a Pharisee in Matthew's Gospel, a scribe in Mark's Gospel, and a lawyer in Luke's Gospel – all scholars of the law (Stuhlmueller 1968, n. 102). The encounter between Jesus and the rich young ruler takes on a similar theme about what does it mean to love God above all things (Mark 10:17–31, Matthew 19:16–30 and Luke 18:18–30). But the parable which Jesus uses to answer the spokesman to illustrate what it means to love one's neighbor is found only in the Gospel of Luke, Chapter 10:25–37:

> On one occasion a lawyer stood up to test Jesus: "Teacher, what must I do to inherit everlasting life?" Jesus answered him: "What is written in the law? How do you read it?" The man replied: "You shall love the Lord your God with all your heart, with all your soul, with all your strength, and with all your mind; and your neighbor as yourself." Jesus said, "You have answered correctly. Do this and you shall live." But because he wished to justify himself he said to Jesus, "And who is my neighbor?"
>
> Jesus answered, "A certain man was going down from Jerusalem to Jericho, and he fell prey to robbers. They stripped him and beat him, and departed, leaving him half dead. By chance a certain priest was going down that same way. When he saw him, he passed by on the other side. In the same way a Levite also, when he came to the place, and saw him, passed by on the other side. But a certain Samaritan, as he travelled, came where he was. When he saw him, he was moved with compassion at the sight, came to him, and bound up his wounds, pouring on oil and wine. He then hoisted him onto his own animal, brought him to an inn, and took care of him. The next day, before he departed, he took out two [silver coins], and gave them to the innkeeper with the request: 'Take care of him. Whatever you spend beyond that, I will repay you when I return.' Now which of these three, in your opinion, was neighbor to the man who fell among the robbers?" The answer came, "The one who treated him with compassion." Then Jesus said to him, "Go and do the same." (NAB 1991)

*He was moved with compassion* (verse 33). The original Greek word for compassion is *splagchnizomai* (σπλαγχνίζομαι, or ἐσπλαγχνίσθη as it is used here in its verb form): literally, he was moved to the depths of his bowels, to the depths of his inmost self, in other words – much like we refer to the heart today as the seat of love and compassion (Aland et al. 1968, p. 254). In the Greek New Testament, various forms of this word are used only twelve times – five times by Matthew, four times by Mark and three times by Luke.[3] In nine of these instances, this same word was used to describe the response of Jesus to people in sorrow, hunger, possession, or disease – all desperate circumstances. The word was used once by Jesus in Matthew to describe in a parable the response of a Master toward his indebted slave; and twice by Jesus in Luke – once to describe in parable the response of the Father to the return of his Lost Son, and once in this parable of the Good Samaritan. The word illustrates what it really *means* to be merciful. As one scholar points out, every time the word is used the result of this kind of compassion is not just a detached concern, pity, or kind word, but *involvement* and *action* (Keathley 2004).

So, who was this Samaritan man who was moved with such compassion – or mercy like the Father? In Jesus' day and culture, Samaritans and Jews generally despised each other. They both considered the other to be foreigners, strangers; certainly *not* neighbors. To the Jew, a Samaritan was a "half-breed" – descendants of those Jews who stayed behind after the Assyrians conquered the northern Kingdom in 722 BCE, mixing with their captors and becoming syncretic in their belief system (Word in Life 1993, pp. 340–341). There were over a million Samaritans at the time of the Roman occupation in Jesus' day – so, interaction between Jews and Samaritans was probably necessary, though tenuous.

Jesus tells the parable in response to the question, "and who is my neighbor?" The inquirer knew full well what the law said. The Book of Leviticus (19:18) says, "Take no revenge and cherish no grudge against your fellow countrymen. You shall love your neighbor as yourself" (NAB 1991). But notice that Jesus does not tell us who the man in the parable was who was beaten and left for half-dead – it just says that he was a traveler. It doesn't say whether he was a Jew or a *foreigner*. So Jesus flips around the question about who is my neighbor – according to Leviticus, my fellow countryman – and instead asks who was *acting as a neighbor* to the man who was beaten, robbed, and left for half dead. Portraying a Samaritan in such positive light would have come as a shock to Jesus' audience (Funk et al. 1993, pp. 271–400).

Some scholars have spent an inordinate amount of energy trying to get into the minds of each of the characters in the parable. Recall from the story of Jesus raising Jairus' daughter from the dead, that in Jesus' culture, contact with a dead body made one unclean. Priests, in particular, were enjoined to avoid such defilement. It has been suggested, for example, that the priest and Levite in the story may have assumed that the fallen traveler was dead and so would have wanted to avoid him to keep themselves ritually clean (Vermes 2004, pp. 152–54). But other commentators

[3] Matthew 9:36, 14:14, 15:32, 18:27, and 20:34; Mark 1:41, 6:34, 8:2, and 9:22; Luke 7:13, 10:33, and 15:20.

point out that they were *leaving* Jerusalem, which makes this interpretation of the parable unlikely (Green 1997, p. 430; Forbes 2000, pp. 63–4). Equally important as leaving Jerusalem may be the fact that Jerusalem is about 2500 ft above sea level and Jericho is roughly 800 ft *below* sea level – conveying, perhaps, the idea that in true love of neighbor, one must be willing to descend into the depths of human misery, setting aside legal questions of ritual purity.

The Jewish New Testament scholar Amy-Jill Levine agrees that since the Levite and the priest were both *leaving* Jerusalem, there was no real or apparent concern in the parable with ritual impurity (Perper on Levine 2011). She dismisses such interpretations as missing the point of the story. The scholar of the law with whom Jesus spoke would have known that saving a life – loving your neighbor as yourself – always trumps any concern about ritual impurity. In the final analysis, Levine argues, the point of the parable is that Jesus is teaching us to have compassion for our enemies. Remember the word *splagchnizomai*: compassion that goes beyond mere sympathy or pity; compassion that goes into action; compassion that is not reserved only for one's friends or fellow-believers; or, as we learn from the other parable in Luke, compassion like the Father. This strikes at the heart of the flaw in the Health and Human Services' rule that defined a religious employer as that organization which "serves primarily persons who share the religious tenets of the organization" (Carlson-Theis 2012). That is precisely *not* what our faith and not what this parable teaches us or calls us to do!

As far back as St. Ignatius, the first century bishop of Antioch in Asia Minor (martyred at Rome in the year 107 AD), church authorities and biblical scholars typically taught that there were four different senses in which one should read the Scriptures: (1) the literal sense – or what the author meant to convey; (2) the spiritual or allegorical sense, especially as an archetype of Christ; (3) the moral sense, and (4) the anagogical or mystical sense, or a foreshadowing of heaven. Keep in mind that historical criticism only began in the seventeenth century, gaining popular recognition in the nineteenth and twentieth centuries, leading to a variety of methodologies today that were practically unknown to biblical scholars during the first sixteen centuries of Christianity.

Beginning with the early Church Fathers, the allegorical interpretation of this biblical parable was favored over the rest, with the Good Samaritan being interpreted as representing Jesus, who came to save the lost (Caird 1980, p. 165). Brigham Young University scholar John Welch points out that this allegorical reading was taught not only by ancient followers of Jesus, but it was virtually universal throughout early Christianity (Welch 2007, pp. 26–33). It continues to be the traditional interpretation of the Orthodox Churches (Schönborn 2008, p. 16). The parable of the Good Samaritan was known throughout the ancient and Medieval Catholic Christian world as a model for understanding Christ's compassion for us – and therefore a model of compassion that believers have felt compelled to imitate. The concept of the Good Samaritan continues to strike a powerful metaphor for people of many faiths today – so much so, that we typically refer to tort laws that give legal immunity from liability to those who help strangers as "Good Samaritan Laws" or "Good Samaritan Rules."

## 2.3 From Scripture to Practice: Historical Foundations

I have recounted here two powerful models – thickly layered stories of Jesus' healings and his method of teaching by way of parable: showing us how to believe and how to love with compassion. So, how did this play out in the centuries that followed and how does it play out for us today? The first two centuries of Christianity are quite remarkable. By the end of the second century, Christian communities had sprung up all over the Roman empire, and into Mesopotamia, Bactria (now Afghanistan), Persia and southern India. In the late nineteenth century, the German scholar von Harnack was probably the first modern historian to recognize that Christianity's early rapid growth and success were due in large part to its strong emphasis on healing and for caring for the sick, the poor and the downtrodden (von Harnack 1892, pp. 37–152). In our own time, Risse, Porterfield and Stark have all researched extensively this phenomenon of the rapid expansion of Christianity in the second and third centuries (Risse 1999, pp. 69–116; Porterfield 2005, pp. 43–65; Stark 1996). These scholars all have presented convincing historical evidence that it was not so much miraculous cures or compelling preaching or the promise of a happy afterlife that attracted so many adherents to the Christian faith in its early centuries; rather, it was the remarkable way that Christians cared for each other and for poor and vulnerable persons.

Take one example. It was not an unusual practice in Roman culture to expose one's own child to the elements if it was a girl when one had too many girls, or if it was sick, or was undesirable for whatever reason (Harris 1982). Christians became known for collecting such abandoned babies and caring for them. In the "Apostolic Constitutions" or early rules of the Church in the second and third centuries – a hundred years before the legalization of Christianity by Constantine – it was the established practice in Christian communities from Syria to Rome, for bishops and deacons to be bound by rules to collect alms and donations from wealthy Christians for redistribution to the poor (Risse 1999, p. 20). The Roman Apostolic Constitution of the year 215 obligated bishops to seek out the sick and care for them in their own homes.

During times of plague and public health crises, when the wealthy would typically abandon cities until the disease passed over, they were often surprised on their return to find the sick and the dying being cared for by Christians, who subjected themselves to great risk for the sake of love of God and neighbor and of the practice of compassion (Stark 1996, pp. 3–28). During one such plague epidemic in Alexandria, Egypt, during the reign of the Emperor Gallienus in the mid third century (259–268), the local bishop Dionysius led an extensive door-to-door relief campaign (Risse 1999, p. 80). Such dramatic caring behavior was noted with great interest by the general populous and with concern and a certain alarm by authorities and chroniclers throughout the Empire. It was interpreted by many in power as subversive activity, and persecutions intensified.

The pace of conversions grew rapidly after the emperor Constantine proclaimed the edict of toleration in 313, and after Christianity was established as the official religion of the Empire by Theodosius in 381. These events of the fourth century brought several important developments in Christianity for its impact on the larger culture. The political endorsement of Christianity brought in wealth, influence, and support for new institutions and organizational structures (Porterfield 2005, p. 44; Risse 1999, pp. 70–79). In many places, bishops began taking over the traditional social welfare function of the aristocracy.

In the East, the Emperor, seen as God's representative on earth, was understood as having special responsibility as the lead patron of charitable institutions, including what we would identify today as shelters, hospitals and hospices (Risse 1999, p. 81; Horden 1985; Partlagean 1977). Monasteries also began to emerge during this period of time, largely as a rebellious response to the perceived watering down of Christian ideals brought about by its official recognition by the state. In both the East and the West, the infirmaries of these monasteries took on an increasingly important role. The usual and preferred place of treatment always remained in the home, but such hospitals, hospices and infirmaries remained important social institutions for the poor and vulnerable throughout the Middle Ages (Porterfield 2005, p. 48; Barrett-Lennard 2011, pp. 9–41, 85–86).

## 2.4 Constitutive Features of the Catholic Ministry of Healing

So what can we conclude about the healing and teaching of Jesus? What can we conclude from the practices and teachings of the Church in the early centuries into the Middle Ages? What do they tell us for our work of healing and caring for the sick and the dying today? There are many constitutive features we can identity, which are explored elsewhere in this book, for example: *solidarity with the poor; reverence and love for the inherent sacredness and dignity of human life; treating persons holistically – recognizing they are a body-spirit unity*; *hungering and thirsting for justice; commitment to the common good*; *hospitality for the foreigner or stranger*. These six constitutive features are evidenced by the healing stories and teaching I have examined here, in this broad sweep of early Christian history, and in the traditions and stories and commitments of the founders of many Catholic healing ministries.

But I would suggest that there are *four* constitutive features of the Gospel healing narratives and of our living Tradition that also represent distinguishing characteristics of the Catholic healing ministry. These are not discussed with the same frequency and care in the literature, but could be equally emphasized as hallmarks of Catholic healthcare, and as inspiration for the ministry of palliative care. Whether we embody and live out these constitutive features *well*, depends on reflection, prayer, vision, leadership, courage, stewardship, determination, humility, friendship and *community*.

### 2.4.1 First, the Healing Ministry of Jesus Is Incarnational

In carrying out the healing ministry of Christ we don't reach out to heal the poor and vulnerable, the sick and the downtrodden simply because of a disinterested sympathy. Rather, we believe we are reaching out to Christ when we reach out to the sick and poor, and that we ourselves are the body of Christ in the world – groaning with the Spirit of God for its renewal and restoration. In his 1995 encyclical, *The Gospel of Life*, Saint John Paul II wrote:

> As disciples of Jesus, we are called to become neighbors to everyone (cf. Lk 10:29–37), and to show special favor to those who are poorest, most alone and most in need. In helping the hungry, the thirsty, the foreigner, the naked, the sick, the imprisoned – as well as the child in the womb and the old person who is suffering or near death – we have the opportunity to serve Jesus. He himself said: "As you did it to one of the least of these my brethren, you did it to me" (Mt 25:40). Hence we cannot but feel called to account and judged by the ever relevant words of Saint John Chrysostom: "Do you wish to honor the body of Christ? Do not neglect it when you find it naked. Do not do it homage here in the church with silk fabrics only to neglect it outside where it suffers cold and nakedness." (John Paul II 1995, n. 87)

### 2.4.2 Second, the Healing Ministry of Jesus Is the Work of Evangelization

By "evangelization," I do *not* mean that we reach out to the poor and vulnerable for the sake of converting them or for proselytization. Rather, the healing ministry of Christ brings about the *Reign of God*: restoring persons to their communities; loving and caring, and *showing* people they are loved and cared for. *Love* must be our motive. Nothing else will sustain us. As I wrote earlier, many of our founders still use this as their motto today: *The love of Christ compels us* (2 Corinthians 5:14). The good news of God's love and care for us goes hand in hand with caring for poor and vulnerable persons and for the sick and frightened whom we serve. Again, in the encyclical, *The Gospel of Life*, we read:

> The mission of Jesus, with the many healings he performed, shows God's great concern even for [our] bodily life. Jesus, as "the physician of the body and of the spirit," was sent by the Father to proclaim the good news to the poor and to heal the brokenhearted (cf. Lk 4:18; Is 61:1). Later, when he sends his disciples into the world, he gives them a mission, a mission in which healing the sick goes hand in hand with the proclamation of the Gospel: "And preach as you go, saying, 'The kingdom of heaven is at hand.' Heal the sick, raise the dead, cleanse lepers, cast out demons" (Mt 10:7–8; cf. Mk 6:13; 16:18). (John Paul II 1995, n. 47)

### 2.4.3 *Third, the Healing Ministry of Jesus Is the Work of Compassion*

Compassion is connected closely with reverence and love for the inherent sacredness and dignity of life, but is distinctive. I am referring here not only to *cum-pati,* the Latin root of our word compassion, which means "to suffer with," but to the Greek word *splagchnizomai*: compassion like Jesus, like the Father in the parable, like the good Samaritan; compassion that goes beyond mere sympathy or pity; being moved to the depth of one's being; being moved into action; compassion that is not reserved only for one's friends or fellow-believers, but is especially reserved for those who are marginalized, most vulnerable, forgotten or abandoned in any way – whose suffering cries out for a response. Suffering can be especially unbearable if the person believes they will lose all control, or will be abandoned or left to languish, whether existentially or physically. Again, we read in the encyclical, *The Gospel of Life*:

> In the sick person the sense of anguish, of severe discomfort, and even of desperation brought on by intense and prolonged suffering can be a decisive factor. Such a situation can threaten the already fragile equilibrium of an individual's personal and family life, with the result that, on the one hand, the sick person, despite the help of increasingly effective medical and social assistance, risks feeling overwhelmed by his or her own frailty; and on the other hand, those close to the sick person can be moved by an understandable even if *misplaced* [kind of] compassion. (John Paul II 1995, n. 15)

*Authentic compassion* is the antidote to despair. As it says in John's Gospel (13:35): "This is how all will know that you are my disciples, if you have love for one another" (NAB 1991). John's account has Jesus speaking these words immediately after he washed his disciples' feet at the last supper, telling them that he was leaving them an example of what it means to love one another. This is what especially drew people to the Christian faith in its early centuries – not great preaching; not great miracles; not rational arguments or intricate philosophies or theologies; but *seeing* and *experiencing* how Christians loved and cared not only for each other, but for the stranger, the forgotten, the abandoned. Wherever and whenever this is *not* the witness of Christians, there is scandal and distrust.

### 2.4.4 *Fourth, the Healing Ministry of Jesus Is the Work of Restoration*

By this, I mean to say that when we reach out to heal, to comfort and to care, even when there is no cure, we are restoring people to community – this is a constitutive feature of the Reign of God. Not only do we restore people to community, but we form community among ourselves by the very activity in which we participate. We are thus called to be *peacemakers*. When we restore people to their relationship with God and to their relationship with their families, their loved ones, and their communities, we are making peace. In a real sense, we are helping to restore them to themselves. Sickness and death separate people. But sickness and death can also bring people together if, through that suffering and death, they experience your caring hands, your competent treatment, your compassionate care, your tender voice, your attentive listening – and so experience the love and compassion of God.

If we have great clarity around these ten constitutive features of Catholic health care, then our contribution to the field and practice of end of life care, hospice and palliative care will truly be distinctive. Certainly, not all of these features will speak to people of every faith or spiritualty. But the more we know *who we are* and *whose* we are, and the rich Tradition that has been handed on to us and that we bring into practice, the easier it will be for us to engage in learning conversations with both confidence and humility. We hold a treasure. We need to protect it and to cultivate it, but most of all, we need to share it.

With these thoughts, I invite you to reflect now on words often attributed to St. Theresa of Avila. Whether or not that attribution is factual, the words are no less inspiring for helping us to recall and envision our vocation and the meaning of our Communion with God:

Christ has no body but yours,
No hands, no feet on earth but yours,
Yours are the eyes with which he looks
Compassion on this world,
Yours are the feet with which he walks to do good,
Yours are the hands, with which he blesses all the world.
Yours are the hands, yours are the feet,
Yours are the eyes, you are his body.
Christ has no body now but yours,
No hands, no feet on earth but yours,
Yours are the eyes with which he looks
Compassion on this world.
Christ has no body now on earth but yours.

## References

Aland, Kurt, et al. 1968. *The greek new testament*. 2nd ed. New York: American Bible Society.

Barrett-Lennard, R.J.S. 2011. *Christian healing after the new testament: Some approaches to illness in the second, third and fourth centuries*. Lanham: Rowman & Littlefield.

Caird, G.B. 1980. *The language and imagery of the bible*. London: Gerald Duckworth & Co Ltd.

Carlson-Theis, Stanley. 2012. Which religious organizations count as religious? The religious employer exemption of the health insurance law's contraceptives mandate. *The Federalist Society* 13: 2. http://www.fed-soc.org/publications/detail/which-religious-organizations-count-as-religious-the-religious-employer-exemption-of-the-health-insurance-laws-contraceptives-mandate. Accessed 9 July 2017.

Forbes, Greg W. 2000. *The god of old: The role of the lukan parables in the purpose of Luke's gospel*. London: Bloomsbury Publishing PLC.

Funk, Robert W., Roy W. Hoover, and the Jesus Seminar. 1993. *The five gospels*. San Francisco: Harper.

Galloway, Lewis. 2012. Taking Jesus seriously: Mark 5:21–43. Http://day1.org/3937-taking_jesus_seriously. Accessed 8 July 2017.
Green, Joel B. 1997. *The gospel of Luke*. Grand Rapids: Eerdmans.
von Harnack, Adolph. 1892. Medicinisches aus der altesten kirchengeschichte, texte und unterscuchungen zur geschichte der altchristlichen literature (TUGAL) 8, n. 4.
Harris, William V. 1982. The theoretical possibility of extensive infanticide in the graeco-roman world. *Classical Quarterly* 32 (1): 114–116.
Horden, P. 1985. The byzantine welfare state: Image and reality. *The Society for Social History of Medicine Bulletin* 37: 7–10.
John Paul II. 1995. *Evangelium vitae*. http://w2.vatican.va/content/john-paul-ii/en/encyclicals/documents/hf_jp-ii_enc_25031995_evangelium-vitae.html. Accessed 9 July 2017.
Keathley IV, Hampton. 2004. Raising the widow's son from nain. In the series, *The miracles of Jesus*. http://bible.org/seriespage/raising-widow%E2%80%99s-son-nain. Accessed 9 July 2017.
Mally, Edward J. 1968. The gospel according to Mark. In *The Jerome biblical commentary*, ed. Raymond E. Brown, Joseph A. Fitzmyer, and Roland E. Murphy. Englewood Cliffs: Prentice Hall.
New American Bible (NAB). 1991. New York: Catholic Book Publishing Co.
Partlagean, Evelyne. 1977. *Pauvrété économique et pauvréét sociale à Byzance*. Paris: Moutin.
Perper, Emily. 2011. Levine: Good samaritan parable teaches compassion for the enemy. *Chautauquan Daily*. http://chqdaily.com/2011/08/17/levine-good-samaritan-parable-teaches-compassion-for-the-enemy. Accessed 9 July 2017.
Porterfield, Amanda. 2005. Healing in early christianity. In *Healing in the history of christianity*. New York: Oxford University Press.
Risse, Guenter B. 1999. Christian hospitality: Shelters and infirmaries. In *Mending bodies, saving souls: A history of hospitals*. New York: Oxford University Press.
Scalia, Elizabeth. 2013. Jesus and the hemorrhagic woman: Accountability and thanksgiving. *The anchoress*. http://www.patheos.com/blogs/theanchoress/2013/02/05/jesus-and-the-hemorrhagic-woman. Accessed 8 July 2017.
Schaff, Philip. 2017. Ireneus, book five, preface, against the heresies. In *Ante-nicene fathers*, ed. Alexander Roberts and James Donaldson, vol. 1. Scotts Valley: CreateSpace Publishing (Amazon).
Schönborn, O.P., Christoph Cardinal (tr. Henry Taylor). 2008. *Jesus, the divine physician: Reflections on the gospel during the year of Luke*. San Francisco: Ignatius Press.
Stark, Rodney. 1996. *The rise of christianity: How the obscure, marginal, Jesus movement became the dominant religious force in the western world in a few centuries*. Princeton: Princeton University Press.
Stuhlmueller, Carroll. 1968. The gospel according to Luke. In *The Jerome biblical commentary*, ed. Raymond E. Brown, Joseph A. Fitzmyer, and Roland E. Murphy. Englewood Cliffs: Prentice Hall.
Word in Life. 1993. *The word in life study bible, new testament edition*. Nashville: Thomas Nelson Publishers.
United States Conference of Catholic Bishops (USCCB). 2018. *Ethical and religious directives for catholic health care services*. 6th ed. Washington, DC: USCCB.
Vermes, Geza. 2004. *The authentic gospel of Jesus*. London: Penguin Books.
Welch, John W. 2007. *The good samaritan: Forgotten symbols*. Salt Lake City: Liahona.

# Chapter 3
# Catholic Moral Teaching and Tradition on Palliative Care and Hospice

**Peter J. Cataldo**

In his *On the Christian Meaning of Human Suffering* (*Salvifici Doloris*), St. John Paul II provides an eloquent and profound reflection on how, in Christ Jesus, human suffering has been "linked to love," a "love which creates good" (1984, n. 18). Indeed, the whole theme of his Apostolic Letter may be read as a reflection on how human suffering has been conquered by love. That theme also describes Catholic moral teaching and tradition on palliative care and hospice. This chapter will show how Catholic moral teaching and tradition provide substantive answers to the many challenges associated with care for serious chronic illness and care at the end of life with a message of love, compassion, and respect for human dignity parallel to what is found in palliative care.

The chapter has three objectives: first, to explain the central principles and historical foundation of Catholic moral tradition and teaching regarding care at the end of life; second, to explain Catholic teaching on palliative care and hospice; and third, to show some important parallels between palliative care and Catholic moral teaching and tradition. The work of palliative care is critically important for those with serious progressive illness or who are at the end of life. Catholic moral teaching and tradition as they relate to palliative care and hospice have a vital, positive role to play for both the care that individuals receive and for society at large.

P. J. Cataldo (✉)
Providence St. Joseph Health, Renton, WA, USA
e-mail: peter.cataldo@providence.org



P. J. Cataldo, D. O'Brien (eds.), *Palliative Care and Catholic Health Care*, Philosophy and Medicine 130, https://doi.org/10.1007/978-3-030-05005-4_3

## 3.1 Catholic Teaching and Tradition on End-of-Life Care

Catholic moral teaching and tradition have identified three touchstones for evaluating our moral obligations regarding life-sustaining treatment and care. The first two entail avoiding the opposite extremes of euthanasia/physician-assisted suicide on the one hand, and the burdensome prolongation of life on the other. The third marker is the pursuit of the mean between these extremes in caring for life.

As early as St. Cyprian in the third century and St. Basil the Great in the fourth century, we find the concept that there is a limit to the particular means that an individual is morally expected to use for conserving life. During the time of the plague in Carthage, St. Cyprian wrote: "How absurd it is and how perverse that, while we ask that the will of God be done, when God calls us and summons us from this world, we do not at once obey the command of His will. We struggle in opposition and resist [a death that is inevitable]" (Amundsen 1996, p. 76). Saint Basil, who had medical knowledge and was known for tending the sick wrote, "Whatever requires an undue amount of thought or trouble or involves a large expenditure of effort and causes our whole life to revolve, as it were, around solicitude for the flesh must be avoided by Christians" (Amundsen 1996, p. 76). In a similar fashion, St. Thomas Aquinas (1225–1274) wrote that "every man has it instilled in him by nature to love his own life and whatever is directed thereto; and to do so in due measure, that is, to love these things not as placing his end therein, but as things to be used for the sake of his last end" (1947, II–II, q. 126, a. 1). The fact that medicine during the times of Saints Cyprian, Basil, and Aquinas was primitive compared with modern medicine does not negate the legitimacy of their views, which imply that the ethical obligation to care for life and the means used to fulfill that obligation are ethically distinct. Even though the obligation to care for life applies equally to all, this fact does not entail that the particular means to fulfill this obligation must be the same in every case, including at the end of life.

Catholic moral tradition and teaching recognize that a positive moral obligation, unlike a negative obligation, can be fulfilled in a variety of ways, given the variability of circumstances and conditions in which positive obligations are carried out in the concrete world.[1] As the accounts of these saints indicate, there are some situations, which are not at all rare, in which the use of all means possible to care for life is not morally required. In particular, for Saint Thomas, the obligation to love one's own life is inextricably linked to a fulfillment of that obligation with means that are in "due measure." If the means are not in due measure, they need not be applied. For Aquinas, "due measure" is an integral component of the universal obligation to care for life.[2] Circumstances that are in due measure entail that the "how," "what,"

[1] The classical distinction is that negative precepts of the natural law are always valid and binding in every case (*lex negativa semper et ad semper*), whereas positive precepts of the natural law are always valid, but with means that are not always applicable in every concrete case (*lex affirmativa obligat semper sed non ad (pro) semper*). See Cathrein (1905, n. 151, p. 130).

[2] For an explanation of the conceptual distinction between the universal obligation to care for life and its ethical distinction from the means used to fulfill the obligation see Cataldo (2004, 1992).

"where," "when," etc. of the means used to fulfill the obligation are proportionate or commensurate with the condition of the life being cared for and with the fact of mortality.[3] This distinction between the obligation to care for life and due measure in the way it is fulfilled is also evident in the work of subsequent theologians.

In particular, the moral theologians of the classical period further refined the Church's understanding of the boundaries of our moral obligation to care for life. Francisco de Vitoria, O.P. (1492–1546) argued that "it is in no way lawful [i.e. morally acceptable] to shorten one's life. But it should be taken into account that . . . it is one thing to shorten life and another thing not to prolong it. Second, it should be noted that although a man is obliged not to shorten his life, he is not however obliged to seek all means, even lawful means, to lengthen it" (1997, p. 103). At a time when food was one of the best options available for the treatment of disease, Thomas Sanchez, S.J. (1550–1610), made similar distinctions regarding the obligation to care for life. He wrote that "one must suppose that it is one thing not to prolong life and it is another to shorten life . . . no one is held to prolong life; indeed neither is he held to conserve it by using the best and most delicate foods" (Cronin 1989, p. 43). In fact, the moralists recognized that in some cases all food may be forgone, as when de Vitoria explained that food is excused ". . . if the depression of spirit is so low and there is present such consternation in the appetitive power that only with the greatest of effort and though by means of a certain torture, can the sick man take food" (Cronin 1989, p. 35). Juan de Lugo, S.J. (1583–1660) argued that while appropriate care is always required, this obligation does not include the undue prolongation of life:

> The 'bonum' [good] of his life is not of such great moment, however, that its conservation must be effected with extraordinary diligence: it is one thing not to neglect and rashly throw it away, to which a man is bound: it is another however, to seek after it and retain it by exquisite means as it is escaping away from him, to which he is not held. (Cronin 1989, p. 53)

In identifying our moral obligation to conserve and care for life as lying between the extremes of suicide and the burdensome prolongation of life, the Catholic moral tradition clearly held that caring for life was not the same thing as the quantitative extension of life. For example, de Vitoria writes that ""someone is not obliged to use every means to preserve his life, but it is enough to use those means which are of themselves ordered and fitting for this. . . . [W]hen someone is sick without any hope of life, granted that some expensive drug could prolong his life a few hours, or even days, he is not obliged to buy it, but it is enough to use common remedies . . ." (Cronin 1989, p. 33). When we consider that medicines used in the seventeenth century had a disproportionate risk of harm, Sanchez argued that a sick person is not morally bound to use "medicines to prolong life even where there would be probable danger of death, such as taking a drug for many years to avoid fevers, etc." (Cronin 1989, p. 33). Even though the circumstances of medicine have changed since the days of the classical moralists, what these authors established is that the

[3] For Aquinas' general account of circumstances in the *Summa Theologica*, see Aquinas (1947, I–II, q. 7).

duty to care for life is not to be equated with the use of all means in all situations. The fact that modern medicine is far advanced over what medicine was during the times of Cyprian, Aquinas, and subsequent theologians, does not negate the non-obligation, established in earlier times, to use all possible means to care for life.

The articulation of this theological tradition regarding the obligation to care for human life continued throughout the eighteenth and nineteenth centuries down to the present time. For example, in the 1950s Marcelino Zalba, S.J., held that "there is no obligation to use extraordinary means or extraordinary diligence" and that "one would not be bound 'to undergo a very dangerous operation or a very troublesome convalescence" (Cronin 1989, p. 75). In commenting on the Catholic moral tradition regarding care for life, Gerald Kelly, S.J., explained that "to preserve one's life is to do good; and the duty of doing good is usually circumscribed by certain limits". "The moralists," he said, "set out to make a prudent estimate of the limits of this duty" (1958, p. 132). In 1957, Pope Pius XII gave an important address to an international congress of anesthesiologists in which he stated the following principle in answer to the ethical question of forgoing or withdrawing a respirator:

> . . . normally one is held to use only ordinary means—according to circumstances of persons, places, times, and culture—that is to say, means that do not involve any grave burden for oneself or another. A more strict obligation would be too burdensome for most men and would render the attainment of the higher, more important good too difficult. Life, health, all temporal activities, are in fact subordinated to spiritual ends. On the other hand, one is not forbidden to take more than the strictly necessary steps to preserve life and health, as long as he does not fail in some more serious duty. (1957)

There are three points that the Holy Father made that reflect the moral tradition and warrant highlighting. First, the moral judgment of the means used to fulfill the obligation to care for life must always be assessed in relation to the particular circumstances of the patient and the place, time, and culture in which the treatment or care is provided. Second, means that are gravely burdensome are not morally required. Third, for the patient who is conscious and capable of pursuing the spiritual ends of life, the means used should not place a burden that prevents the person from pursuing those spiritual ends.[4] The Vatican Congregation for the Doctrine of the Faith has also maintained the distinction found in the Catholic moral tradition between appropriate treatment and care on the one hand and both suicide and the burdensome prolongation of life on the other:

> . . . one cannot impose on anyone the obligation to have recourse to a technique which is already in use but which carries a risk or is burdensome. Such a refusal is not the equivalent of suicide; on the contrary, it should be considered as an acceptance of the human condition, or a wish to avoid the application of a medical procedure disproportionate to the results that can be expected, or a desire not to impose excessive expense on the family or the community. (1980, n. IV).

Two other more recent sources of teaching on the obligation to care for life are St. John Paul's encyclical, *The Gospel of Life*, and the *Catechism of the Catholic Church*. St. John Paul II emphasizes how the ethical obligation to care for life is one

---

[4] For an expanded explanation of this third point, see Cataldo (2004, pp. 535–536).

thing and its fulfillment is another. Fulfillment of the obligation is dependent on the particular circumstances of the patient, as he states:

> Certainly there is a moral obligation to care for oneself and to allow oneself to be cared for, but this duty must take account of concrete circumstances. It needs to be determined whether the means of treatment available are objectively proportionate to the prospects for improvement. To forego extraordinary or disproportionate means is not the equivalent of suicide or euthanasia; it rather expresses acceptance of the human condition in the face of death. (1995, n. 65)

Finally, the *Catechism* makes clear why there is no moral obligation to use all means in all circumstances when attempting to be good stewards of the gift of life:

> Discontinuing medical procedures that are burdensome, dangerous, extraordinary, or disproportionate to the expected outcome can be legitimate; it is the refusal of "over-zealous" treatment. Here one does not will to cause death; one's inability to impede it is merely accepted. (1997, n. 2278)

The essential points that we may learn from Catholic moral teaching and tradition on the obligation to care for life are that first, an adequate understanding of this obligation begins with the acknowledgment that life is a gift from God and that each person is called to be a good steward of God's gift. Second, stewardship of life recognizes that the receiver of the gift of life ultimately does not have absolute control over it, and this means that the exercise of stewardship steers a middle course between the deliberate and direct taking of life in an attempt to end suffering and the use of any and all means for the quantitative extension of life. St. John Paul II insightfully linked this middle way to the recognition of the fact that the human person does not have absolute control over the gift of life: "Both the artificial extension of human life and the hastening of death, although they stem from different principles, conceal the same assumption: the conviction that life and death are realities entrusted to human beings to be disposed of at will. This false vision must be overcome. It must be made clear again that life is a gift to be responsibly led in God's sight" (1998). The opposite extremes of deliberately hastening death and over-zealous treatment are essentially linked by the denial that life is a gift from God and that responsibility for life is a stewardship on behalf of the giver in the face of concrete circumstances. The third essential point to learn from Catholic moral teaching and tradition on treatment and care at the end of life is that the obligation to care for life (as a matter of principle) is not the same thing as its fulfillment in any particular case. There is one universal obligation always to care for life, but many different ways to fulfill it depending upon – as Pope Pius XII put it – circumstances of persons, places, times, and culture. This means that no one type of treatment or care is always obligatory in every case.[5]

[5] The Catholic moral tradition on the obligation to care for life has culminated in what is known as the principle of ethically proportionate and disproportionate means of sustaining life. This principle is designed to give guidance in judging one's stewardship obligations regarding the use of life-sustaining treatment and care. The principle states that a means that has a reasonable hope of benefit for a patient *and* will not likely cause an excessive burden for the patient, her or his family, or the community, is morally obligatory or ethically proportionate. A means that has no reasonable

## 3.2 Catholic Teaching on Palliative Care and Hospice

Given Catholic moral teaching and tradition on the obligation to care for human life, is there a place for palliative care and hospice in that teaching and tradition? Are palliative care and hospice compatible with the Catholic faith, the Catholic vision of health care, and the stewardship of life? To answer these questions, we need to begin with general definitions of palliative care and hospice. The *Dying in America* report describes the interdisciplinary nature of palliative care in this way:

> In addition to palliative medicine specialists, palliative nurses, social workers, and chaplains, team members may include, for example, pharmacists, dietitians or nutritionists, physical therapists, occupational therapists, psychotherapists, speech-language pathologists, and others such as art or music therapists and child life specialists. (IOM 2015, p. 71)[6]

Such an interdisciplinary team can address the spectrum of needs that the patient with serious, progressive chronic disease or the dying patient has. As the Report points out, these needs include: "patients' needs involving medication management, loss of appetite, functional limitations, depression, difficulties in swallowing and communicating, spiritual guidance, and other problems arising, perhaps for the patient's first time, during an advanced stage of illness" (IOM 2015, p. 71). *Dying in America* also highlights the fact that spiritual care parallels "the goals of palliative care, with its attention to body, mind, and spirit, and of patient-centeredness, which encompasses 'compassion, empathy, and responsiveness to the needs, values, and expressed preferences of the individual patient'" (IOM 2015, p. 163).

In order to examine Catholic teaching with respect to palliative care and hospice, clarity is also needed about how palliative care and hospice are related but distinct. Hospice is a subset of palliative care, that is, hospice is palliative care as the sole focus of care in the final stages of illness and life. The National Hospice and Palliative Care Organization describes hospice as providing "expert medical care, pain management, and emotional and spiritual support expressly tailored to the patient's needs and wishes. Support is provided to the patient's loved ones as well. . . . Hospice services are available to patients with any terminal illness or of any age,

hope of benefit for a patient *or* will likely cause an excessive burden for the patient, or an excessive financial burden for the family or community, is called ethically disproportionate; in other words, it is morally optional. It is morally optional up to the point of causing significant harm to the patient, at which point there may be an obligation not to provide it.

[6] The interdisciplinary nature of palliative care also allows for better coordination of care in addition to better quality improvement and pain management: "The interdisciplinary team approach that typically distinguishes palliative care contributes to the development and implementation of comprehensive plans of care, helps ensure coordination of care, enhances the anticipation and remediation of problems that arise during transitions and crises, facilitates quality improvement, and contributes to good pain management" (IOM 2015, p. 71). See IOM (2015, p. 27) for distinctions drawn between the definitions for "basic palliative care," "specialty palliative care," and "palliative care."

religion, or race" (2017, p. 2).[7] Since palliative care may be provided as the sole focus of care, as in the case of hospice, or may be delivered concurrently with disease modifying treatment apart from terminal illness, references to palliative care throughout this chapter are inclusive of both modes of delivering palliative care.

Corresponding to the development of palliative as a medical specialty, recent Catholic teaching has explicitly recognized and encouraged palliative care. The *Catechism of the Catholic Church* affirms the use of palliative care and teaches that "palliative care is a special form of disinterested charity [that is, it is free from self-interest]. As such it should be encouraged" (1997, n. 2279). St. John Paul's understanding of palliative care essentially matches how palliative care is defined by the profession itself. In the *Gospel of Life*, he stated that "in modern medicine, increased attention is being given to what are called 'methods of palliative care,' which seek to make suffering more bearable in the final stages of illness and to ensure that the patient is supported and accompanied in his or her ordeal" (1995, n. 64). Later, in 2004, he expanded his explanation of palliative care with this definition:

> . . . palliative care aims, especially in the case of patients with terminal diseases, at alleviating a vast gamut of symptoms of physical, psychological and mental suffering; hence, it requires the intervention of a team of specialists with medical, psychological and religious qualifications who will work together to support the patient in critical stages. (2004)

It is important to note that in this statement St. John Paul II implicitly recognizes that palliative care is not restricted to patients with terminal conditions but can be appropriate for those with serious chronic diseases.

St. John Paul II so highly regarded palliative care that he stated we are morally required to use it rather than treatment that has no reasonable hope of benefit or is excessively burdensome. "Particularly in the stages of illness when proportionate and effective treatment is no longer possible," he said, "while it is necessary to avoid every kind of persistent or aggressive treatment, methods of 'palliative care' are required" (2004). Indeed, Pope Benedict XVI went so far as to describe palliative care as a "right":

> There is a need to promote policies which create conditions where human beings can bear even incurable illnesses and death in a dignified manner. Here it is necessary to stress once again the need for more palliative care centres which provide integral care, offering the sick the human assistance and spiritual accompaniment they need. This is a right belonging to every human being, one which we must all be committed to defend. (2006)

---

[7] Hospice also has an interdisciplinary approach: "The hospice team develops a care plan that meets each patient's individual needs for pain management and symptom control. This interdisciplinary team . . . usually consists of the patient's personal physician, hospice physician or medical director, nurses, hospice aides, social workers, bereavement counselors, clergy or other spiritual counselors, trained volunteers, and speech, physical, and occupational therapists, if needed" (National Hospice and Palliative Care Organization 2017, p. 2). See also IOM (2015, p. 60): "For people with a terminal illness or at high risk of dying in the near future, hospice is a comprehensive, socially supportive, pain-reducing, and comforting alternative to technologically elaborate, medically centered interventions. It therefore has many features in common with palliative care . . . ."

Pope Francis has continued papal support and encouragement of palliative care on several points. In a speech to the Pontifical Academy for Life, he gave a strong endorsement of palliative care:

> Palliative care is an expression of the truly human attitude of taking care of one another, especially of those who suffer. It is a testimony that the human person is always precious, even if marked by illness and old age. Indeed, the person, under any circumstances, is an asset to him/herself and to others and is loved by God. This is why, when their life becomes very fragile and the end of their earthly existence approaches, we feel the responsibility to assist and accompany them in the best way. (2015)

Elsewhere in this speech, Pope Francis recognizes that palliative care is not restricted to the dying; "today," he notes, "there is a great variety of diseases characterized by chronic progressive deterioration, often linked to old age, which can benefit from this type of assistance." He also accurately identifies the essence of palliative care when he stated that even though it is not designed to save life, "palliative care accomplishes something equally important: it values the person" (2015). Finally, Pope Francis shows his support for efforts to make palliative care more accessible and for individuals to specialize in this field: "Thus I appreciate your scientific and cultural commitment to ensuring that palliative care may reach all those who need it. I encourage professionals and students to specialize in this type of assistance . . ." (2015).

In addition to explicitly approving and encouraging palliative care, these samples of contemporary Catholic teaching on palliative care indicate the same moral premise regarding life-sustaining treatment that has existed in the Church's moral tradition for centuries. This is the moral truth that the obligation to care for life is distinguishable from, yet inextricably tied to, the concrete circumstances in which it is fulfilled. This fact means that palliative care is often the best means possible to fulfill our duty to care for those with serious, progressive illness or for those with terminal disease. The Church's teaching and tradition on the moral obligation to care for life can be demonstrated through curative treatment that is provided along with palliative care, but also through the use of palliative care alone. In either case, palliative care exemplifies care for life that achieves the moral mean between the extremes of deliberately hastening death and overzealous treatment, as articulated in Catholic moral teaching and tradition. To see just how closely palliative care and hospice reflect Catholic moral teaching and tradition, it is helpful to compare the major domains of palliative care with Catholic moral and social teaching and tradition, not only with respect to the obligation to care for life, but also with respect to other aspects of the dignity of the human person and the family.

## 3.3 Parallels Between the Eight Domains of Palliative Care and Catholic Moral Teaching and Tradition

There are eight domains of palliative care that are widely recognized as constituting its goals.[8] All of the domains of palliative care presume that the patient cannot be reduced to his or her disease and that the patient has a holistic dignity that must be respected. "To a person who is living with an incurable illness," Dr. Ira Byock writes, "life is more than a set of medical problems to be treated. The fundamental nature of illness is not medical; it is personal" (2012, p. 25). Thus, the first domain, on the structure and processes of palliative care, focuses on the need for teams that are well trained and are able to address the physical, psychological, social and spiritual aspects of the patient in an interdisciplinary manner, and in a way that respects the values and preferences of the patient and the patient's family. This approach is most consistent with the understanding – articulated in Catholic teaching – that the human person is a body-soul unity: "The unity of soul and body is so profound that one has to consider the soul to be the 'form' of the body: i.e., it is because of its spiritual soul that the body made of matter becomes a living, human body; spirit and matter, in man, are not two natures united, but rather their union forms a single nature" (Catechism 1997, n. 365).

The ontological unity of the human being is also summarized well in *Donum Vitae*: "By virtue of its substantial union with a spiritual soul, the human body cannot be considered as a mere complex of tissues, organs and functions, nor can it be evaluated in the same way as the body of animals; rather it is a constitutive part of the person who manifests and expresses himself through it" (Congregation for the Doctrine of the Faith 1987, I, n. 3). In order to highlight the ramifications of this view of the unity of the human individual for personal dignity, *Donum Vitae* quotes St. John Paul II: "Each human person, in his absolutely unique singularity, is constituted not only by his spirit, but by his body as well. Thus, in the body and through the body, one touches the person himself in his concrete reality. To respect the dignity of man consequently amounts to safeguarding this identity of the man '*corpore et anima unus'*, as the Second Vatican Council says (*Gaudium et Spes*, 14, par.1)" (Congregation for the Doctrine of the Faith 1987, I, n. 3).[9] Given the unity of the human person and its inherent dignity articulated in Catholic teaching, to address bodily disease and physical symptoms in medical care is to treat the person in her or his physical and spiritual dimensions as an individual, unified being. This is a fun-

[8] See NCP (2018); see also IOM (2015, pp. 85–87) for a table of twelve components similar to the eight domains and an explanation.

[9] Pope Benedict XVI summarized the teaching in his encyclical, *Charity in Truth* (2009, n. 76): ". . . the human person is a 'unity of body and soul', born of God's creative love and destined for eternal life."

damental agreement between palliative care and the Catholic view of the human person.

The second domain of palliative care addresses the physical aspects of care. Expert assessment and management of symptoms, including pain, is central to the compassionate care provided in palliative medicine. Extra attention is given to tailoring treatment plans to the particular needs and circumstances of the patient. This includes ensuring patient and family understanding of the disease, its symptoms, side effects, functional impairment, and treatments. Here there is a consistency between the Catholic focus on fulfilling our obligations to care based on the particular circumstances of the patient and the approach of palliative care. There is also a complete compatibility between Catholic teaching on pain management and the treatment of pain in palliative care. In order to appreciate this compatibility, it is important to clear away the misconceptions about Catholic teaching on the use of painkillers.

The Catholic moral tradition long ago affirmed the moral acceptability of an individual to take measures to prevent pain, even if that meant forgoing painful treatment or surgery altogether during times when effective analgesics and anesthesia were not available. For example, Dominic Soto, O.P. (1494–1560) wrote that ". . . no one can be forced to bear the tremendous pain in the amputation of a member or in an incision into the body: because no one is held to preserve his life with such torture" (Cronin 1989, p. 38). Similarly, Dominic Banez, O.P. (1528–1604) wrote, ". . . although a man is held to conserve his own life, he is not bound . . . to a certain extraordinary and horrible pain . . . " (Cronin 1989, p. 42).

In modern times, beginning with Pope Pius XII, and continuing with the teaching of Saint John Paul and the *Catechism*, Catholic teaching on this subject has been quite explicit. St. John Paul addressed the morality of "using various types of painkillers and sedatives for relieving the patient's pain when this involves the risk of shortening life" (1995, n. 65). He wrote that

> [w]hile praise may be due to the person who voluntarily accepts suffering by forgoing treatment with pain-killers in order to remain fully lucid and, if a believer, to share consciously in the Lord's Passion, such 'heroic' behaviour cannot be considered the duty of everyone. Pius XII affirmed that it is licit to relieve pain by narcotics, even when the result is decreased consciousness and a shortening of life, 'if no other means exist, and if, in the given circumstances, this does not prevent the carrying out of other religious and moral duties.' (1995, n. 65)

The *Catechism* echoes this teaching when it states that the "use of painkillers to alleviate the sufferings of the dying, even at the risk of shortening their days, can be morally in conformity with human dignity if death is not willed as either an end or a means, but only foreseen and tolerated as inevitable" (1997, n. 2279).[10]

[10] Morally, this theoretical shortening of life is justified because the dose is therapeutic (i.e., it directly treats the pain and does not have death as its direct effect), the therapeutic effect is what is intended, the bad effects of administering pain medications (including a possible shortened life) do not cause the good effect of pain relief, and because the good effect is proportionate to the bad effects. Moreover, from a spiritual point of view, the use of pain medication may actually assist a

It is important to know the Catholic moral justification for the use of pain medication with respect to the question of the shortening of life as a theoretical possibility. However, the reality is that the clinical need for this justification is becoming rare, if non-existent, in modern clinical settings. Many studies have shown that the use of opioids and sedatives do not affect the survival of patients receiving palliative care and hospice at the end of life. In fact, for some patients, the use of morphine equivalent drugs for pain actually increases their survival time. In one study, the mean survival was 29 days longer for hospice patients compared with non-hospice patients (Connor et al. 2007).[11] In general, studies show that the use of sedatives for intractable pain or restlessness in the last hours of life is directed at relieving symptoms, not unconsciousness, and there is no evidence that these drugs precipitate death (Sykes and Thorns 2003).

Palliative care teams also address the psychological and psychiatric aspects of care using the "best available evidence to maximize patient and family coping and quality of life" (NCP 2013). This is the third domain of palliative care. Palliative care provides continuing psychological and psychiatric assessment and intervention for patients that addresses various syndromes such as anxiety, depression, delirium, hopelessness, and suicidal ideation, and includes patient and family education about the disease and coping strategies. This domain also includes grief and bereavement services for patients, family, and support staff that are respectful of cultural and religious preferences. In a particular way, this domain of palliative care affirms the holistic view of the human person espoused by the Catholic faith. The psychological dimension of the person is integral to who the patient is as a human person and as a result, palliative care addresses psychological suffering with a focus equal to treatment of physical symptoms. This recognition of the centrality of caring for the psychological well-being of the patient in palliative care is paralleled in the Catholic view of the human person, as can be seen in this statement from St. John Paul II:

> In fact, illness and suffering are not experiences which concern only man's physical substance, but man in his entirety and in his somatic-spiritual unity. For that matter, it is known how often the illness which is manifested in the body has its origins and its true cause in the recesses of the human psyche. (1985)

Months before he died, St. John Paul II, reflecting on the term "health," wrote that "the word is intended to refer to all the dimensions of the person, in their harmony and reciprocal unity: the *physical, the psychological,* and *the spiritual and moral* dimensions" (2005). Attention to the psychological and emotional suffering of patients in palliative care has also proven to be critically important for responding to requests from patients to hasten death or for physician-assisted suicide. A

---

patient's spiritual journey by helping the patient to sustain lucidity as opposed to writhing in pain. Pain for the sake of pain is not required to experience the redemptive meaning of suffering, but rather having a level of lucidity in order to enter into the spiritual journey. As Ira Byock has stated (2012, p. 160): "when pain is carefully treated and closely monitored, people's minds usually remain clear."

[11] See also, Good et al. (2005), Bercovitch et al. (1999), Morita (2001), and Sykes and Thorns (2003).

significant percentage of these patients make such requests because they are either depressed or suffer from hopelessness (Breitbart et al. 2000; Jones et al. 2003; Maytal and Stern 2006; Wilson et al. 2007; Rodin et al. 2007, 2009; Shim and Hahm 2011). Taking such needs of dying patients seriously is both the goal of palliative care and is an approach that is fully consistent with the Catholic view of health.

The fourth domain of palliative care pertains to the social aspects of care. In this domain, the interdisciplinary palliative care team assesses and addresses patient-family needs, promotes patient-family goals, and works to maximize patient-family strengths and well-being (NCP 2018, p. 26). It also involves a comprehensive, person-centered assessment of factors such as the family structure and function, the patient's and family's perception about caregiving, and access to community resources. This family-centered and community-centered approach to care is fully compatible with Catholic teaching on the nature of the human person as a social being. Catholic social teaching affirms that the "human person is essentially a social being because God, who created humanity, willed it so. . . . God did not create man as a 'solitary being' but wished him to be a 'social being'. Social life therefore is not exterior to man: he can only grow and realize his vocation in relation with others" (Pontifical Council for Peace and Justice 2004). The social nature of the human person is fundamentally developed within the family, and the family is the context in which love as the gift of self is nourished and fostered. According to Catholic teaching, "*the family is present as the place where communion . . . is brought about. It is the place where an authentic community of persons develops and grows, thanks to the endless dynamism of love . . .*" (Pontifical Council for Justice and Peace 2004). Thus, the high priority in palliative care given to social relations affecting both patient care and the patient's family goes to the heart of Catholic teaching on who the patient is as a human person.

Palliative care also places high importance on the spiritual, religious, and personal dimension of the patient, and recognizes that this is a fundamental component of compassionate, patient and family centered care. This aspect of care forms the fifth domain of palliative care. This domain includes a spiritual assessment that identifies religious or spiritual background, preferences, beliefs, rituals, and practices of the patient and family. It seeks to respond to an individual's desire for "meaning, purpose, and trancendence" (NCP 2018, p. 32). The interdisciplinary team includes professional chaplains—board certified when possible—and facilitates requests for clergy. Palliative care also encourages the spiritual self-care and self-reflection of its staff.

This fifth domain of palliative care finds a parallel in Catholic teaching on the spiritual dimension of physical suffering. The words of St. John Paul II on this subject show this parallel very well:

> The concept of health that we find in Christian thought is quite the opposite of the vision that reduces it to a purely psycho-physical balance. Such a vision of health disregards the spiritual dimensions of the human person and would end by harming his true good. For the believer . . . health "strives to achieve a fuller harmony and healthy balance on the physical, psychological, spiritual and social level". This is the teaching and witness of Jesus, who was so sensitive to human suffering. With his help, we too must endeavor to be close to

> people today, to treat them and, cure them, if possible, without forgetting the requirements of the spirit. (2002, n.2)

Christianity is a religion of healing. At the core of the Catholic faith is the healing ministry of Jesus, who heals the sick in both spirit and body. Palliative care is fully receptive to the spiritual component of sickness and healing found in the Catholic faith.

The sixth domain of palliative care encompasses the cultural aspects of care. This domain seeks to ensure that palliative care teams deliver "care that respects patient and family cultural beliefs, values, traditional practices, language, and communication preferences . . ." (NCP 2018, p. 38). Attention to cultural sensitivity in palliative care includes the identification of cultural strengths, concerns, and needs for patients and families and makes respect for culture a high priority. Among other things, this means communicating with patients and families in a language and manner that is appropriate and sensitive to their cultural experience. The Catholic Church affirms the critical place of culture in society in general and the role of cultural considerations in the rights and life of minorities in particular: "The *Magisterium affirms that minorities constitute groups with precise rights and duties, most of all, the right to exist*. . . . Moreover, minorities have the right to maintain their culture, including their language, and to maintain their religious beliefs, including worship services" (Pontifical Council for Justice and Peace 2004). The cultural aspects of palliative care are also in agreement with Catholic teaching on human dignity and justice. The Second Vatican Council taught that man's dignity, as created in the image of God, is rooted in the powers of intellect and free will: ". . . man's dignity demands that he act according to a knowing and free choice that is personally motivated and prompted from within, not under blind internal impulse nor by mere external pressure" (Second Vatican Council 1965). Providing care that respects culture enables patients and families to make better informed choices in accordance with their human dignity as knowing and free persons.[12]

The seventh and eighth domains of palliative care encompass care of the patient at the end of life and the ethical and legal aspects of care respectively (NCP 2018, pp. 45–59). Palliative care at the end of life seeks to be attentive to all of a patient's needs both physical and spiritual and it proactively addresses the patient's and family's concerns, fears, and hopes during these phases.[13] The ethical and legal aspects of palliative care include understanding and respecting patient and surrogate goals, preferences, and choices, educating patient and family about advance care planning, and following a patient's advance directive documents. The ethics of palliative care also involves the prevention, identification, and resolution of ethical dilemmas or concerns, such as the forgoing or withdrawal of life-sustaining treatment and care,

---

[12] We are also obligated in justice to give what is due to God and neighbor. This moral obligation includes treating others according to their human dignity, which includes appropriate respect for their cultural beliefs.

[13] As Ira Byock (2012, p. 44) has put it: "The time it takes palliative clinicians to effectively listen, convey information, and respond to questions and concerns is not time taken from our medical practice; it *is* the practice."

or the use of pain medication and sedatives. Here, the philosophy of palliative care is consistent with the Catholic moral principle of ethically proportionate and disproportionate means of sustaining life explained earlier. Moreover, no matter what the proponents of so-called "aid in dying" or "the right to die" might argue or wish for, euthanasia and physician assisted suicide are not part of the intrinsic nature of palliative care, nor are they appropriate options among others in the continuum of care at the end of life.

The essential reason why physician-assisted suicide and euthanasia are incompatible with palliative care is that palliative care targets the suffering of the patient, not the very life of the patient. This truth about palliative care is fully consistent with the story of the Good Samaritan in Saint Luke's gospel, which is the model of palliative care for Catholic health care. The Good Samaritan did not treat the robbery victim left half-dead along the road by taking his life; rather, he supported the life of this ill person and responded to his suffering. The fact is that euthanasia and physician-assisted suicide are not truly designed to prevent or end suffering. In order for suffering to be prevented or ended, there must be a person who is the subject of the action and who survives the transition from suffering to relief of suffering as the substrate of this change. In euthanasia and physician-assisted suicide, there is no subject who is the recipient of this change, because these actions are designed to eliminate the subject. (Cataldo 2008, pp. 151–152; New Catholic Encyclopedia, Supplement 2012–2013) While there may be similarities between palliative care and physician-assisted suicide with respect to some of the circumstances in each case and with respect the ultimate subjective intensions of the health professionals, such similarities do not make the actions morally equivalent. The circumstances may be similar insofar as in both cases there is a terminal or serious chronic disease, pain and suffering, and critical physician involvement. Similar also is the ultimate or remote intention of the physicians, which is to relieve suffering. Yet, despite these similarities, the actions in both situations remain radically different because the immediate or proximate intentions and the nature of the actions are radically different. In the case of physician-assisted suicide, the immediate or proximate intention of the physician is to provide a dose of a drug that is designed to cause death, and the nature of the act of ingesting this drug is lethal by design. In the case of withdrawing or forgoing life-sustaining treatment in palliative care, or in the case of administering pain medication in palliative care, the immediate intention is only to provide that care which has a reasonable hope of benefit or does not cause an excessive burden on the patient. Moreover, the nature of such acts is therapeutic. They are not designed to cause death but to directly treat symptoms of a life-impinging illness. Death is foreseen as an outcome of the disease process, but is not willed either as an end or means in itself. The essential difference between physician-assisted suicide and palliative care is aptly summarized by Leon Kass and Ira Byock. Kass made the simple but undeniable point that there is "no benefit without a beneficiary" (Kass 2002, p. 34), and Byock has echoed this truth by pointing out that "*alleviating suffering* and *eliminating the sufferer* are very different acts" (2014, p. 164).

## 3.4 The Need for Palliative Care and Hospice

Part of the ethical justification for palliative care and hospice is that they are successful. That is, they have been proven to provide what patients need who have chronic disease, serious progressive illnesses, or who are dying. In the words of Ira Byock, they provide "the best care possible" for these patients. For example, Jennifer Temel's significant study in *The New England Journal of Medicine* compared the early integration of palliative care in standard cancer therapy for metastatic non-small-cell lung cancer with patients who received standard cancer therapy alone (2010).[14] As compared to the group that received standard cancer therapy alone, the patients who received palliative care had a greater survival by 2 months, better quality of life and mood, and better documentation of preferences regarding resuscitation. I have already cited other similar studies above. There may be several reasons for these outcomes. Dr. Byock provides a very good description of what happens:

> . . . we help a lot of people with advanced cancer to tolerate treatments that are effective against their tumors but have difficult side effects. In alleviating their symptoms and optimizing their ability to eat and drink, be active and rest, people with cancer are able to stay in the fight longer. If a time comes when chemotherapy and radiation prove more toxic than therapeutic, we can still help people live as well and as long as possible. Patients with late-stage cancers commonly find that when they are finally free of the side effects of treatment, they feel better and stronger. In *living with,* rather than relentlessly fighting their cancer, they ultimately live longer. (2014, p. 99)

In order for patients and their families to ensure that they receive this type of care, there must be change on the policy level and change with respect to how patients, families, and health professionals communicate about end of life decision-making. This latter this problem deserves some consideration.

There are many recognized challenges to discussion about end of life matters (Adams et al. 2011; Barclay et al. 2011; Hwang et al. 2003). There is a difficulty determining the best time to initiate the conversation due to an uncertainty about the prognosis, especially if a patient is hospitalized. There are fears about removing all hope and the emotional challenge of stopping curative treatments. Many health care professionals are also not adequately trained to conduct conversations about end of life decision-making. Other problems include time pressures, health professionals' perceptions of medical failure, fear of upsetting patients, and general discomfort in discussing death and dying.

Palliative care and hospice strive to prevent such communication barriers and in this regard they serve as a model for all of health care. Studies have shown that good communication among patients, their families, and health professionals has many beneficial results, such as greater understanding of illness, the reduction of anxiety and depression, building of trust, greater understanding of the patient's values, pref-

---

[14] See also Greer et al. (2013).

erences and needs, and more involvement in treatment decisions (Lown et al. 2011; Levinson et al. 2010; Wright 2008).

A patient will not reap all of the benefits of palliative care or hospice unless there is good communication among patient, family, and health professionals that begins early. Studies have shown that only 18–36% of the general American population has completed an advance health care directive document (HHS 2008). However, among elderly Americans near the end of life, a recent study found that 67% had an advance directive (Silveira et al. 2010, p. 1217). We all have a reluctance to discuss and complete advance health care directives, but as palliative care professionals point out, "it's always too soon until it's too late" to address these matters (Byock 2014, p. 172). Ideally, this communication should take place at a time when a patient is in relative good health and before the patient lacks capacity to make health care decisions. It should begin with a conversation or a series of conversations between patient and physician about the patient's condition, prognosis, and options and goals of care relative to the patient's health. The patient should make known his or her values, religious beliefs, and preferences that could affect health care decision-making. Equipped with the physician's information about the patient's condition and options, the patient should also discuss the matter with family or close friends. This is the beginning of a process known as advance care planning, which may culminate in the completion of an advance health care directive document that preferably names an individual who will act as the patient's health care agent. This person should make health care decisions *as the patient would make them* based on the patient's stated values and preferences.

Advance directives that name a health care agent can be very helpful because they are better able to ensure that treatment decisions for an incapacitated patient will be made with the free and informed consent of the patient consistent with human dignity. However, setting up an advance directive and making decisions in accordance with it presuppose a reciprocal relationship between patient and health professional. This relationship is acknowledged in the *Ethical and Religious Directives for Catholic Health Care Services*:

> A person in need of health care and the professional health care provider who accepts that person as a patient enter into a relationship that requires, among other things, mutual respect, trust, honesty, and appropriate confidentiality. . . . Neither the health care professional nor the patient acts independently of the other; both participate in the healing process. (United States Conference of Catholic Bishops 2018, Part Three, Introduction)

The advance planning process should emerge from this reciprocal relationship. In addition to ensuring that an incapacitated person near the end of life will receive the best health care possible, advance directives are also about love for others. Having an advance directive can be a great help to family and others if they are required to make critical decisions when their loved one cannot make them (Byock 2012, p. 172).

## 3.5 Suffering Conquered by Love

Loving care is the foundation and infrastructure of palliative care and hospice. Love is the point from which and toward which this care is provided. The eight domains of palliative care concretely represent love for the patient both as a human person and as this individual with personal needs and a story. The loving care provided in palliative care and hospice includes treating the symptoms of disease with competence, thoroughness, and continuity. However, as Byock has pointed out, it also goes the extra mile to love the patient in little ways, whether this might be a volunteer with a patient laughing over a newspaper story, or a nurse gently humming while she baths a patient, or assistance given to a family to celebrate the life of their loved one (2014, pp. 284 and 287).

Palliative care's reason for being is to alleviate suffering with loving care. This mission is fully compatible with the essential connection that is drawn between suffering and love in the Christian faith. This relationship between suffering and love is recognized even from a non-religious perspective of palliative care. For example, Byock comments that for patients who are helpless and hopeless, "love is the answer . . . loving care opens up a full range of possibilities that are not seen through the problem-based filters of medicine" (2014, p. 284). These opportunities for love are in direct response to suffering. This aspect of palliative care and hospice is compatible with the Christian meaning of suffering explained by St. John Paul II:

> Human suffering has reached its culmination in the Passion of Christ. And at the same time it has entered into a completely new dimension and a new order: it has been linked to love, to that love of which Christ spoke to Nicodemus, to that love which creates good, drawing it out by means of suffering . . . . In the messianic programme of Christ, which is at the same time the programme of the *Kingdom of God*, suffering is present in the world in order to release love, in order to give birth to works of love towards neighbour, in order to transform the whole of human civilization into a "civilization of love." (1984, ns. 18 and 30)

Catholic moral teaching tradition represents a rich and robust font for stewardship at the end of life, which steers a middle course between euthanasia and physician-assisted suicide on the one hand, and over-zealous treatment on the other. Evident also from Catholic teaching and tradition is that palliative care represents a concrete, loving way in which this stewardship may be realized (whether concurrent with curative treatment or at the end of life), and is a form of care that is fully compatible with the Catholic vision of human dignity.

## References


Adams, Judith A., Donald E. Bailey Jr., Ruth A. Anderson, and Sharron L. Docherty. 2011. Nursing roles and strategies in end-of-life decision making in acute care: A systematic review of the literature. *Nursing Research & Practice*. https://doi.org/10.1155/2011/527834.

Amundsen, Darrel W. 1996. *Medicine, society, and faith in the ancient and medieval worlds*. Baltimore: Johns Hopkins University Press.

Aquinas, Thomas. 1947. *Summa theologica*. Trans. Fathers of the English Dominican Province. New York: Benziger Bros. http://dhspriory.org/thomas/summa/index.html. Accessed 16 Dec 2015.

Barclay, Stephen, et al. 2011. End of life care conversations with heart failure patients: A systematic literature review and narrative synthesis. *British Journal of General Practice* 61 (582): e49–e62.

Bercovitch, Michaela, Alexander Waller, and Abraham Adunsky. 1999. High dose morphine use in the hospice setting: A database survey of patient characteristics and effect on life expectancy. *Cancer* 86 (5): 871–877.

Breitbart, W., Barry Rosenfeld, Hayley Pessin, Monique Kaim, Julie Funesti-Esch, Michele Galietta, Christian J. Nelson, and Robert Brescia. 2000. Depression, hopelessness, and desire for hastened death in terminally ill patients with cancer. *Journal of the American Medical Association* 284: 2907–2911.

Byock, Ira. 2012. *The best care possible: A physician's quest to transform care through the end of life*. New York: Avery.

Cataldo, Peter J. 1992. To conserve human life. *Ethics & medics: A commentary of the national catholic bioethics center on health care and the life sciences* 17 (12): 2–4.

———. 2004. Pope John Paul II on nutrition and hydration: A change of catholic teaching? *The National Catholic Bioethics Quarterly* 4: 513–536.

———. 2008. The ethics of Pope John Paul's allocution on care of the pvs patient: A response to J.L.A. Garcia. In *Artificial nutrition and hydration: The new catholic debate*, ed. Christopher Tollefsen, 141–161. New York: Springer.

Catechism of the Catholic Church. 1997. 2nd ed. Vatican. Libreria Editrice Vaticana.

Cathrein, Victore S.J. 1905. *Moralis in usum scholarum*. Freiburg: Herder. https://ia801408.us.archive.org/0/items/philosophiamoral00cath/philosophiamoral00cath.pdf. Accessed 17 Dec 2015.

Congregation for the Doctrine of the Faith. 1987. *Instruction on respect for human life in its origin and on the dignity of procreation*. http://www.vatican.va/roman_curia/congregations/cfaith/documents/rc_con_cfaith_doc_19870222_respect-for-human-life_en.html. Accessed 17 Dec 2015.

Connor, Stephen R., Bruce Pyenson, Kathryn Fitch, Carol Spence, and Kosuke Iwasaki. 2007. Comparing hospice and nonhospice patient survival among patients who die within a three-year window. *Journal of Pain and Symptom Management* 33 (3): 238–246.

Cronin, Daniel A. 1989. Conserving human life. In *Conserving human life*, ed. Russell E. Smith, 3–145. Braintree: Pope John XXIII Medical-Moral Research and Education Center.

Good, P.D., P.J. Ravenscroft, and J. Cavenagh. 2005. Effects of opioids and sedatives on survival in an Australian inpatient palliative care population. *Internal Medicine Journal* 35: 512–517.

Greer, Joseph A., Vicki A. Jackson, Diane E. Meier, and Jennifer S. Temel. 2013. Early integration of palliative care services with standard oncology care for patients with advanced cancer. *CA: A Cancer Journal for Clinicians* 63: 349–363.

Hwang, Shirley S., Victor T. Chang, Janet Cogswell, and Shanthi Srinivas. 2003. Knowledge and attitudes toward end-of-life care in veterans with symptomatic metastatic cancer. *Palliative & Supportive Care* 1: 221–230.

Institute of Medicine of the National Academies (IOM). 2015. *Dying in America: Improving quality and honoring individual preferences near the end of life*. Washington, DC: The National Academies Press.

John Paul II. 1984. *The Christian meaning of human suffering*. http://w2.vatican.va/content/john-paul-ii/en/apost_letters/1984/documents/hf_jp-ii_apl_11021984_salvifici-doloris.html. Accessed 16 Dec 2015.

———. 1985. Apostolic letter, "motu proprio", *Dolentium Hominum*, Establishing Pontifical Commission for the Apostolate of Health Care Workers. http://w2.vatican.va/content/john-paul-ii/en/motu_proprio/documents/hf_jp-ii_motu-proprio_11021985_dolentium-hominum.html. Accessed 15 Dec 2015.

———. 1995. *The gospel of life*. http://w2.vatican.va/content/john-paul-ii/en/encyclicals/documents/hf_jp-ii_enc_25031995_evangelium-vitae.html. Accessed 17 Dec 2015.

———. 1998. Address at the Rennweg hospice in Vienna. http://w2.vatican.va/content/john-paul-ii/en/speeches/1998/june/documents/hf_jp-ii_spe_19980621_austria-infermi.html. Accessed 15 June 2015.

———. 2002. Address to the world organization of gastro-enterology. http://w2.vatican.va/content/john-paul-ii/en/speeches/2002/march/documents/hf_jp-ii_spe_20020323_congr-gastro-enterologia.html. Accessed 10 Nov 2015.

———. 2004. Address to the participants in the 19th international conference of the pontifical council for health pastoral care. http://w2.vatican.va/content/john-paul-ii/en/speeches/2004/november/documents/hf_jp-ii_spe_20041112_pc-hlthwork.html. Accessed 10 Nov 2015.

———. 2005. Letter to the president of the pontifical academy for life on the occasion of a study congress on 'quality of life and ethics of health. https://w2.vatican.va/content/john-paul-ii/en/letters/2005/documents/hf_jp-ii_let_20050219_pont-acad-life.html. Accessed 10 Dec 2015.

Jones, Jennifer M., Mary Anne Huggins, Anne C. Rydall, and Gary M. Rodin. 2003. Symptomatic distress, hopelessness, and the desire for hastened death in hospitalized cancer patients. *Journal of Psychosomatic Research* 55: 411–418.

Kass, Leon R. 2002. "I will give no deadly drug": Why doctors must not kill. In *The case against assisted suicide for the right to end-of-life care*, ed. M.D. Kathleen Foley and Herbert Hendin, 17–40. Baltimore: The Johns Hopkins University Press.

Kelly Gerald, S.J. 1958. *Medico-moral problems*. St. Louis: The Catholic Hospital Association of the United States and Canada.

Levinson, Wendy, Cara S. Lesser, and Ronald M. Epstein. 2010. Developing physician communication skills for patient-centered care. *Health Affairs* 29 (7): 1310–1318.

Lown, Beth A., Julie Rosen, and John Martilla. 2011. An agenda for improving compassionate care: A survey shows about half of patients say such care is missing. *Health Affairs* 30 (9): 1772–1778.

Maytal, G., and T.A. Stern. 2006. The desire for death in the setting of terminal illness: A case discussion. *Primary Care Companion Journal of Clinical Psychiatry* 8 (5): 299–305.

Morita, Tatsuya. 2001. Effects of high dose opioids and sedatives on survival in terminally ill cancer patients. *Journal of Pain and Symptom Management* 21 (43): 282–289.

National Consensus Project for Quality Palliative Care (NCP). 2018. *Clinical Practice Guidelines for Quality Palliative Care*. 4th ed. Richmond, VA: National Coalition for Hospice and Palliative Care.

National Hospice and Palliative Care Organization. 2017. *NHPCO's facts and figures hospice care in America*. Alexandria: National Hospice and Palliative Care Organization.

New Catholic Encyclopedia, Supplement. 2012–2013. *Ethics and philosophy*, vol. 2, s.v., "euthanasia."

Pontifical Council for Justice and Peace. 2004. *Compendium of the social doctrine of the church*, Libreria Editrice Vaticana. Washington, DC: United States Conference of Catholic Bishops.

Pope Benedict XVI. 2009. *Charity in truth*. http://w2.vatican.va/content/benedict-xvi/en/encyclicals/documents/hf_ben-xvi_enc_20090629_caritas-in-veritate.html#_edn156. Accessed 17 Dec 2015.

Pope Francis. 2015. Address of his holiness Pope Francis to participants in the plenary of the pontifical academy for life. https://w2.vatican.va/content/francesco/en/speeches/2015/march/documents/papa-francesco_20150305_pontificia-accademia-vita.html. Accessed 15 Sept 2015.

Pope Pius XII. 1957. Address to an international group of physicians (February 24, 1957).
Rodin, Gary, Camilla Zimmermann, Anne Rydall, Jennifer Jones, Frances A. Shepherd, Malcolm Moore, Martin Fruh, Allan Donner, and Lucia Gagliese. 2007. The desire for hastened death in patients with metastatic cancer. *Journal of Pain and Symptom Management* 33: 661–675.
Rodin, G., C. Lo, M. Mikulincer, A. Donner, L. Gagliese, and C. Zimmermann. 2009. Pathways to distress: The multiple determinants of depression, hopelessness, and the desire for hastened death in metastatic cancer patients. *Social Science & Medicine* 68: 562–569.
Sacred Congregation for the Doctrine of the Faith. 1980. *Declaration on euthanasia*. http://www.vatican.va/roman_curia/congregations/cfaith/documents/rc_con_cfaith_doc_19800505_euthanasia_en.html. Accessed 17 Dec 2015.
Shim, E.J., and B.J. Hahm. 2011. Anxiety, helplessness/hopelessness and 'desire for hastened death' in Korean cancer patients. *European Journal of Cancer Care* 20: 395–402.
Silveira, Maria J., S.Y. Kim, and K.M. Langa. 2010. Advance directives and outcomes of surrogate decision making before death. *New England Journal of Medicine* 362 (13): 1211–1218.
Sykes, Nigel, and Andrew Thorns. 2003. Sedative use in the last week of life and the implications for end-of-life decision making. *Archives of Internal Medicine* 163: 344.
Temel, Jennifer S., Joseph A. Greer, Alona Muzikansky, Emily R. Gallagher, Sonal Admane, Vicki A. Jackson, Constance M. Dahlin, Craig D. Blinderman, Juliet Jacobsen, William F. Pirl, J. Andrew Billings, and Thomas J. Lynch. 2010. Early palliative care for patients with metastatic non-small-cell lung cancer. *New England Journal of Medicine* 368 (8): 733–742.
The National Consensus Project Clinical Practice Guidelines for Quality Palliative Care (NCP). 2013. 3rd ed. http://www.nationalcoalitionhpc.org/ncp-guidelines-2013/. Accessed 10 Dec 2015.
U.S. Department of Health and Human Services Assistant Secretary for Planning and Evaluation Office of Disability, Aging and Long-Term Care Policy (HHS). 2008. *Advance directives and advance care planning: Report to congress*. https://aspe.hhs.gov/basic-report/advance-directives-and-advance-care-planning-report-congress. Accessed 18 Sept 2016.
United States Conference of Catholic Bishops. 2018. *Ethical and religious directives for catholic health care services*. 6th ed. Washington, DC: United States Conference of Catholic Bishops.
Wilson, Keith G., Harvey Max Chochinov, Christine J. McPherson, Merika Graham Skirko, Pierre Allard, Srini Chary, Pierre R. Gagnon, Karen Macmillan, Marina De Luca, Fiona O'Shea, David Kuhl, Robin L. Fainsinger, Andrea M. Karam, and Jennifer J. Clinch. 2007. Desire for euthanasia or physician-assisted suicide in palliative cancer care. *Health Psychology* 26 (3): 314–323.
Wright, Alexi A. 2008. Associations between end-of-life discussions, patient mental health, medical care near death, and caregiver bereavement adjustment. *Journal of the American Medical Association* 300 (14): 1665–1673.

# Chapter 4
# Meeting Mortality: Palliative Care and the *Ars Moriendi*

**Charles Bouchard**

Our preoccupations change from one generation and one age to the next. Today, our preoccupations do not include mortality; in fact, there has been a great deal of writing about our disdain for illness and death, our eagerness to overlook or ignore it, and our ability to entertain the possibility of endless vitality and strength. Pitches for senior living always show happy, active adults who do not seem to be encumbered by the aches and pains we feel, literally, in our bones as we grow older. Meador and Henson talk about the same reality in terms of a "therapeutic culture." "The excesses of medical therapeutic modernity," they say, "have frequently perpetuated the self-deception that death can be avoided if we work hard enough and sufficiently trust our rational scientific abilities" (2003, p. 90). Atul Gawande in his marvelous reflection on medicine and death, *Being Mortal*, says, "you don't have to spend much time with the elderly or those with terminal illness to see how often medicine fails the people it is supposed to help. Our reluctance to honestly examine the experience of aging and dying has increased the harm we inflict on people and denied them the basic comforts they most need" (2014, p. 9).

C. Bouchard (✉)
Theology and Ethics, The Catholic Health Association of the United States,
St. Louis, MO, USA
e-mail: cbouchard@chausa.org



P. J. Cataldo, D. O'Brien (eds.), *Palliative Care and Catholic Health Care*,
Philosophy and Medicine 130, https://doi.org/10.1007/978-3-030-05005-4_4

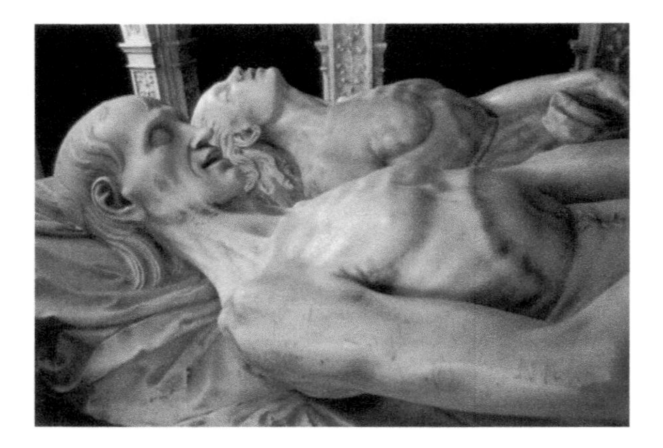

Other ages were different. I remember visiting the Cathedral of St. Denis outside of Paris some years ago. It was filled with the tombs of dead nobility and other important persons, and I recall being shocked by one sarcophagus that was adorned by the prone figures of King Louis XII and his queen, almost naked, quite obviously dead. The sculptures showed even the embalmer's stitches to make it clear that these formerly privileged persons had, like everyone else, succumbed to death and decay. I gasped and thought to myself that the fifteenth century must have had a very different relationship to death than we do today.

The reason for this different and graphic portrayal of death was they had a much more intimate relationship with death, especially during the plague that killed as much as a third of Europe's population between 1348 and 1351. Death was quick, gruesome and widespread. It was made worse by the fact that no one understood how it was spread. There was no knowledge of disease vectors, so it could just as easily have been a sign of God's pleasure or displeasure if you were infected or spared.

There were generally no hospitals or medical care of any kind to speak of during the fourteenth century. After the worst of the plague, there were few priests because many of them had also died. Death was vivid and cruel physically. It was made worse by the fact that there would probably be no one to attend to you or to hear your confession in your dying days. The fear of physical death was exacerbated by the fear of spiritual death. In fact, death was for most of history a domestic and spiritual event rather than a medical and scientific event (Vogt 2004).[1] As late as the 1940s, most people died at home rather than in a hospital.

---

[1] Vogt (2004) notes that much later, during the Puritan period in America, the same thing was true: "The life expectancy of an English nobleman for those toward the end of the seventeenth century was 35 years, and even [less] for those of humble birth…Medicine could do little to prevent or cure illness or to extend life. …Dying was not a medical event so much as a familiar, communal and religious one. People often acted almost as presiders over their own death, generally dying at home with friends and family close at home." Dying often had a rapid onset and progressed to an equally rapid conclusion [54].

The Church responded to this pastoral crisis at the Council of Constance (1415) when it commissioned the creation of a handbook on "the art of dying well."[2] This handbook was based on an earlier version written by an anonymous Dominican friar and supplemented by the work of Jean Gerson, the chancellor of the University of Paris. Gerson had written a pastoral handbook on dying that contained three parts: the ten commandments, confession, and "Scientia mortis," or the knowledge of death (Verhey 2011, pp. 85–86).

Despite his academic duties at the University, Gerson appears to have been an eminently practical theologian. He saw his books as resources for the laity rather than clergy and his desire was to enable a practical, "non-elitist" mysticism which would strengthen the dying and enable them to take full advantage of the Church's spiritual tradition. This view seems predictive of the more egalitarian movements that led to the Reformation in the sixteenth century. Mysticism, Gerson maintained, was not reserved to professionals; he thought it should be accessible to any faithful Christian, regardless of status, sex or education, regarding it as the responsibility of the pastor to nurture this knowledge in the laity (Verhey 2011).[3]

Amy Appleford says in her recent study of death in fourteenth century London, "*Scientia mortis* is addressed solely to those who are not priests, but neighbors, hospital workers and others in active life, who undertook to perform their merciful work of sick bed attendance in the interstices of the regular rites and sacraments for the dying" (2014, p. 158). This accompaniment in dying was seen as an obligation of friendship rather than of ecclesial office, and the role of family and friends was emphasized in the *ars moriendi* ("the art of dying"; hereafter AM) literature.

Appleford says that this cultivation of a relationship with death extended beyond merely personal spirituality. She argues that

> the schooled awareness of mortality was a vital aspect of civic culture, critical not only to the individual's experience of interiority and the management of families and households, but also to the practices of cultural memory, institution building and the government of the city itself. At a time when increasingly laicized religiosity coexisted with an ambitious program of urban renewal and cultural enrichment, and sometimes with violent political change, having an educated attitude to death was understood as essential to good living. (2014, p. 1)

Christopher Vogt makes the same point from a different perspective when he says he will "rely on the work of social historians who have shown that the experience people had of dying centuries ago was significantly eased by the intellectual, cultural and religious traditions of those times" (2004, p. 16).

Things are far different today. Ethicist Daniel Callahan says that in our own time "we live in a society increasingly scant in those cultural resources necessary to

---

[2] Many of the early editions were entitled *Speculum Artis Bene Moriendi* or *Tractatis Artis Bene Moriendi*; an English edition by William Caxton in 1490 was entitled "The Arte and Craft to Know Well to Dye."

[3] Verhey says subsequent versions of these treatises "bore the marks of Gerson's nonelitist mystical theology, nurturing the union of the dying person with God in spiritual affection for God and conforming the will to God's will" (2011, pp. 85–86).

sustain our interior life as we make sense of our endings" (1993, p. 25). The sentiment is echoed by John Stannard in his study of Puritan death when he says that contemporary Americans no longer possess the conceptual resources for giving believable or acceptable meaning to death (1977).

## 4.1 Structure of the *Ars Moriendi* Literature

The AM literature was a series of self-help booklets that proved to be enormously popular, going through hundreds of editions in multiple languages over most of two centuries. Although the books varied somewhat from one country and time to another, they generally stuck to the same general five-part schema (Verhey 2011, pp. 77–173):

1. **Commendation of death.** This was a common feature in the AM literature, designed not to "recommend" death, but to acknowledge that it is a feature of life and that it can be accepted and done well.[4] This commendation was bolstered for Christians by hope in the Resurrection, even though some versions did not make much reference to this and sounded more like "suck it up and die like a man," or "look forward to the day when you will leave your burdensome body behind." This omission has led some historians to suggest that some of these manuals were more influenced by Stoic philosophy or Platonism than by Christian hope of the Resurrection.
2. **Temptations confronted by the dying.** Another common feature of the literature was to recite temptations that commonly afflict the dying, and sometimes to juxtapose them with corresponding virtues, to which we will return later. Different versions cited temptations to lose faith, to despair, impatience, pride, or avarice. The descriptions were often accompanied by illustrations, replete with demonic tempters to make it clear what was at stake. One illustration of impatience shows "Moriens," the dying man, overturning his food tray and kicking the doctor (Fig. 4.1).[5]
3. **Instruction on repentance and assurance of God's forgiveness.** Because eternal life was in jeopardy for the dying, there were frequent exhortations to repentance and admonitions to remember God's mercy. There was a question-and-answer format, designed to elicit as much participation as possible from the dying person. The questions reflected an early list from St. Anselm, which are summarized by Vogt (2004, p. 158):

    - Are you glad to die in the faith of Christ?

[4] This approach emerged early, as we can see from St. Ambrose's *De Bono Mortis* ("On the Good of Death") which may have inspired subsequent authors.

[5] Illustration of *Moriens* from English *Ars Moriendi*, fifteenth century, found in Verhey (2011, p. 121). Many illustrations in subsequent editions were adapted from these.

**Fig. 4.1** Moriens behaving badly (Ars Moriendi, 15th c)

- Do you acknowledge that you have not done what you should and that you have done what you should not?
- Do you repent of these sins?
- Would you amend your life should you survive?
- Do you believe that our Lord Jesus Christ, God's son, died for you?
- Do you thank him with all of your heart?
- Do you believe that you can be saved only by Christ's passion and death?

4. **Imitation of the dying Christ and prayers for use by the dying.** This is an example of very traditional Christian spirituality rooted in the Christian Scriptures, especially St. Paul, who developed a strong theology of what we

might call "Christo-identification" in his letters.[6] It may also reflect the "Imitation of Christ" spirituality associated with Thomas à Kempis. These later became part of popular spirituality in which the believer "offers up" his suffering, or prays that it be united to the redemptive suffering of Christ.

5. **Prayers to be said by those ministering to the dying person.** The AM literature was intended to be used in the absence of a priest, so there is a strong emphasis on the role of the ministering community.[7] Long before the medicalization of death, the role of friends and family was paramount. Today, the disconnect between patient and family during serious illness and dying is one of the key problems that hospice care has tried to address. Friends and family gathered around the dying person had a special responsibility to encourage the patient, to challenge him, and to bring him to the point of confessing his sins. In many illustrations that accompanied the written text of AM literature, this family was extended to include even the saints and others who had died, creating a "communion of saints" that transcended earthly life.

## 4.2 *Ars Moriendi* and Palliative Care

Of course, palliative care is not the same thing as care for the dying. In fact, one of our challenges has been to make it clear that palliative care is *not* hospice care. Hospice is actually a subset of palliative care, where attention to spiritual and physical suffering, and symptom management become the sole focus of care during the terminal phase of an illness. Palliative care can also be delivered concurrently with disease modifying treatment in order to manage symptoms and improve quality of life when cure is not possible and death is not imminent. The art of dying is relevant for both applications of palliative care.

St. Benedict in his Rule advises his monks to keep death ever before their eyes (1904, Chapter IV, n. 47).[8] That is good advice for all of us, yet fear and medicalization of death have affected the dying process and our ability to deal with terminal illness as well as our attitudes about any chronic, incurable disease. This is especially true in a society where "our only response to suffering is to strive for its eradication" (Vogt 2004, p. 72), an attitude that is clearly a factor in the increasing societal acceptance of euthanasia and physician-assisted suicide. It seems that many would rather be dead than suffer.

---

[6] See for example, Galatians 2:20, "I have been crucified with Christ, and I no longer live, but Christ lives in me"; or Philippians 3:10–11, where Paul sees himself "becoming like him (Christ) in his death" so that he "may attain the resurrection from the dead."

[7] The word "ministering" is an anachronism here. The word minister would not have been used in a generic sense at that time, and it certainly would not have applied to prayers by lay people. However, since they were filling in for the absent sacramental minister it seems appropriate to describe what they did here as a kind of ministry.

[8] *Mortem quotidie ante oculos suspectam habere.*

Both Robert Bellarmine (d. 1621) and Erasmus (d. 1536), in their books on death said that we die as we live, and that it is too late to prepare for death when we have already reached our deathbeds.[9] Christian spirituality urges us to be mindful of the reality of death, and union with God, as the ultimate purpose of our lives. In that way we prepare ourselves for the final journey. This is all the more true for those who are suffering from serious illness, who are carrying the death of the Lord in their very bodies (2 Cor 4:10).

This life-long intentionality about death requires certain virtues. These are not deathbed virtues but habitual moral qualities that can be built up over a lifetime through diligence. These habits begin to pay off when we first start to experience diminishment, when we have to discipline ourselves by exercise, diet, medications, periodical checkups and lab tests. These ways of dealing with chronic illness make us acutely aware, every day, of our frailty. Think for example of the diabetic who can be hours away from coma and death if blood sugar gets out of control. Or think of the patient with COPD for whom even walking across the street can result in exertion; or of someone with controlled, but not cured, leukemia.

Palliative care can be a response to the "little deaths" that we experience with greater regularity as we begin to age. William Perkins' *Salve for the Sicke Man* (1595) notes this fact long before we had palliative care or any way of treating chronic illness:

> He that would be able to bear the cross of all crosses, namely death itself, must first of all learn to bear small crosses, as sicknesses in body and troubles in mind, with losses of goods and of friends, of good name, which I may fitly term little deaths. We must first of all acquaint ourselves with these little deaths before we can well be able to bear the greatest death of all. (Atkinson 1992, p. 141)

So palliative care deals with the "little deaths" that come in the form of incurable illness, chronic pain, loss of mobility, or diminishment. Palliative care also seeks to manage suffering – whether mental, physical or spiritual; integration of these dimensions of the person is one of the hallmarks of quality palliative care. It treats the person rather than just the illness (Cassell 2010, p. 129).[10] So it seems fair to say that even though hospice and palliative care have different purposes – preparing for death on the one hand and managing the symptoms of incurable illnesses on the other – they are both part of the process of dying. They both require "remembrance of death" as part of living.

---

[9] The full title of Bellarmine's book is "The Art of Dying Well, or how to be a Saint, Now and Forever." The instructions he provides – prayer, sobriety, faithfulness to one's baptismal promises, detachment, reverence and fasting are clearly life-long spiritual habits. Erasmus' "Preparing for Death" was published in 1553.

[10] Eric Cassell (2010) makes three important points about suffering: (a) it is experienced by persons, and persons are not merely minds, or spiritual; (b) Suffering occurs when an impending destruction of the person is perceived – it continues until the threat of disintegration has passed or until the integrity of the person can be restored; (c) Suffering can occur in relations to any aspect of the person whether it is in the realm of social roles, group identification, the relations with self, body or family, or the relation with a transpersonal, transcendent source of meaning.

## 4.3 Palliative Care and the Life of Virtue

In his little book *Preparing for Death*, the philosopher Erasmus says:

> The preparation for death must be practiced through our whole life and the spark of faith must be continually fanned so that it grows and gains strength. Love, joined to it, will attract hope, which gives no cause for shame. None of these things, however, comes from us; rather they are gifts of God to be sought by continuous prayers and petitions if we lack them; if they should be present they must be strengthened so that they grow. The stronger is our faith, accompanied by love and hope, the more diminished is our fear ... An action continually repeated will become part of a habit, the habit a state, and the state a part of your nature (1998, p. 398, 421).[11]

Erasmus here speaks of some of the virtues – faith, hope and charity – that are essential to the Christian life and that nourish us when we first experience sustained illness. Let us take a moment to explore some of the ordinary virtues that become extraordinarily important as we cope with chronic illness.

Most of the AM resources cite loss of faith as the first temptation the dying person faces. After faith itself – which is the special knowledge of God and God's mysteries, not available to reason – hope is probably the most important virtue in illness. As we age and get sick we have a natural tendency to shift our focus from the future to the past. In older age we have already achieved many of the goals and hopes we had when we were younger (or the opportunity to do so has passed and then we may have regret). Whatever we set out to do as a life project, at least in terms of material achievements, is probably completed. Age and illness often present us with an opportunity to reflect on "what it all meant," to weave the strands of our hopes, disappointments, successes and failures, into a coherent narrative of our lives. But for the person receiving palliative care, life is not over. Therefore, part of the spiritual challenge of palliative care is to remain hopeful. Hope pulls us into the future; it helps us focus on God's plan for us and our eternal destiny. "Hope implies seeking and pursuing," St. Thomas Aquinas says in *Disputationes* (1982, 203). Hope is a dynamic energy that pulls us forth even when we might be tempted to wallow in a morbid preoccupation with our past failures or our current limitations. In an Advent meditation Ladislas Orsy is even more direct when he says: "Embedded as we are in the present, we are creatures of the future. Our whole being tends toward what we hope for – a much-desired favorable event. Take away our hope, and we find ourselves in a void worse than Dante's inferno" (2005). Remaining hopeful – even when our bodily condition, our medications and our weakness no longer sustain the hopefulness of youth – is a basic spiritual discipline.

**Courage, or fortitude**, is the will to pursue the good in spite of adversity. Trappist Father Thomas Keating says that the virtue of fortitude, which we can attain to some extent by our own efforts, enables us to pursue the difficult good, but the *Gift* of fortitude, which comes from God, "gives energy to overcome major obstacles in the way of spiritual growth" (2000, p. 49). The kind of courage that is

[11] Quoted by Vogt (2004, p. 255).

needed as we enter into palliative care is not a one-shot deal. It is sustained courage that keeps us involved in our relationships, connected to friends, experiencing art and music and even aspects of nature around us. The temptation is to allow chronic illness to discolor everything so that nothing looks fresh and appealing anymore. Depression is a clinical version of lack of courage, when we cannot bring ourselves to step out and enjoy life. Everything seems pointless and uninteresting. I remember a friend who told me she realized how depressed she was one day at the grocery store when she stood in front of the shelves of breakfast cereal and could not make a choice. Courage gives us the strength to move ahead, and sometimes to seek the help of others in dealing with our limits. We often see courage as a private thing, but as Stanley Hauerwas notes, "there is no such thing as private courage" (1997, p. 227). The tendency to see assistance from others as a weakness or lack of courage is exactly wrong, and has contributed to untold suffering and even suicide. This is disproportionately true for men, especially returning veterans whose tragic stories are well known to us.

Courage is also necessary in order to let go. We spend our whole lives accumulating, growing, acquiring, advancing, becoming more knowledgeable and skilled. Chronic illness and the need for palliative care are often the time to begin divesting, leaving behind everything that is non-essential. Richard Rohr built a whole book around this idea (2013). Invoking John of the Cross he says that the second half of life is kind of a "luminous darkness" which is the simultaneous existence of deep suffering and joy. In this second half of life, "one has less and less need or interest in eliminating the negative or fearful, making again those old rash judgments, holding on to old hurts, or feeling any need to punish other people." Our view of God changes, and we become more mellow and more attuned to what is essential. "There is a gravitas in the second half of life," he says, "but it is now held up by a much deeper lightness or 'okayness.' Our mature years are characterized by a kind of bright sadness and sober happiness" (2013, xiii). This ability may be most helpful as we face death, but it is a Christian virtue even if we are just facing chronic illness that requires palliative care.

Charles Pinches (2003) says this spiritual maturity is marked by the virtue of delight, the result of learning to "travel light," and by the virtue of simplicity. He says the elderly (and perhaps also those with chronic illness) must

> live every day with the awareness that the bodily powers are diminishing, which forces them to live with a constant sense of their own limitation. This may not sound much like a virtue, and for our culture it decidedly is not. Indeed…there is nothing that horrifies our culture more than the thought that we might have to live within the limits of what we a truly able to do… Christians, however, see how limitedness can appear not only as a gift but also as a call. If lived well, it can be called a virtue. (p. 212)

**Solidarity** is another important virtue for those in need of palliative care. Sometimes palliative care is appropriate because of chronic pain. It is easy to let that isolate me, as though I am the only person in the world to have such misfortune. Solidarity with Christ and with others who suffer is important. It enables us to feel close to those who are geographically, spiritually or socially different, but who share our human condition with us. On my best days, I can look at someone who is clearly in pain and say a prayer for them; at that moment, I am able to allow my own suffering to be a pathway to someone else. My suffering is the suffering of others. If the moment is graced, I am able to allow that suffering to be a pathway to Christ.

Although we often think of prayer as an activity, it is also a virtue, or an acquired habit that orients us to God and to reality. Two kinds of prayer are important for the spiritual life of those receiving palliative care. First, personal prayer (contemplative, or quiet meditative prayer) is the most important. Regular contemplative prayer keeps us focused on what is important, and allows us to cultivate gratitude. Lack of contemplative prayer creates space where weeds of resentment, anger and discouragement take root. Contemplative prayer enables us to discern God's will, to keep our illness in perspective and to chase away illusions. It allows us to "name grace" in our actual circumstances and allows us to be open to the possibility of future grace (Hilkert 1997).[12]

The second kind of prayer is liturgical or communal prayer, which is prayer shared with others, in public, through ritual and bodily involvement. This kind of prayer is characteristic of an ecclesial spirituality, one that is not solitary but actively communal. Liturgical prayer like the Eucharist is a reenactment of the mysteries of faith. It communicates to me and to the people with whom I am praying, our mystical connection with Christ and our real connection with one another. Unfortunately, many parish masses are lifeless; they can hardly bear their own weight, much less the weight of the transcendent. Finding a worshipping community that is committed to beautiful liturgy, good preaching and strong mission can be a great aid to those beginning to experience loss of physical integrity. It also helps us remember that salvation comes to a people, not just to individuals, and makes us mindful of our dependence on others. So in some ways, common prayer is an act of humility in which we say "I can't do this on my own; I need God's grace and I need the faith of others to sustain me." This can be difficult especially if you are, like me, someone who tends to want to hide when in pain.

## 4.4 Palliative Care and Intentional Living

In the days before modern medicine, death – and chronic illness – were primarily spiritual events. Then, with the Enlightenment, came the "clinic" where the objective scientific gaze was normative, where the disease, not the patient, was the object of interest (Sulmasy 2006).[13] Today, when illness and dying have been thoroughly

[12] I am indebted to Mary Catherine Hilkert's idea of "naming grace" as the basis for a Catholic theology of preaching. When we preach, we don't "pull grace down" from some heaven that is incompatible with creation, but rather try to "name grace" in the world around us. Being able to name grace is a great gift from God, and it becomes especially important when we are tempted to focus on pain and loss.

[13] Sulmasy (2006, p. xiii) says the clinic "came to eschew the mystical. It became blind to the mystery of the person within and blind to the mystery that lay beyond the range of its scientific gaze."

medicalized (and medicalization is without doubt a great blessing and benefit), our challenge is to recover the spiritual dimension of illness and put it more fully into the context of our lives, of God's call to us, and of our vocation. As Daniel Sulmasy says, people today are looking for what the Enlightenment clinic failed to deliver: "They want a form of medicine that does not abandon science, but also does not eschew the mystical. Often they seek this ideal in alternative forms of medical care. Most, however, want soul medicine and scientific medicine at the same time" [xiii]. Bringing these two together is the challenge of holistic palliative care.

Palliative care and the medical conditions that it treats require that we grow in the ability to *pay attention*. We have to watch our symptoms, our medications, our state of mind. But we also have to pay attention to the fundamental direction and value of our lives. In his version of AM, St. Robert Bellarmine (2005) maintains that living well is the best preparation for dying well. The initial chapters of his book suggest virtues and practices that sustain us and help us grow toward dying:

- Strive to live well
- Learn to die to the world
- Persevere in faith, hope and charity
- Be ever ready to meet Christ
- Remain detached from worldly possessions
- Be sober, just and pious
- Pray fervently
- Practice fasting
- Give generously of the gifts God has given you
- Be faithful to your baptismal promises
- Use the gifts you have received in Confirmation
- Receive the Eucharist worthily
- Confess your sins with true contrition
- Revere the sacred
- Love and honor your spouse and your children
- Guard your senses against sin, and finally, *resolve to die well.*

None of these instructions is inconsistent with what the AM literature recommended, but none of them is particular to it, either. They are human and Christian virtues, the hallmarks of holiness. They should mark our living and our dying.

## 4.5 A Spirituality for Palliative Care

What I have tried to do here is to use aspects of the *ars moriendi* tradition to build a framework for a spirituality of palliative care. The Center to Advance Palliative Care (2014) says that palliative care is a "new and better paradigm... [one that] provides interdisciplinary coordination and team-driven continuity of care that best responds to the episodic and long-term nature of serious, multifaceted disease." This new paradigm presents Catholic health care with a unique opportunity to integrate clinical and spiritual dimensions of care so that spirituality and formation for spiritual maturity will not be just an add-on or an afterthought, but an essential aspect of creating new habits of life that acknowledge new health circumstances. This paradigm should allow us to ask our patients not just how they feel or how they are responding to medication, but "What does this condition, this 'little death' mean to you spiritually?"

Patients who enter into palliative care don't need these practices of holiness more than other Christians, but they have a unique opportunity to know about them and to embrace them as part of their own spiritual journeys. I hope this brief essay will make a small contribution to the evolution of a spirituality suited to growing numbers of us who face serious illness and want to use everything we can from medicine and from our rich spiritual tradition to help us face it in confident faith.

## References

Appleford, Amy. 2014. *Learning to die in London, 1380–1540*. Philadelphia: University of Pennsylvania Press.

Aquinas, Thomas, St. 1982. *Disputationes de spe*, 1 ad 1. In *Thomas Aquinas: Theological texts*. Trans. T. Gilby. Durham: The Labyrinth Press.

Atkinson, David William. 1992. *The English ars moriendi*. New York: Peter Lang.

Bellarmine, Robert, St. 2005. *The art of dying well, or how to be a saint, now and forever*. Manchester: Sophia Institute Press.

Benedict of Nursa. 1904. *The rule of St. Benedict*. Ed. and Trans. D. Oswald Hunter Blair, 2nd ed. London: Sands & Co.

Callahan, Daniel. 1993. *The troubled dream of life: In search of a peaceful death*. New York: Simon and Schuster.

Cassell, Eric. 2010. The nature of suffering and the goals of medicine. In *Palliative care: Transforming the care of serious illness*, 125–136. San Francisco: Josey-Bass.

Center for the Advancement of Palliative Care (CAPC). 2014. *The case for hospital palliative care*. https://media.capc.org/filer_public/06/90/069053fe-12bf-4485-b973-d290f7c2ecbf/thecaseforhospitalpalliativecare_capc.pdf. Accessed 31 Aug 2017.

Erasmus, Desiderius. 1998. Preparing for death. In *Collected works of Erasmus: Spiritualia and Pastoralia*, ed. J.W. O'Malley, 392–450. Toronto: University of Toronto Press.

Gawande, Atul. 2014. *Being mortal: Medicine and what matters in the end*. New York: Henry Holt.

Hauerwas, S. 1997. Courage exemplified. In *Christians among the virtues*, ed. S. Hauerwas and C. Pinches. Notre Dame: University of Notre Dame Press.

Hilkert, Mary Catherine. 1997. *Naming grace: Preaching and the sacramental imagination*. New York: Bloomsbury Academic.

Keating, Thomas. 2000. *The fruits and gifts of the Holy Spirit*. New York: Lantern Books.

Meador, Keith, and Shawn Henson. 2003. Growing old in a therapeutic culture. In *Growing old in Christ*, ed. Carole B. Stoneking, David Cloutier, Keith G. Meador, and Stanley Hauerwas, 90–111. Grand Rapids: Eerdmans.

Orsy, Ladislas. 2005. The virtue of hope. *America*: December 5 2005. https://www.americamagazine.org/issue/553/faith-focus/virtue-hope. Accessed 2 July 2017.

Pinches, Charles. 2003. The virtues of aging. In *Growing old in Christ*, ed. Carole B. Stoneking, David Cloutier, Keith G. Meador, and Stanley Hauerwas, 202–225. Grand Rapids: Eerdmans.

Rohr, Richard. 2013. *Falling upward: Spirituality for the two halves of life*. San Francisco: Josey-Bass.

Stannard, John. 1977. *The Puritan way of death*. New York: Oxford University Press.
Sulmasy, Daniel. 2006. *The rebirth of the clinic*. Washington, DC: Georgetown.
Verhey, Allen. 2011. *The Christian art of dying: Learning from Jesus*. Grand Rapids: Eerdmans.
Vogt, Christopher. 2004. *Patience, compassion and hope: The Christian art of dying well*. Lantham: Rowman and Littlefield.

# Chapter 5
# Palliative Care: Euthanasia by Another Name?

**Ron Hamel**

Is palliative care simply euthanasia by another name, or "stealth euthanasia," as some have called it? Is there any legitimacy to the concerns these individuals raise and to the observations they make that lead them to this conclusion, a conclusion that unfortunately can plant seeds of doubt about and even undermine one of the desired hallmarks of Catholic health care and even imperil good, holistic care of sick and dying persons? Are their claims the equivalent of the canary in the coalmine? Are they signaling significant dangers in palliative care and hospice that call for vigilance and possible corrective actions? In what follows, I will examine the claims being made about palliative care and hospice being stealth euthanasia, the legitimacy of these claims, and their implications for palliative care and hospice in Catholic health care.

## 5.1 Claims of Stealth Euthanasia

One of the first challenges to palliative care and hospice to appear in a Catholic journal was a 2006 article by Romanus Cessario, O.P. titled "Catholic Considerations on Palliative Care" (2006)[1]. While he does not refer to palliative care and hospice as stealth euthanasia, Cessario does raise doubts about their moral acceptability. Commenting on the National Hospice and Palliative Care Organization's mission statement, which draws on the WHO definition for palliative care, Cessario observes:

> While the WHO [World Health Organization] definition of palliative care contains no provision that contravenes the *Catechism of the Catholic Church*, n. 2279, one should not

[1] In addition to the three sources discussed in this section, there are several others that discuss stealth euthanasia. See Helmueller (2005), Mallon (2009), Abbott (2013), Dial (2003), and Isajiw (2015).

R. Hamel (✉)
SSM Health Ministries and SSM Health Board of Directors, St. Louis, MO, USA



P. J. Cataldo, D. O'Brien (eds.), *Palliative Care and Catholic Health Care*, Philosophy and Medicine 130, https://doi.org/10.1007/978-3-030-05005-4_5

> assume that the implementation of these provisions will remain 'unambiguously pro-life'…. As it now stands, the NHPCO mission statement leaves itself open to allowing palliative care that would contravene the moral law, especially inasmuch as terms like 'quality of life,' 'values,' and 'decisions' invite judgments ordered toward assisted suicide or euthanasia. (Cessario 2006, pp. 642, 643–44, 645)

Cessario finds there to be an absence of specific, concrete moral norms in the NHPCO's mission statement as well as any indication that freedom of choice (as stated in *Veritatis splendor*, n. 72) should be in "*conformity with man's true good,*" the person's ultimate end, which is God (2006, p. 643). In fact, Cessario wonders whether dame Cecily Saunders was influenced by what *Veritatis splendor* calls "'the modern concern for the claims of autonomy'" (John Paul II 1993, n. 36). This being the case, palliative care can easily be influenced by a culture of death and "prevailing political ideologies that threaten the true dignity of human existence" (Cessario 2006, p. 644) and "affords an easy venue" for various groups that promote assisted suicide and euthanasia to exert their influence on what it means to "die with dignity" [p. 645].

In the face of these challenges, Cessario insists that what is required of Catholic health care professionals is a virtuous mean between avoiding the extremes of high-technology interventions on the one hand and assisted suicide or euthanasia (passive or active) on the other. He urges careful examination of the *Catechism's* teaching on medical interventions at the end of life (n. 2279) and specifically examines directive 61 of the 2001 edition of the United States Conference of Catholic Bishops' *Ethical and Religious Directives for Catholic Health Care Services.*

Based on this directive, Cessario makes four observations: (1) even when death is imminent, ordinary care (e.g., nutrition and hydration) should continue to be provided; (2) in the administration of analgesics, blunted consciousness and the willing of death either as an end or as a means by the one who administers the pain medication or the one who receives it raise serious moral concerns, especially the willing of death; (3) palliative care is a form of "disinterested charity" (*Catechism*, n. 2279) that allows for no counterfeit; (4) palliative care, as a form of disinterested charity should be encouraged, but this requires that palliative care protocols "conform to the truth about the human person, and that the services offered in Catholic institutions do not confuse misplaced compassion for true God-inspired charity" (2006, pp. 647–648).

After enumerating five proposals to guide palliative care in Catholic health care organizations, Cessario concludes by saying:

> The initiative that Dame Cecily Saunders instigated in the late 1960s occurred at a time when the "problems of the contemporary world" began to show signs of a new and disturbing moral disequilibrium. Catholics will want to ensure that the tradition of palliative care that finds its true origin and source in Christ's own love for his people will not become a locus for violating a fundamental truth that distinguishes Christian existence in our period: respect for human life. (2006, p. 650)

Although Cessario acknowledges that palliative care and hospice do good and can be forms of disinterested charity, he also raises the possibility that they can fall victim to the culture of death and employ measures that violate human dignity and

Catholic moral teaching. As a whole, the article plants seeds of doubt about the moral legitimacy of palliative care.

One of the first, if not the first, uses of the term "stealth euthanasia" occurs in the title of a 2011 book by Ron Panzer, LPN a former hospice nurse and founder of Hospice Patients Alliance (a patient advocacy organization). The book, *Stealth Euthanasia: Health Care Tyranny in America,* is a blistering attack (or perhaps more appropriately, a tirade) on hospice and palliative care in the U.S., as well as the health care system, health care professionals, payers, the media, politicians, government and more (whom he sees as complicit in the abuses of palliative care and hospice). Much of the book is based on first hand reports from families and clinicians. Panzer claims that there is widespread abuse in the hospice industry by "rogue hospices" that, in the quest for profits, deprive patients of needed services leading to their premature death and, worse, hasten the death of many, many patients. "In a rogue hospice, you may have patients dying the day or week they are admitted, even when they are not clinically close to death" (Panzer 2011). Furthermore, many hospices are admitting patients who are not terminal in order to be able to bill for them for a longer period of time. When these hospices can no longer bill, the patients' lives are medically ended.

This is accomplished by overly sedating patients, not feeding them, giving medications that are not needed, withholding needed medications, over-dosing with opioids, and palliative or terminal sedation. These are what constitute stealth euthanasia under the guise of palliative care. In these situations, Panzer claims, families are lied to about what is actually happening (2011, Part VII, Sections 2, 3, 5, 6, 8, 11, 12, 13).

All of this, according to Panzer, is due to the "culture of death" infiltrating American culture and the hospice movement, combined with a "quality of life" mentality (i.e. some lives are not worth living—the very sick, disabled and elderly) and utilitarianism. In addition, Panzer maintains that the National Hospice and Palliative Care Organization (NHPCO) has its roots in organizations that have promoted assisted suicide and euthanasia:

> Even better: "Partnership for Caring" (2001). That sounds like we're all working together and caring, nothing about euthanasia in the name. "Last Acts Partnership (2004)?" That's totally unrecognizable as to what it's about. And lastly, the NHPCO's "Caring Connections" (2004). That's as far from "Euthanasia Society of America" as you can get. But the National Hospice and Palliative Care Organization's "Caring Connections" project is the successor of all these organizations! … the final successor organization to the Euthanasia Society of America. (Panzer 2011, Part III, Section 4, "Euthanasia Society: Covert Operations in the Health Care & Hospice Industry")

Elsewhere, Panzer claims that 80% of hospices in the US are members of the NHPCO and that the NHPCO's website is listed as a "right-to-die" website by the World Federation of Right-to-Die Societies (2011, Part III, Section 8, "End-Run Around Right-to-Life: Hospice No Longer Safe is Safe Alternative to Euthanasia & Assisted Suicide").

It is precisely these euthanasia supporting and promoting organizations, states Panzer, that have also promoted living wills and advance directives that are "an **incremental step** toward open euthanasia" (2011, Part I, Section 5, "'Palliative Care' & Its Approach to End-of-Life Care"). He also attributes POLST to them.

Furthermore, Panzer claims that "those who are true believers in euthanasia and assisted suicide manage many of these hospices or palliative care units. They run the National Hospice and Palliative Care Organization. They're on the board of directors. They train the staff, perpetuating twisted clinical practice and interpretations of the hospice and palliative care mission. Hospice has become a playground for the right-to-die zealot!" (Panzer 2011). Singled out by Panzer is Ira Byock, MD (a renowned palliative care physician known for opposing assisted suicide and euthanasia) who, he maintains, "is one of those promoting the use of palliative sedation to intentionally end life within hospice." Byock, he says,

> Has encouraged the hospice industry to widely implement terminal sedation as a way to end life within the hospice setting. Terminal or palliative sedation does not strictly come under the prohibition of open euthanasia and it's not a direct agent to cause death that a patient might take, so it avoids both the legal prohibitions against euthanasia and those against assisted-suicide. It's a right-to-die advocate's dream solution: clever, devious, and legally "do-able" throughout the health care system, especially in "hospice". (Panzer 2011, Part VII, Section 12, "Hospice's Third Way: Quill and Byock Promote Palliative Sedation to Hasten Death")

In Panzer's view, "While 'palliative care' in its purest form is made possible by advances in medical science and practice and is applied to relieve suffering at any stage of life, there are palliative care leaders who seek to use this newer specialty within medical practice to further an agenda that takes our society further away from affirming the value of human life and 'doing no harm'" (2011, Part I, Section 5, "'Palliative Care' & Its Approach to end-of-Life Care"). The seeds of that agenda were planted in the hospice movement, Panzer claims, by Florence Wald, the founder of hospice in the United States. Wald explicitly supported physician assisted suicide (2011, Part I, Section 2, "Three Hospice 'Giants'").

Since the publication of Panzer's book, an article appeared in the National Catholic Bioethics Center's *Ethics and Medics* titled, "The Rise of Stealth Euthanasia" (Capone et al. 2013). It echoes many of the same themes and concerns as the Panzer book. Stealth euthanasia is occurring, the authors contend, because physicians in palliative and end-of-life care settings are increasingly intending to kill patients through the administration of opioids and palliative sedation, while appealing to the principle of double effect. "It is horrifying that health care professionals… intentionally hasten death while pretending to be providing appropriate end-of-life care. That this is a *pretense* is becoming more and more evident to patients and families" (Capone et al. 2013, p. 2). When using opioids to cause death, the authors claim, health professionals "usually cloak their actions by telling families that the signs of approaching death being observed are due to a terminal illness, not to the adverse effects of a clinical overdose, which conveniently mirror some of the signs of the end-stage active phase of dying" (Capone et al. 2013, p.1).

Like Panzer (who is one of the co-authors of this article), the authors maintain that patients who are not terminal are being admitted to hospice, some chronically ill with conditions such as Alzheimer's and brain damage, only to die from dehydration. They contend that withholding nutrition and hydration is "often done with the intention that the patient die" and that "physicians who seek to continue providing food and fluids are often pressured not to do so." Indeed, they state, "the culture of death has deeply infiltrated the hospice and palliative care industry" (Capone et al. 2013, p. 3). The authors further claim that those involved in these practices maintain that their only intent is to do good—to relieve suffering, or a poor quality of life, or not to have a patient linger when their time has come, or to honor patient autonomy. The patient in these cases is thought to be "better off dead."

Echoing Panzer, the authors claim that the NHPCO "is the actual legal and corporate successor to the Euthanasia Society of America" (Capone et al. 2013, p. 2) which they believe "explains the contradiction between the publically stated hospice mission and the reality in too many clinical settings." This is compounded by their belief that many hospice and palliative care physicians "are urging, and actually performing, euthanasia by stealth" (Capone et al. 2013, p. 2). Here they particularly single out Joanne Lynn, Timothy Quill, and Ira Byock.

Not all palliative care and hospice programs, however, are engaged in stealth euthanasia, according to the authors. When "the founding principles of hospice—to maintain dignity, to increase quality of life, and to provide comfort and pain control—are followed, hospice is a safe haven for patients in need of expert end-of-life care" (Capone et al. 2013, p. 1).

## 5.2 What Is to Be Made of These Claims?

What are we to make of the claims in these sources that palliative care and hospice are, in very many instances, nothing more than stealth euthanasia? The claims seem to center around organizations, people, tools and practices. We will consider each in turn.

First, regarding the National Hospice and Palliative Care Organization's historical connections to euthanasia-promoting organizations, this does seem to be the case. On the website of "CaringInfo," a program of the NHPCO that provides information to help people make decisions about end-of-life care and services with a focus on advance directives, there is a statement that suggests the connection. "… for all intents and purposes, NPHCO became a successor organization of Last Acts Partnership (and its predecessor entities)" (CaringInfo).

A timeline on the same webpage explains the connection. In 1998, Choice in Dying became Partnership for Caring. Choice in Dying came about in 1991 from the merger of two right-to-die organizations—Society for the Right to Die and Concern for Dying. Society for the Right to Die was the new name given to the Euthanasia Society in 1974. In 2000, Partnership for Caring (formerly Choice in Dying) became the national program office of Last Acts. In January 2004, Partnership for Caring and Last Acts merged to become Last Acts Partnership. Finally, later in 2004, Last Acts Partnership ceased operations and the Robert Wood Johnson Foundation funded NHPCO to transition the work of Last Acts Partnership and begin Caring Connections. "… NHPCO acquired virtually all of the physical and intellectual assets of the Last Acts Partnership." CaringInfo is the offspring of Caring Connections.

However, while NHPCO seems to have distant roots in euthanasia-promoting organizations, these past connections have not resulted in NHPCO becoming a euthanasia organization. It is important to note that more recent incarnations of the earlier organizations shifted their focus from advocating for euthanasia to the promotion of living wills and advance directives, and improving end-of-life care. But for some, the very connection raises suspicion about palliative care and hospice.[2]

It is also interesting to note the NHPCO's position statements on physician assisted suicide and palliative sedation. In both instances, the organization takes a "pro-life" stance. With regard to physicians assisted suicide, the NHPCO resolution states:

> RESOLVED, That the National Hospice and Palliative Care Organization reaffirms its commitment to the value of life and to the optimization of the quality of life for all people at the end of life.
>
> RESOLVED, That the National Hospice and Palliative Care Organization supports improved knowledge of and access to hospice and palliative care for terminally ill people and their families, regardless of individuals' views, decisions, or preferences regarding physician assisted suicide.
>
> RESOLVED, That the National Hospice and Palliative Care Organization does not support the legalization of physician assisted suicide.

The resolution is preceded by a commentary in which two guiding principles are articulated. The second has to do with the need for a variety of efforts to improve knowledge about and access to high quality end-of-life care. The first is an affirmation of the value of life and respect for human dignity:

> **NHPCO Values Life**: The philosophical constructs and evolving practices of hospice/palliative care are concerned foremost with the dignity of persons throughout the trajectory of life-limiting illness. When symptoms or circumstances become intolerable to a patient, effective therapies are now available to assure relief from almost all forms of distress during the terminal phase of an illness without purposefully hastening death as the means to that end. These modalities and the means to safely administer them must be within the expertise of and available from all hospice/palliative care providers as an alternative to PAS. (NHPCO 2005)

With regard to palliative sedation, the NHPCO's statement is also very balanced. Only a small portion of it can be noted here:

> For the small number of imminently dying patients whose suffering is intolerable and *refractory*, NHPCO supports making the option of palliative sedation, delivered by highly trained health care professionals, available to patients. … Properly administered, palliative sedation of patients who are imminently dying is not the proximate cause of patient death, nor is death a means to achieve symptom relief in palliative sedation. As such, palliative sedation is categorically distinct from euthanasia and assisted suicide. (Kirk and McMahon 2010, pp. 915, 916)

Hence, while the NHPCO might have a loose and remote association with euthanasia advocating organizations, the mission of the NHPCO and its immediate pre-

[2] The NHPCO is not (at least currently) listed as a member of the World Federation of Right to Die Societies as claimed by Panzer. See, for the listing: http://www.worldrtd.net/member-organizations

decessor organizations is dramatically different from the Euthanasia Society of America. Furthermore, there is nothing in position statements of the NHPCO on critical ethical issues (including medically administered nutrition and hydration) that, at least on their face, would be morally problematic.[3]

Second, regarding people in palliative care and hospice who have been associated with physician assisted suicide and euthanasia, there is also some truth to the claims made by those who view palliative care as stealth euthanasia. Florence Wald, in fact, was a supporter of assisted suicide. In a 1999 interview that appeared in JAMA, Wald said: "I know that I differ from Cicely Saunders, who is very much against assisted suicide. I disagree with her view on the basis that there are cases in which either the pain or the debilitation the patient is experiencing is more than can be borne, whether it be economically, physically, emotionally, or socially. For this reason, I feel a range of options should be available to the patient, and this should include assisted suicide" (Freidrich 1999, p.1684).[4]

Another individual cited by those who claim stealth euthanasia is Timothy Quill. Quill is probably best known for his article about "Diane," that appeared in the *New England Journal of Medicine* (1991), in which he described prescribing barbiturates to his patient with leukemia so that she could end her life. Five years later, on April 29, 1996, Quill offered testimony at an Oversight Hearing on "Assisted Suicide in the United States," before the Subcommittee on the Constitution of the House of Representatives Committee. In his remarks, "A physician's position on physician-assisted suicide," subsequently published in the *Bulletin of the New York Academy of Medicine* (1997), Quill supported physician assisted suicide, but as a last resort, to be used only if good palliative care and hospice have been unsuccessful in addressing the patient's issues. "Assisted death should never be an alternative to good palliative care. Physician-assisted death should be restricted to those relatively few patients for whom hospice care ceases to be effective and suffering is so intolerable that death is their only answer" (Quill 1997, pp. 115–116). Quill, board certified in palliative care and the Georgia and Thomas Gosnell Distinguished Professor of Palliative Care at the University of Rochester School of Medicine, is recognized as a leader in the palliative care and hospice movement. In 2013, he was named by members of the American Academy of Hospice and Palliative Medicine as one of 20 top visionaries in the field. These two examples illustrate the association of physician assisted suicide with significant individuals in the field of palliative care, feeding suspicions of stealth euthanasia in palliative care and hospice.[5]

---

[3] Cessario might have done well to examine NHPCO position statements and their Code of Ethics and not just their mission statement. He might have found in these some guiding ethical principles that he found absent in the mission statement.

[4] See also Nelson (2013) who discusses Wald's "pro-euthanasia ideology."

[5] Capone et al. (2013) accuse Ira Byock of supporting the use of palliative sedation to hasten death. This is a false accusation. In an article co-authored with Timothy Quill, Byock states: "The purpose of the medications is to render the patient unconscious to relieve suffering, not to intentionally end his or her life" (2000, p. 410). He also refers to palliative sedation as a "last resort clinical response to extreme, unrelieved physical suffering" (ibid.).

A third area that has given rise to suspicions consists in the tools that relate to end-of-life care, namely living wills, durable powers of attorney, and, more recently POLST. Suffice it to say here, that the living will (developed by attorney Louis Kutner in 1967 and published in the *Indiana Law Journal* in 1969 in an article titled "Due process of euthanasia: the living will, a proposal") was widely distributed by the Euthanasia Society of America (renamed Society for the Right to Die in 1974). The association with euthanasia is present, even though the living will did not advocate either assisted suicide or euthanasia. POLST was developed in Oregon, beginning with the formation of a Task Force in 1991. Of course, Oregon was the first state in the United States to legalize assisted suicide, which occurred in October of 1997 with the passage of the Oregon Death with Dignity Act. While there is no connection between POLST and the Death with Dignity Act, it might be said by some that the culture in Oregon that led to the latter also influenced the former. Furthermore, one of the concerns raised by opponents of POLST is that the form lends itself to the practice of assisted suicide and even euthanasia. Again, associations (though remote) with physician assisted suicide and euthanasia raise suspicions about advance directives and POLST, which are frequently used is providing palliative care and in hospice.

Fourth, and finally, there are *practices*, particularly overdosing patients with opioids and palliative sedation to cause death that are claimed to be a part of palliative care. Panzer and the authors of "The Rise of Stealth Euthanasia" claim that these (and other) practices are widespread in palliative care and hospice.

> Traumatized families are reporting the hastened deaths of loved ones, and hospice and palliative care providers are warning that euthanasia and stealth euthanasia are sometimes being performed in end-of-life care settings. … Many in the field of hospice and palliative care… confirm that there is a clear trend toward hastening deaths of patients. Oncologists and primary care practitioners are shocked when their patients, who have chronic or terminal illnesses but are not in the active phase of dying and are not expected to die suddenly, die within days or weeks of entering hospice. Internationally known hospice and palliative care leaders confirm these reports. These professionals cannot *all* be wrong or ignorant". (Capone et al. 2013, p. 2)

Similar statements occur in Panzer (2011).

The claims being made here are very serious. Their serious nature requires more than anecdotal evidence, however. Hard evidence is required, as difficult as it may be to acquire. While some of the reporting may be factually correct, some may be the result of lack of information and/or lack of understanding or misunderstanding, especially by family members. For example, it is not uncommon for patients in the U.S. to be enrolled or admitted into hospice very late in their dying process, which results in patients dying *in* hospice very soon after admission. This can look suspicious to a family that has not truly understood that their loved one was already imminently dying when they entered into hospice. This misunderstanding is compounded when the attending physician doesn't recognize or report that the patient is imminently dying, leading to the family's perception that death was unexpected. Notwithstanding such cases, if there are, in fact, widespread abuses in palliative care and hospice that violate palliative care and hospice mission statements, this needs to be independently verified based on examination of medical records and rigorous interviews with the principal parties. In any case, these anecdotal reports generate suspicion of all palliative care and hospice programs, an unfortunate development that undermines the good work that most palliative care and hospice programs do.

Having said this, it would not be entirely surprising if the abusive practices noted do in fact occur, especially since requests for hastened death are not uncommon among hospice patients. For example, a 2004 study that appeared in the *Journal of Pain and Symptom Management* examined hospice social workers' perceptions of requests by patients or family members for a hastened death (Arnold et al. 27 [6] June 2004: 523–32). Slightly more than half of the 73 respondents indicated that they had a patient who requested a hastened death. Only a bit more than a quarter experienced the same from a family member. The study concludes that requests for a hastened death are not uncommon from hospice patients.

A 2014 study of the use of terminal sedation (Continuous Sedation until Death, or CSD) in nursing homes in Flanders, Belgium indicated that it is frequently used to hasten death (Rys et al. 2014). In a recent article in *Health Services Insights* titled, "Palliative Care and Patient Autonomy: Moving Beyond Prohibitions Against Hastening Death," the authors write:

> The National Hospice and Palliative Care Organization (NHPCO) upholds policies prohibiting practices that deliberately hasten death. We find these policies overly restrictive and unreasonable. We argue that under some circumstances, namely, those patients who have a terminal illness and whose life expectancy is less than six months, it is both reasonable and morally sound to allow for treatments that may deliberately hasten death; these treatments should be part of the NHPCO guidelines. Broadening such policies would be more consistent with the gold standard of bioethical principles, ie, respecting the autonomy of competent adults. … We will show how including VSED [voluntary stopping of eating and drinking], CSD upon request, and PAS (where legal) as part of the official NHPCO guidelines will better serve the needs of dying patients by expanding options that are both morally sound and reasonable. (LiPuma and DeMarco 2016, p. 37)

In an earlier article, the same authors propose that "decisionally capable, terminally ill patients who have a life expectancy of less than 6 months may request CSD before being subjected to the refractory suffering of a treatment of 'last resort'" (LiPuma and DeMarco 2015, p. 121). It is also the case that in the extensive debate about terminal or palliative sedation, many authors argued that there was actually no difference between palliative sedation and euthanasia so that, in fact, the use of palliative sedation in palliative care and hospice was surreptitious euthanasia (Billings and Brock 1996; Rady and Verheijde 2010; LiPuma 2011; Raus et al. 2011).

What this small sampling of citations illustrates is that hastening death by various means in palliative care and hospice is at minimum "in the air" and is actively being proposed and being performed by some. Together with what has been said above, it is no wonder that some are very concerned that death is being hastened in palliative care and hospice programs.

The challenge when it comes to "practices" is determining which are morally legitimate and which are not. In addition to what is actually done and the circumstances of the act, intention is crucial when it comes to the administration of opioids, palliative sedation, and the withdrawal of medically administered nutrition and hydration. And intention can be difficult to determine from the outside. Be that as it may, double effect reasoning, which is critical to decisions about these practices, needs to be correctly understood and applied. As Capone et al. observe:

> The principle of double effect is used to assess a good action that has both an intended good (primary) effect and an unintended bad (secondary) effect. Invoking the principle of double effect to justify speeding the dying process is disingenuous. In some palliative and end-of-life care settings, death from palliative sedation or use of opioids is not a secondary effect, but either the *intended* primary goal or the *unintended* result of failure to properly educate clinical staff. … Double-effect reasoning should not be used to justify inappropriate use of opioids and sedation. At best, it is inaccurate to use the principle of double effect to justify an a priori intention to kill. At worst, it is a misleading attempt to justify an evil action. (2013, p.1)

## 5.3 Responding to the Threat of Stealth Euthanasia

What are we to make of these concerns overall? First, when one considers the associations of the NPHCO, prominent palliative care practitioners, the living will and POLST, along with anecdotes, the literature on various practices that can be used to hasten death, and a culture that is increasingly accepting of physician assisted suicide, it is understandable why some see palliative care and hospice programs as an opportunity for stealth euthanasia. Furthermore, it is quite likely that at least some of the anecdotes are true, that some physicians do in fact hasten the death of their patients out of concern and in an effort to spare their patients additional intolerable and uncontrollable pain and suffering. Hence, the claims of stealth euthanasia may be seen as "the canary in the coal mine." They probably should not be dismissed out of hand, even though many are based on hearsay and inference.

However, it also needs to be emphasized that it is far from clear that stealth euthanasia is a pervasive problem in palliative care and hospice. What is needed here is not inferences, suspicion, and hearsay, but hard evidence as to what is actually going on and how extensive it might be. It would be unfortunate and unfair, at minimum, to impugn an entire movement on the basis of suppositions. Just because the NHPCO has some historical connections to organizations that supported or promoted physician assisted suicide and euthanasia doesn't mean that the NHPCO itself does. Just because some prominent palliative care physicians and proponents have advocated for the legitimacy of PAS does not mean that most of those working in palliative care and hospice currently do. And just because the living will was originally connected with a euthanasia-promoting organization and POLST originated in the first state to legalize PAS do not mean that these tools are currently being used to hasten death. The focus must be on what is actually happening in palliative care and hospice. Also, the fact that there may be some abuse by some clinicians does not vitiate all of palliative care and hospice. Nor does it mean that all palliative care and hospice have been infiltrated by a culture of death.

Second, several major studies have shown that end of life care in this country has opportunities for considerable improvement (Connors et al. 1995; Field and Cassel 1997; IOM 2015). Much progress has been made over the years in making hospice known and used, and in initiating palliative care programs in acute and long term care facilities across the country. These have benefitted innumerable individuals with chronic, life-threatening and terminal illnesses in a holistic way, along with their families. As these and other studies suggest, however, there is still a long way to go. It would be terribly unfortunate if progress were halted or undermined by claims that palliative care and hospice are simply euthanasia by another name. Not only would this undermine progress in addressing unmet needs of sick and dying persons, it could also undermine the confidence of patients, families and clinicians in the benefits of palliative care and a willingness to make use of it. While attempting to bring attention to possible abuses in palliative and end-of-life care, those questioning palliative care are likely to deal a blow to the best hope for improved end-of-life care that currently exists. In trying to save the lives of some dying patients, they could contribute to undermining good care for tens of thousands of other dying patients. The reality is, there is no other alternative. If palliative care and hospice fail, there is no alternative for caring for the sick and dying, there is nothing else to accompany them in their journey, to alleviate their symptoms and their pain and suffering. Actually, there is an alternative—PAS and euthanasia. Undermining palliative care and hospice plays directly into the hands of those advocating for hastening of death by self or others.

Third, palliative care is both supported and encouraged by the three most recent popes. John Paul II mentioned palliative care in his encyclical *Evangelium Vitae*, in the context of his discussion of decisions to forgo disproportionate medical treatments (1995, par. 65). Many years later, he was more explicit in his support:

> Particularly in the stages of illness when proportionate and effective treatment is no longer possible, while it is necessary to avoid every kind of persistent or aggressive treatment, methods of "palliative care" are required. As the encyclical *Evangelium Vitae* affirms, they must "seek to make suffering more bearable in the final stages of illness and to ensure that the patient is supported and accompanied in his or her ordeal" (n. 65). In fact, palliative care aims, especially in the case of patients with terminal diseases, at alleviating a vast gamut of symptoms of physical, psychological and mental suffering; hence, it requires the intervention of a team of specialists with medical, psychological and religious qualifications who will work together to support the patient in critical stages. (John Paul II 2004, par. 5)

This statement, as the one from *Evangelium Vitae*, is followed by a word of caution on the licit use of pain killers and the need to avoid the administration of "massive doses of a sedative for the purpose of causing death."

Pope Benedict XVI endorsed palliative care on several occasions, even referring to it as a right. In his message for the 15th World Day of the Sick, for example, Benedict stated:

> There is a need to promote policies which create conditions where human beings can bear even incurable illnesses and death in a dignified manner. Here it is necessary to stress once again the need for more palliative care centers which provide integral care, offering the sick the human assistance and spiritual accompaniment they need. This is a *right* belonging to every human being, one which we must all be committed to defend (2005) [emphasis added]

In 2009, while visiting the Hospice Foundation of Rome, Pope Benedict again referred to palliative care:

> Whoever has a sense of human dignity knows instead that they must be respected and supported while they face the difficulties and sufferings linked with their health conditions. Toward this end, today one takes recourse more and more to the use of palliative care, which is able to soothe pain that comes from the illness and to help infirm persons to get through it with dignity. Nevertheless, together with the indispensable palliative care clinics, it is necessary to offer concrete gestures of love, of nearness and Christian solidarity to the sick… (2009)

Finally, Pope Francis, in a 2015 "Address to the Pontifical Academy of Life," said this about palliative care:

> Palliative care is an expression of a proper human attitude to take care of one another, especially of those who suffer. They give testimony that the human person is always precious, even if marked by old age and sickness. The person is, in fact, regardless of the circumstance, a good in itself and for others and is loved by God. Therefore, when his life becomes very fragile and the conclusion of his earthly existence approaches, we feel the responsibility to assist and support him in the best way. … Therefore, I appreciate your scientific and cultural commitment to ensure that palliative care reaches all those in need of it. I encourage professionals and students to specialize in this type of assistance, which does not have less value because of the fact that it does "not save life." Palliative care does something that is equally important: it appreciates the person. (2015)

In the same address, however, Francis does raise a word of caution. He urges all those who practice palliative care to keep "integral the spirit of service" and to remember that medical knowledge can never be achieved "'against' a person's life and dignity." "It is in this capacity of service to life and to the dignity of the sick person," he says, "…which measures the true progress of medicine and of the whole society." Francis concludes by quoting John Paul II's appeal: "'Respect, defend, love and serve life, every human life'" (*Evangelium Vitae*, n. 5).

Of course, we ought not to forget that compassionate care for the sick and dying is integral to the Gospel and to Christianity. Care for the sick and dying have been integral to Christianity from its beginnings, and hospital-like facilities have played an important role in Christianity from at least the early fourth century onward (Risse 1999; Ferngren 2009). While palliative care may be a new discipline in health care, what it stands for and what is does is not new. The Gospels are replete with examples of Jesus caring for the sick, the infirm and the dying. Early and later Christianity was known for its care of the sick—Christians' care for victims of plagues, monasteries establishing hospices, the establishment of hospitals and of religious congregations devoted to care of the sick. Historian Henry Sigerist writes that the Christian faith introduced

> … the most revolutionary and decisive change in the attitude of society toward the sick. Christianity came into the world as the religion of healing, as the joyful Gospel of the redeemer and of redemption. It addressed itself to the disinherited, to the sick and the afflicted, and promised them healing, a restoration both spiritual and physical. … It became the duty of the Christian to attend to the sick and poor of the community. … The social position of the sick man thus became fundamentally different from what it had been before. He assumed a preferential position which has been his ever since. (1943, pp. 69–70)

Caring for the sick and dying in a holistic manner is not optional for Catholic Christians or for Catholic health care. It is essential to living out the Gospel, irrespective of the faith or beliefs of the recipients of that care. So even if palliative care and hospice did not formally exist, what they do would have to exist, even by some other name.

Fourth, and finally, Catholic health care must be vigilant and must take steps to ensure that palliative and hospice care delivered in its facilities truly respect human life and the dignity of the human person. To this end, those working in these disciplines need to be informed about and be helped to understand the Church's teaching on end-of-life matters (i.e., the use of opioids, the principle of double effect, and the withdrawal of medically administered nutrition and hydration), the meaning of "allowing someone to die," the prohibitions against assisted suicide and euthanasia and how these differ from allowing to die, and all of this in the broader context of respect for the human person and the tradition's approach to the meaning of illness, suffering and death.

Vigilance must also be exercised regarding the practices occurring in Catholic-sponsored palliative and hospice care programs to ensure that nothing is done intentionally or in ignorance that would violate respect for human life and dignity. Any suspected abuses need to be investigated and addressed.

## 5.4 Conclusion

There is too much at stake here for Catholic health care and practitioners of palliative and hospice care to assume a passive attitude. If a culture that permits the hastening of death for the sick, the elderly and the dying infects Catholic-sponsored palliative and hospice care programs, this will be their demise. The same is true even if there is only a suspicion of this occurring. Palliative and hospice care in Catholic facilities must be known for the quality of their care of patients and for the profound respect and reverence they embody for the persons in their care.

Furthermore, excellence in palliative and hospice care is Catholic health care's and the Church's only real defense against PAS. The fight against PAS is virtually lost in a society that overvalues and misunderstands autonomy. The only real alternative is palliative and hospice care—the care of the whole person, the effective alleviation of symptoms, and accompanying the sick, the elderly and the dying in their journey with all its struggles.

Palliative and hospice care were never intended to be stealth euthanasia. The chance of their becoming so is a perversion of their original mission. Those who have written about stealth euthanasia may well have sounded an alarm about a grave threat to that mission and to the practice of palliative care and hospice, even if the evidence is inconclusive. In light of this, Catholic health care needs to double its efforts to ensure that palliative care remains, as Pope Francis has stated, "an expression of a proper human attitude to take care of one another, especially of those who suffer."

# References

Abbott, Matt C. 2013. The tragic reality of "covert euthanasia". *Renew America*. http://www.renewamerica.com/columns/abbott/130509. Accessed 14 Dec 2016.

Arnold, E.M., K.A. Artin, J.L. Person, and D.L. Griffith. 2004. Consideration of hastening death among hospice patients and their families. *Journal of Pain and Symptom Management* 27 (6): 523–532.

Benedict XVI. 2005. *Message of his holiness Benedict XVI for the 14th world day of the sick*. http://w2.vatican.va/content/benedict-xvi/en/messages/sick/documents/hf_ben-xvi_mes_20051208_world-day-of-the-sick-2006.html. Accessed 14 Dec 2016.

———. 2009. *Address of his holiness Benedict XVI on the occasion of his visit to the hospice foundation of Rome*. http://w2.vatican.va/content/benedict-xvi/en/speeches/2009/december/documents/hf_ben-xvi_spe_20091213_hospice.html. Accessed 14 Dec 2016.

Billings, J.A., and S.D. Brock. 1996. Slow euthanasia. *Journal of Palliative Care* 12: 21–30.

Capone, Ralph A., Kenneth R. Stevens, Julie Grimstad, and Ron Panzer. 2013. The rise of stealth euthanasia. *Ethics & Medics* 38 (6): 1–4.

CaringInfo. Caring connections timeline. http://www.caringinfo.org/i4a/pages/index.cfm?pageid=3402. Accessed 19 Jan 2017.

Catechism of the Catholic Church. 1994. http://www.vatican.va/archive/ENG0015/_INDEX.HTM. Accessed 14 Dec 2016.

Cessario, Romanus. 2006. Catholic considerations on palliative care. *National Catholic Bioethics Quarterly* 6 (4): 639–650.

Connors, Alfred F., Jr., Neal V. Dawson, Norman A. Desbiens, et al. 1995. A controlled trial to improve care for seriously ill hospitalized patients. The study to understand prognoses and preferences for outcomes and risks of treatments (SUPPORT). *JAMA* 274 (20): 1591–1598.

Dial, Kathy. 2003. Are euthanasia advocates taking over America's hospice industry? LifeNews.com. http://www.lifenews.com/2003/12/19/bio-186/. Accessed 14 Dec 2016.

Ferngren, Gary B. 2009. *Medicine and health care in early Christianity*. Baltimore: John Hopkins University Press.

Field, Marilyn J., and Christine Cassel, eds. 1997. *Approaching death: Improving care at the end of life*. Washington, DC: National Academy Press.

Freidrich, M.J. 1999. Hospice care in the United States: A conversation with Florence S. Wald. *JAMA* 281 (18): 1683–1685.

Helmueller, Mary Therese. 2005. Warning—Are you being targeted for euthanasia? Rense.com. http://www.rense.com/general63/euth.htm. Accessed 14 Dec 2016.

Institute of Medicine (IOM). 2015. *Dying in America: Improving quality and honoring individual preferences near the end of life*. Washington, DC: The National Academies Press.

Isajiw, George. 2015. "Medical education in the shadow of 'stealth euthanasia'" among Catholics: Are we fighting secularism or heresy? *Linacre Quarterly* 83 (3): 210–216.

John Paul II. 1993. *Veritatis splendor*. http://w2.vatican.va/content/john-paul-ii/en/encyclicals/documents/hf_jp-ii_enc_06081993_veritatis-splendor.html. Accessed 14 Dec 2016.

———. 1995. *Evangelium vitae*. http://w2.vatican.va/content/john-paul-ii/en/encyclicals/documents/hf_jp-ii_enc_25031995_evangelium-vitae.html. Accessed 14 Dec 2016.

———. 2004. *Address to the participants in the 19th international conference of the Pontifical Council for Health Pastoral Care*. Retrieved from http://w2.vatican.va/content/john-paul-ii/en/speeches/2004/november/documents/hf_jp-ii_spe_20041112_pc-hlthwork.html. Accessed 14 Dec 2016.

Kirk, Timothy W., and Margaret M. McMahon. 2010. NHPCO position statement and commentary on the use of palliative sedation in imminently dying terminally ill patients. *Journal of Pain and Symptom Management* 39 (5): 914–923.

Kutner, Louis. 1969. Due process of euthanasia: the living will, a proposal. *Indiana Law Journal* 44: 539–554.

LiPuma, Samuel H. 2011. The lacking of moral equivalence for continuous sedation and PAS. *American Journal of Bioethics* 11: 48–49.

LiPuma, Samuel H., and Joseph H. DeMarco. 2015. Expanding the use of continuous sedation until death: Moving beyond the last resort for the terminally ill. *Journal of Clinical Ethics* 26 (2): 121–131.

———. 2016. Palliative care and patient autonomy: moving beyond prohibitions against hastening death. *Health Services Insights* 9: 37–42.

Mallon, John. 2009. Palliative care: The new stealth euthanasia. *Celebrate Life Magazine*. http://www.clmagazine.org/article/palliative-care-the-new-stealth-euthanasia/. Accessed 14 Dec 2016.

National Hospice and Palliative Care Organization. 2005. *Commentary and resolution on physician assisted suicide*. http://www.nhpco.org/sites/default/files/public/PAS_Resolution_Commentary.pdf. Accessed 7 Feb 2017.

Nelson, Kelleigh. 2013. Killing us softly—the pro-euthanasia ideology of American hospice founder Florence Wald. *Freedom Outpost*. http://freedomoutpost.com/killing-us-softly/. Accessed 14 Dec 2016.

Panzer, Ron. 2011–2013. *Stealth Euthanasia: Health Care Tyranny in America*. Rockford: Hospice Patients Alliance. HTML/Web version. http://www.hospicepatients.org/this-thing-called-hospice.html. Accessed 28 Dec 2016.

Pope Francis. 2015. Address to participants in the plenary of the Pontifical Academy for Life. http://w2.vatican.va/content/francesco/en/speeches/2015/march/documents/papa-francesco_20150305_pontificia-accademia-vita.html. Accessed 14 Dec 2016.

Quill, Timothy E. 1991. Death and dignity—a case of individualized decision making. *New England Journal of Medicine* 324: 691–694.

———. 1997. A physician's position on physician-assisted suicide. *Bulletin of the New York Academy of Medicine* 74 (1): 114–118.

Quill, Timothy E., and Ira R. Byock. 2000. Responding to intractable terminal suffering: The role of terminal sedation and voluntary refusal of food and fluids. *Annals of Internal Medicine* 132: 408–414.

Rady, M.Y., and J.L. Verheijde. 2010. Continuous deep sedation until death: Palliation or physician-assisted suicide? *American Journal of Hospice and Palliative Care* 27: 205–214.

Raus, K., S. Sterckx, and F. Mortier. 2011. Is continuous sedation at the end of life an ethically preferable alternative to physician-assisted suicide? *American Journal of Bioethics* 11: 32–40.

Risse, Guenter B. 1999. *Mending bodies, saving souls*. New York: Oxford University Press.

Rys, S., R. Deschepper, F. Mortier, L. Deliens, and J. Bilsen. 2014. Continuous sedation until death with or without the intention to hasten death—A nationwide study in nursing homes in Flanders, Belgium. *Journal of the American Medical Directors Association* 15 (8): 570–575.

Sigerist, Henry. 1943. *Civilization and disease*. Ithaca: Cornell University Press.

# Part II
# Body, Mind, and Spirit in Palliative Care

# Chapter 6
# The Catholic Moral Tradition on Pain and Symptom Management

G. Kevin Donovan

## 6.1 Human Dignity and Palliative Care

The treatment of pain is a fundamental necessity in the practice of medicine. In fact, the adjective "patient" as meaning "able to accept problems or suffering" is also the noun for a person under medical treatment (Oxford English Dictionary 2010). Medicine has been served by the Hippocratic tradition spanning two and a half millennia, in which we have been taught to do good and avoid harm to the patient. In addition, for the past 2000 years, the profession has benefited from the insights of the Christian tradition in which the sufferer is seen as reflecting the sufferings of Christ, and serving the needs of that sufferer is our most directly accessible path of service to Christ. It is no wonder that this Catholic Christian tradition has much to say about pain, suffering, and the proper moral as well as medical response to it.

When we consider the role of palliative care, we must appreciate its proper focus on pain <u>and</u> suffering. According to the World Health Organization (WHO) definition, "Palliative care is an approach that improves the quality of life of patients and their families facing the problems associated with life threatening illness, through the prevention and relief of suffering by means of early identification and impeccable assessment of pain and other problems, physical, psychosocial, and spiritual" (WHO 2017). This definition has much to be said for it, especially in its recognition of the link of suffering to pain and other problematic conditions. There are areas in which it could be improved upon, however. We should recognize that even in palliative care, pain is a problem for more than just those "life-threatening" illnesses, remembering the focus of palliative care extends beyond those facing the imminent end of life. Palliative care is not dependent on prognosis, but can be and often is

G. K. Donovan (✉)
Pellegrino Center for Clinical Bioethics/ Georgetown Univeristy Medical Center, Washington, DC, USA
e-mail: G.Kevin.Donovan@georgetown.edu



P. J. Cataldo, D. O'Brien (eds.), *Palliative Care and Catholic Health Care*, Philosophy and Medicine 130, https://doi.org/10.1007/978-3-030-05005-4_6

offered in conjunction with curative therapies. Properly understood, it is intrinsic to all good medical care, aiming for relief from troublesome symptoms. Therefore, it is not merely part of hospice care for the dying, even though their goals may often overlap.

Moreover, the WHO definition refers to the aim of improving the "*quality of life*" of patients. Although a laudable goal, we must be cautious in its application. When the term quality-of-life appears, we must be certain to avoid any use which may diminish the status of the patient to whom it is being applied. Too often, it is the observer's quality-of-life assessment as a third-party which is being introduced into the discussion. In such situations, it is too frequently applied to the most vulnerable patients whose own views cannot be expressed, such as infants, children, the frail elderly, and those with mental limitations both congenital and acquired. They may be seen to be in such a state that "no one would want to live that way." Edmund Pellegrino reminds us:

> Quality-of-life, however, is an infinitely malleable term. No two persons have the same definition of a satisfactory life. No one is qualified to make a quality-of-life decision for another, especially for an infant or child who has had no opportunity to experience life. In Christian charity and morality, there is no such thing as *Lebens unwertes lebens* nor metaphysically wrongful existence. Quality-of-life for those who cannot assess that quality for themselves is not a consideration in a Christian view of life, which bestows dignity on every human being, regardless of physical or intellectual limitations. Who among us can discern God's intent or provide initial purposes for any human life or for those in whose midst that life may be placed? Whether a person is a "useful" or "contributing" member of society does not affect his or her dignity, or the sanctity of that person's life. Human dignity is intrinsic, conferred by God, and therefore not "lost" or "gained" by human judgments. (Pellegrino 1999)

When seen through the Catholic moral tradition for palliative care, and for medicine generally, the patient who experiences pain and suffering does not also experience a diminishing of their inherent value in the presence of mental or physical disabilities. Such disabilities can in no way affect their human dignity, which is predicated on their relationship with their creator, their status as a child of God. They will be rightly seen as a person with disabilities, but fully a person, fully deserving the respect given to all others. This is in contrast to a more common secular perspective in which those who are unable to fully engage society mentally or physically are often seen as having a lesser worth. It is in reaction to this view that some disabled patients and their advocates insist on being thought of as "differently abled" or "handi-capable." In some cases this reaction has led to an avowed preference for their disability, to the point of insisting that such persons should select for children who would be born with the same disability rather than without it (Blankmeyer Burke 2011). A view in which disabilities can be accepted as part of the human condition, without diminishing or disparaging the patient who has them, leads to a fuller and more acceptable approach, and more in keeping, I think, with the World Health Organization's intended approach to improving the quality of life of patients and their families through palliative care.

In our assessment of the palliative approaches to pain and suffering, we should appreciate that the two are correctly distinguished in the palliative care literature.

Our focus will be on a moral tradition regarding these, but nevertheless must incorporate knowledge of the aspects of medical management, as we shall see. We will first consider theoretical and theological principles guiding pain relief, and later we will look at some of the technical aspects relevant to our discussion. We can then consider the problem of suffering as an issue even more challenging than that of pain, both in its causes and in the attempts to relieve it.

## 6.2 Catholic Teaching on Burdensome Treatment and the Relief of Pain

Pain management is a medical skill requiring knowledge, expertise, and good communication skills. It relies on some guiding principles, including the possibility of pain control (which underlies the philosophy of palliative care), the preference for the maintenance of consciousness, and the balancing of acceptable risks, particularly for those patients who are experiencing pain near the end of their lives.

The church has a long moral tradition regarding the care of the sick and dying, and the moral legitimacy of the means to do so. There are multiple contemporary expressions of this teaching as well. In his address to the International Congress of Anesthesiologists in 1957, Pope Pius XII explained the moral obligations regarding the preservation of human life:

> Natural reason and Christian morals say that man (and whoever is entrusted with the task of taking care of his fellow man) has the right and the duty in case of serious illness to take the necessary treatment for the preservation of life and health. This duty that one has toward himself, toward God, toward the human community, and in most cases, toward certain determined persons, derives from well ordered charity, from submission to the Creator, from social justice, and even from strict justice, as well as from devotion toward one's family. But normally one is held use only ordinary means – according to circumstances of persons, places, times, and culture – that is to say, means that do not involve any grave burden for oneself or another. A more strict obligation would be too burdensome for most men, and would render the attainment of the higher, more important good too difficult. Life, health, all temporal activities, are in fact subordinated to spiritual ends. On the other hand, one is not forbidden to take more than the strictly necessary steps to preserve life and health, as long as he does not fail in some more serious duty. (Pope Pius XII 1957)

Moreover, the *Catechism of the Catholic Church* recapitulates Catholic teaching on care at the end of life:

> Discontinuing medical procedures that are burdensome, dangerous, extraordinary, or disproportionate to the expected outcome can be legitimate; it is the refusal of "overzealous" treatment. Here, one does not will to cause death; one's inability to impede it is merely accepted. The decisions should be made by the patient if he is competent and able or, if not, by those legally entitled to act for the patient, whose reasonable will and legitimate interests must always be respected.
>
> Even if death is thought imminent, the ordinary care owed to a sick person cannot be legitimately interrupted. The use of painkillers to alleviate the sufferings of the dying, even at the risk of shortening their days, can be morally in conformity with human dignity, if death is not willed as either an end or a means, but only foreseen, and tolerated as inevitable.

> Palliative care is a special form of disinterested charity. As such, it should be encouraged. (*CCC*, ns. 2276–2279)

In *Evangelium Vitae,* St. Pope John Paul II encourages the development of programs of palliative care and the provision of proper pain management to patients:

> In modern medicine, increased attention is being given to what are called "methods of palliative care," which seek to make suffering more bearable in the final stages of illness and to ensure that the patient is supported and accompanied in his or her ordeal. Among the questions which arise in this context is that of the licitness of using various types of painkillers and sedatives for relieving the patient's pain when this involves the risk of shortening life. In such a case, death is not willed or sought, even though for reasonable motives, one runs the risk of it: there is simply a desire to ease pain effectively by using the analgesics which medicine provides. All the same," it is not right to deprive the dying person of consciousness without a serious reason" as they approached death. People ought to be able to satisfy their moral and family duties, and above all, they ought to be able to prepare in a fully conscious way for their definitive meeting with God. (1995, n. 65)

## 6.3 The Principle of Double Effect and Pain Management

Both these statements make use of a moral principle to justify the use of pain relieving medications near the end-of-life. This principle or rule is known as the Principle of Double Effect. Its classic version is formulated in the following way:

> A person may licitly perform an action that he or she foresees will produce both a good and bad effect, provided that four conditions are fulfilled at one and the same time:

1) That the action, in itself, by reason of its very object, be good, or at least indifferent.
2) That the good effect and not the evil effect be intended.
3) That the good effect is not produced by means of the evil effect.
4) That there exists a proportionately grave reason for permitting the evil effect. (Mangan 1949)

This moral principle has wide application, but is particularly useful in the provision of pain relieving medication in the care of the dying. It allows those physicians who would never consider euthanasia or assisted suicide to provide sufficient pain relief for a dying patient without violating their conscience or traditional medical morality. In following this rule, it is morally acceptable to use morphine for the provision of effective pain relief, even though there may be a risk of respiratory depression. If pain relief is the direct object and intention of providing morphine, and it is administered under appropriate circumstances, then the morphine dose can be escalated until that relief is achieved. If the escalation should lead to a fatal respiratory depression prior to the relief of pain, that would neither be the intended outcome, the direct means of pain relief, nor a disproportional result for a patient who is already facing death. Therefore, it should be considered a morally acceptable action. Following the guidelines of the Principle of Double Effect allows physicians to aggressively treat pain in these situations with a clear conscience. The importance of the proper application of this principle cannot be over-emphasized. It provides the needed moral underpinnings for the successful pursuit of pain relief for patients near the end-of-life. No patient should have to die in pain, and no doctor should hesitate to provide adequate pain relief.

Given a full understanding of the principle of Double Effect, we might rightly ask, why is it that pain is often poorly controlled by physicians? Is it merely a misunderstanding of the proper application of the principle or is there more involved? Undoubtedly poor education in proper pain control with subsequent poor understanding of proper techniques and approaches can be a factor. It was just this medical reality that gave rise to the needed specialty of palliative care. Nevertheless, there is more to it. Some physicians have underdosed their patients for fear of inducing undesirable narcotic side effects, such as constipation, sedation, or cloudy mentation, in addition to respiratory depression. Some physicians have reported a fear of causing narcotic addiction, a concern that becomes more inexplicable the closer the patient gets to the end of life. Finally, there are those physicians who express a fear of reprisal or censure by governmental and other supervisory agencies for narcotic overuse (Reese 2016). It is this latter justification that is most ethically problematic. It means that a physician, in order to protect himself or herself from imagined or potential inconvenient explanations, would knowingly and willingly allow their patients to suffer from inadequate pain relief. Unlike the other described reasons, this is unlikely to be corrected by matter of education alone. It would constitute a moral failing incompatible with the best principles of the medical profession, particularly that of putting the needs of the patient first.

Even when our focus is on the moral aspects of the relief of pain and suffering, some acknowledgment of the particulars in medical management must occur. It is beyond the scope of this chapter to give a detailed description and analysis of the safe and effective approaches to managing pain. Nevertheless, we can briefly review some of the multiple technical aspects that make pain relief effective. In mild to moderate pain, there is little medical or moral controversy. The means for controlling such pain include a variety of pharmacologic delivery systems, ranging from topical creams and patches, often of lidocaine, which are useful for diminishing pain from injections, immunizations, and providing superficial pain relief, such as in postoperative pain. Added to this are a variety of familiar analgesics and anti-inflammatories, particularly acetaminophen and several non-steroidal anti-inflammatory drugs. Equally important and useful are non-pharmacologic approaches. These cover a broad range, including the proper application of ice and heat, transcutaneous electrical nerve stimulation and other neurologic approaches, oral sucrose for infants and toddlers, as well as basic and time-honored approaches, such as soothing, comforting, and distraction. Acupuncture and movement (yoga, exercise) have some evidence for pain management in this setting. Both children and adults may be helped by mind–body medicine approaches such as meditation, guided imaging, and other relaxation techniques. (Barkin and Barkin 2001)

The mainstays in the treatment of severe pain tend to rely on narcotics and opioids, also given through a variety of routes with a variety of schedules and approaches. They can be effective orally, topically, parenterally, either intramuscular or intravenous, and intrathecal. Guidelines tend to consider opioids comparable and interchangeable in the treatment of chronic pain, but individual responses can vary. The World Health Organization "analgesic ladder" suggests treatment based on the intensity of pain, from no drug in the case of no pain (step zero) to strong opioids (step three) for moderate to severe pain. Adjuvant analgesic drugs should then be added according to guidelines. Morphine and morphine like drugs are often seen as medications with comparable properties and thought to be interchangeable, but in fact they have different pharmacokinetic and pharmacodynamic properties. Other individual factors, including genetics, or liver or renal failure, or treatment with other drugs can alter an individual patient's response. The type of pain can also influence efficacy, particularly neuropathic and breakthrough pain. Experienced clinicians know that the achievement of overall pain reduction with limited side effects is the main goal, but some dynamic aspects over time, including dose changes, switching to other opioids and supplementary analgesics, are similarly important. Not all opioids are identical in pharmacokinetics or pharmacodynamics, so differences in results should be anticipated. Even patients reaching a good response may need frequent adjustments of their therapy schedule. The main objective of physicians for those patients with chronic pain is to relieve the pain over time. To do this, they must adapt the therapy whenever pain is increasing. These adaptations will include schedule changes, the addition of other opioids or adjuvants, or even changing analgesic techniques entirely. An understanding of these many and important variables is necessary for a customized approach leading to the relief of an individual patient's pain. It points out the necessity for a specialist's familiarity with the drugs, their interactions, and their variability in different patients (Corli 2016). When this expertise is lacking, the availability of palliative care consultation becomes a medical necessity. Only good palliative care and pain medicine, which should be able to control pain in almost every case, can serve to blunt the perceived need by some to end their lives prematurely in order to avoid the fear of dying in pain (Stetka 2017; Achenbach 2016).

## 6.4 The Ethics of Palliative Sedation

In rare cases, when the best palliative care is unable to ameliorate severe and intolerable pain at the end of life, the use of sedative medications that have the effect of reducing a person's responsiveness and awareness to varying degrees, are sometimes useful. The drugs employed range from benzodiazepines, barbiturates, and antipsychotics, to anesthetic agents. Their use triggers some ethical questions, particularly in regards to the association between this practice and intended euthanasia. Even though euthanasia remains illegal in the United States, terminal sedation has been used in some cases in manners and methods which appear indistinguishable from intentional euthanasia. We must consider whether sedation to unconsciousness for intractable pain is an acceptable approach at the end-of-life, and if so, how can it be ethically employed without committing euthanasia (Erdek 2015)

The *Ethical and Religious Directives for Catholic Health Care Services* issued by the United States Conference of Catholic Bishops defines euthanasia as "an action or omission that of itself or by intention causes death in order to alleviate suffering" (USCCB 2009, n. 60). The Conference reminds us that neither Catholic healthcare institutions nor faithful Catholic practitioners may condone or participate in euthanasia or assisted suicide in any way. Importantly, we are reminded that "patients should be kept free of pain as possible so that they may die comfortably and with dignity, and in the place where they wish to die. Since a person has the right to prepare for his or her death, while fully conscious, he or she should not be deprived of consciousness without a compelling reason" (USCCB 2009). There is much to be said for the advantages of some preservation of conscious activity near the time of death. John Paul II taught, however, that "while praise may be due to the person who voluntarily accept suffering by forgoing treatment with painkillers in order to remain fully lucid and, if a believer, to share consciously in the Lord's passion, such 'heroic' behavior cannot be considered a duty of everyone" (St. John Paul II 1995, n. 65). Therefore, from a moral as well as a medical point of view, the use of sedatives may be offered when the patient's symptoms or distress are refractory and intolerable, having had no adequate relief with management that avoided reducing the patient's awareness. This is clearly justifiable for intractable pain, but requires closer examination and assessment.

Palliative sedation does not have to be considered an all-or-nothing phenomenon. Both the duration and the depth of sedation should be produced proportionate to the requirements of the patient's pain, as well as the wishes of the patient and family to maintain a level of awareness. Therefore, the administration of sedation should be guided by the intention to relieve suffering rather than of sedation intending to render the patient totally unaware and unresponsive, even though the deepest levels may ultimately be required by the patient's condition. This approach was confirmed in an AMA report from their Council on Ethical and Judicial Affairs, "Sedation to Unconsciousness in End-of-Life Care." They also recommended an approach of sedation proportionate to the requirements of the patient's pain when it proves unresponsive to aggressive, symptom-specific treatments. It is noteworthy that the AMA also recognized that palliative sedation is not an appropriate response to suffering that is primarily existential, and must never be used to intentionally cause patient's death (AMA/CEJA Report 2008 and Opinion 2.201).

The decision to employ palliative sedation is not perforce a decision to mercifully end the patient's suffering by prematurely ending their life. This would constitute a direct and intended euthanasia. Euthanasia may be deliberately produced by treating with larger doses of sedation, painkillers, and anesthetic agents than is required for the control of symptoms. It can also be accomplished by the concurrent withholding of nutrition and hydration from a patient during the period of terminal sedation. It is widely understood and accepted that the requirements for nutrition and hydration are likely to be diminished in a dying patient. Unless the patient's medical condition makes it impossible, or ineffective to offer nutrition and hydration, these should be given to the level physically required and tolerated. Forgoing such life-sustaining treatments should never be considered a routine accompaniment of palliative sedation. Such withholding when they might be otherwise effective would constitute an impermissible euthanasia (Sullivan and Heng 2012).

## 6.5 The Catholic Moral Tradition on Caring for the Suffering

As challenging as the management of chronic pain may appear, it pales in comparison to those challenges the physician must manage in the face of the patient's suffering. While the physician's obligation to relieve pain and suffering stretches back to antiquity, medical education is focused primarily on the relief of pain and curing of disease. This may be due to the overwhelming, amorphous and global nature of suffering. It can be a response triggered by pain, but often is the distress associated with mental, psychological, existential, or spiritual anguish. As Eric Cassell as described, suffering is not confined to physical symptoms. Patients may suffer not only from their disease, but from its treatment, and one cannot anticipate what the patients will see as a source of suffering – they must be asked. Some patients may find distinctively unpleasant side effects to be tolerable when seen as working towards a good end, but sometimes those effects that would seem minor or transient may cause real suffering to an individual patient. Patients may become self-conscious of their appearance to the point of isolating them from others. Other patients may find even temporary physical restrictions may keep them from their most important and self-defining activities. Therefore, patients will see the relief of suffering is one of the primary ends of medicine, a view often not shared by members of the medical profession. Patients do not make a distinction between physical and nonphysical sources of suffering in the same way the doctors do (Cassell 1982).

Although recognizing the presence of suffering and offering treatment for it is as intrinsic to good medical practice as is that of pain treatment, a successful approach is much more difficult, in part due to its variegated nature. Patients experienced not only physical suffering, which medicine expects to encounter and tries to ameliorate. Suffering can also be nonphysical, such as psychological, emotional, spiritual, or existential. These categories can be triggered by post-traumatic stress disorder, past abuses or losses to oneself, or wrongs done to others, as well as failed responsibilities, leading to a weak relationship with God and in some cases even to despair. Finally, and especially in the face of terminal disease, a patient may face an existential crisis questioning the meaning of one's suffering, the meaning of one's life, and the totality of one's existence. Physicians are far less capable of recognizing and managing these other sources of suffering.

The fact that physicians are poorly trained to deal with these issues can be attributed to more than a failure in the medical curriculum. Cassell argues that it goes back to Cartesian dualism, which proposed a split between mind and body. This dualism, Cassell claims, made it possible for science to escape the control of the church by assigning the non-corporeal, spiritual realm to the church, leaving the physical world as the domain of science. "Therefore, so long as the mind- body dichotomy is accepted, suffering is either subjective and is not truly "real" – not within medicine's domain – or identified exclusively with bodily pain." This leads to "an anachronistic division of the human condition into what is medical (having to do with the body) and what is nonmedical (the remainder) [which] has given medicine too narrow a notion of its calling. Because of this division, physicians may, in concentrating on the cure of bodily disease, do things that cause the patient as a person to suffer" (Cassell 1982).

Because pain is primarily a physical phenomenon, shared in common with all the bodies of the animal world, it is seen as comfortably within medicine's purview. Suffering is really different – unlike animals, humans are aware of their own suffering, and in the face of it, wonder why it is happening to them. To answer this question would require more than just medical knowledge, and we do well to look for a moral interpretation from the Catholic Church. A full examination of the problem of human suffering goes well beyond the scope of this chapter, or the capabilities of its author. However, a superficial survey of what has been taught and written may serve to round out a more complete understanding of the moral tradition which informs the medical management of pain and suffering.

When a patient is faced with an illness that entails suffering, requiring palliative care, questions arise about the meaning of that suffering. If we were to look to the church for answers, we see that the church first looks to scripture, and to Christ, who is the center of the gospel. Suffering, like all human things, finds its true meaning in Jesus Christ. It is both a burden and a joy. While the burdensomeness is evident, understanding the joy requires reflection into the mystery of redemption in Jesus Christ. Christ's redemption actively comes through his sacrifice on the cross, through suffering. He became man in order to experience this, and share it with us. Thus our redemption is directly linked to Christ's suffering, and our suffering is linked to our salvation (*Salvifici Doloris* 1984). A rare patient may take to heart St. Paul's words in Colossians 1:24: "Now I rejoice in my sufferings for your sake, and in my flesh I complete what is lacking in Christ's afflictions for the sake of his body, that is the church."

They would understand that Christ's sufferings were perfect and lacking nothing, but we are imperfect. When we act as his body, the church, to unite our sufferings to his, we take our imperfection and our suffering and join it to him, to perfect our own suffering and to benefit the church. Thus, in such a Christian perspective, one's suffering can be seen as a cause for joy, not only distress. Such a patient might even confound his physicians by refusing needed pain medicine, in order to experience fully the suffering that is the cause for spiritual growth and benefit. Such supererogatory virtue will not often be encountered, less often will be understood, but should always be respected by the caregivers.

It is this link between our suffering and our salvation that provokes the most questions and seems hardest to respond to appropriately. Many have struggled with the meaning of suffering for ages, sometimes finding answers that satisfy, sometimes finding solace. I will not attempt to re-examine or recapitulate all of these; I will only suggest a few possible responses to a patient's suffering that can come from physicians and others drawn to palliative care.

Prior to his own death, Edmund Pellegrino had some insightful reflections regarding our understanding of the suffering patient:

> Suffering demands that endurance and dying be personal enterprises. It unmercifully exposes our freedom to fashion a personal response. In that freedom lies the paradoxical possibility of healing even the dying patient. Unlike Job we must respond without seeing God eye-to-eye, without directly confronting his ineffability. To embrace or reject suffering, to make it our experience and not become its passive object, is the final test of that autonomy we hold so dear when we are healthy…. In the modern world, suffering is not very often accepted, simply and submissively, as God's will, as Job first accepted it. Rather, modern man resists openly the affront to pride and self-esteem, that misfortune, disease, and death represent. (Pellegrino 1989)

To modern man, situated in the secular world, suffering and death pose an insult to his dignity. The physician may not be able to help the patient find an explanation or purpose in their suffering. In particular, when a patient is a nonbeliever, religious explanations will not suffice. What can be done, even in a secular and egocentric world, is understanding the patient's perspective that no transcendent values may be recognized, and therefore suffering must be avoided at all costs. This may prompt attempts to create "death with dignity," either through suicide or euthanasia. Are physicians prepared to respond adequately to this existential crisis in their patients? They would seem less prepared and less capable than many other caregivers, such as chaplains, psychologists, and even family and friends. Nevertheless, they can give what they have, they can give of themselves. Suffering can and should elicit compassion, a word whose meaning is "suffering with." The compassionate physician must never minimize or belittle the sentiments of the sick, or take advantage of their vulnerability to proselytize or attempt to convert. They can give them the time they need to reconcile their lives, either with God or with significant others. The patient must always know that their physician will always be there for them, never abandon them, and do what is clearly in medicine's purview – manage the symptoms as best possible. As has been pointed out:

> Patients should be kept as free of pain [and suffering] as possible so that they may die comfortably and with dignity, and in the place where they wish to die. Since a person has the right to prepare for his or her death while fully conscious, he or she should not be deprived of consciousness without a compelling reason. Medicines capable of alleviating or suppressing [symptoms] may be given to a dying person, even if this therapy may indirectly shorten the person's life, so long as the intent is not to cause a death. (The Pontifical Council for Pastoral Assistance to Health Care Workers 1995)

In extreme cases of psychological or emotional anguish, palliative sedation has been offered. This is not without its problems, however. Depriving the dying person of consciousness may truncate the possibility of a truly "good death." The dying person is deprived of the possibility of living his own life by reducing him to a state of unconsciousness, which is unworthy of the dignity of a human being if there is no proportionate reason (The Pontifical Council for Pastoral Assistance to Health Care Workers 1995).

The Catholic moral tradition offers a strong challenge to concepts that see suffering as unredeemable and always needing to be eradicated by any means whatsoever. That challenge is based in the understanding that human dignity is not bestowed externally (by society, by social recognition, by class, by abilities, etc.), nor is it dependent on factors in one's environment. Human dignity is intrinsic to the condition of being human, a result of the dignity bestowed by God. It is neither increased nor decreased by accidental circumstances surrounding one's death. If not necessitated by intractable pain, to cause the loss of consciousness at the end of life in fact impedes the use of the higher faculties. It thereby unjustly eliminates all possibility of the exercise of the autonomy that many prize so dearly. It sacrifices the last opportunity to think and act autonomously at the moment of death, to finish important business, to reach a good death. Moreover, the suffering removed by terminal sedation of the patient may sometimes be the suffering of the medical and family caregivers, rather than the patient's alone. It is difficult to continue to provide compassion, to "suffer with" those patients who demand so much emotionally from the rest of us as they near the end of their journey. These opportunities to reflect and make peace can be overlooked when framed by less robust understandings of the meaning of suffering.

Finally, we must acknowledge that we are living in a society which seeks to avoid the awareness of pain, suffering, illness, and death at all costs. This is true for many physicians as it is for patients. Physicians often view the death of the patient as a defeat, a personal failure. We treat the practice of medicine as a battle with disease, we employ a "therapeutic armamentarium," and we fail to accept or acknowledge that all patients, no matter how expertly cared for, eventually die. The greatest strength and beauty of palliative care may be in its reversal of the usual emphasis, to face these problems with gentleness and sensitivity, to deal with pain and suffering, and most importantly, to help our patients find a way to deal with it. This search may lead patients to a place they did not seek, but perhaps needed to find, and needed to find it on their own. The chorus in *Agamemnon*, a famous play by the Greek playwright Aeschylus, who lived circa 524–456 BC, speaks eloquently to this:

> …he who learns must suffer. And even in our sleep, pain that cannot forget falls drop by drop upon the heart, and in our own despair, against our will, comes wisdom to us by the awful grace of God (Aeschylus 1929)

## References

Achenbach, J. 2016. *An opioid epidemic is what happens when pain is treated only with pills*. https://www.washingtonpost.com/news/to-your-health/wp/2016/12/23/an-opioid-epidemic-is-what-happens-when-pain-is-treated-only-with-pills. Accessed 19 Jan 2017.

Aeschylus. 1929. *Ten greek plays*. Trans. Gilbert Murray et al. New York: Oxford University Press.

American Medical Association. The AMA code of medical ethics' opinions on sedation at the end of life opinion 2.201 sedation to unconsciousness in end-of-life care. *AMA Journal of Ethics*. http://journalofethics.ama-assn.org/2013/05/coet1-1305.html. Accessed 19 Jan 2017.

Barkin, R., and D.L. Barkin. 2001. Pharmacologic management of acute and chronic pain: Focus on drug interactions and patient-specific pharmacotherapeutic selection. *Southern Medical Journal* 94 (8): 756–770.

Blankmeyer Burke, T. 2011. *Quest for a deaf child: Ethics and genetics*. Dissertation. University of New Mexico.

Cassell, E.J. 1982. The nature of suffering and the goals of medicine. *The New England Journal of Medicine* 306 (11): 639–645.
Corli, O. 2016. Are strong opioids equally effective and safe in the treatment of chronic cancer pain? A multicenter randomized phase IV 'real life' trial on the variability of response to opioids. *Annals of Oncology*. https://academic.oup.com/annonc/article/27/6/1107/1741500/Are-strong-opioids-equally-effective-and-safe-in. Accessed 19 Jan 2017.
Erdek, M. 2015. Pain medicine and palliative care as an alternative to euthanasia in end-of-life cancer care. *The Linacre Quarterly* 82 (2): 128–134.
Mangan, J.T. 1949. An historical analysis of the principle of double affect. *Theological Studies* 10: 43.
Oxford English Dictionary. 2010. Oxford University Press. http://www.oed.com/view/Entry/138820?rskey=6vshz6&result=1&isAdvanced=false#eid. Accessed 19 Jan 2017.
Pellegrino, E. 1989. The trials of job: A physician's meditation. *The Linacre Quarterly* 56 (2): 76–88.
———. 1999. Decisions at the end of life: The use and abuse of the concept of futility. The dignity of the dying derson. *Proceedings of the Fifth Assembly of the Pontifical Academy for Life*. Vatican City: Libreria Editrice Vaticana.
Pope Pius XII. 1957. Address to an International Congress of Anesthesiologists. *L'Osservatore Romano*, November 25–26.
Reese, S. 2016. Medscape ethics report 2016: Life, death, and pain. *Medscape*. http://www.medscape.com/features/slideshow/ethics2016-part2. Accessed 19 Jan 2017.
Sacred Congregation for the Doctrine of the Faith. 1980. *Declaration on Euthanasia*, Part II. http://www.vatican.va/roman_curia/congregations/cfaith/documents/rc_con_cfaith_doc_19800505_euthanasia_en.html. Accessed 19 Jan 2017.
St. John Paul II. 1984. *Salvifici Doloris*. https://w2.vatican.va/content/john-paul-ii/en/apost_letters/1984/documents/hf_jp-ii_apl_11021984_salvifici-doloris.html. Accessed 19 Jan 2017.
———. 1995. *Evangelium Vitae* n. 65. http://w2.vatican.va/content/john-paul-ii/en/encyclicals/documents/hf_jp-ii_enc_25031995_evangelium-vitae.html. Accessed 19 Jan 2017.
Stetka, B.S. 2017. The evolving role of opiods in managing chronic pain. *Medscape*. http://www.medscape.com/viewarticle/879009. Accessed 12 Apr 2017.
Sullivan, W.F., and J. Heng. 2012. The use of sedatives in the care of persons who are seriously ill or dying. *National Catholic Bioethics Quarterly* 12 (3): 489.
The Pontifical Council for Pastoral Assistance to Health Care Workers. 1995. *The charter for health care workers*. http://www.vatican.va/roman_curia/pontifical_councils/hlthwork/documents/rc_pc_hlthwork_doc_19950101_charter_en.html. Accessed 19 Jan 2017.
United States Conference of Catholic Bishops. 2009. *Ethical and religious directives for catholic health care services*. 5th ed. http://www.usccb.org/issues-and-action/human-life-and-dignity/health-care/upload/Ethical-Religious-Directives-Catholic-Health-Care-Services-fifth-edition-2009.pdf\. Accessed on 19 Jan 2017.
World Health Organization (WHO). 2017. *Who definition of palliative care*. http://www.who.int/cancer/palliative/definition/en/. Accessed 19 Jan 2017.

# Chapter 7
# Spiritual Care in Palliative Care

**David A. Lichter**

## 7.1 June's Story

The outpatient palliative care team met with June, a woman in her mid-60s and her daughter Maggie, her primary caregiver. The patient had COPD and was going through a "manic" stage where she was very non-compliant with her medications. While one could sense a dynamic of anger and resentment between the patient and her daughter, there also existed a deep mutual love and care for one another. As part of the goals of care, the chaplain coordinated with a parish nurse from a local church to visit regularly the patient and provide additional spiritual support. Two weeks after this coordination, the daughter, Maggie, called the chaplain. She felt guilty because she could no longer care for her mother. When they found an assisted living facility to care for June, Maggie wanted to know if the chaplain would visit her mother every day. The chaplain explained that this was not possible due to distance, but was able to coordinate visits with the local church. The following week a palliative care consult was called in the Intensive Care Unit. Unbeknown to the chaplain, it was for June. She had aspirated at the nursing home, was transferred to the hospital, and was now in the ICU. The family was already familiar with palliative care; they had met with the chaplain several times, as well as spoke via phone. The family expressed gratitude for the chaplain's familiarity with their family and the chaplain's continuity of care. June's daughters, Maggie and Rose, were present, while her son was on speaker phone. They all expressed a desire for their mother to receive comfort care measures based on the physician's report. The siblings agreed that this would be in accord with mom's wishes.

This narrative reveals many human and spiritual needs. Along with June's physical care needs, there exists a need for healing between mother and daughter, and Maggie's need to face the guilt she feels for her inability to provide for June's unmet needs. There is also the need for ongoing spiritual support through a religious representative, as well as the family's need to be prepared to make decisions based on

D. A. Lichter (✉)
National Association of Catholic Chaplains, Milwaukee, WI, USA
e-mail: dLichter@nacc.org



P. J. Cataldo, D. O'Brien (eds.), *Palliative Care and Catholic Health Care*,
Philosophy and Medicine 130, https://doi.org/10.1007/978-3-030-05005-4_7

their mother's values. In this story, spiritual care to meet these human and spiritual needs is integral to the palliative care provided.

## 7.2 Francisco's Story

When I invited a palliative care chaplain to reflect on his spiritual care practice, he offered the following:

> We use a holistic approach, open to all dimensions and not just the religious aspect. We connect with a most profound inner world. In my encounters, I ask, "Seems you have been through a lot. (*pause*) What gives you strength to be alive? Or what helps you to go through all of what you have been through?" Their first answer, in my experience, reveals their journey, their agenda, what they really are processing in their hearts and minds. Usually, their relationship with a Higher Power comes third in their priorities or does not appear at all. It doesn't have to do with a particular faith tradition. The first answer most often relates to a specific human being – a family member, a significant other, a loved one, a group of people. This relationship is what seems to bring hope, meaning, a sense of connection, and especially at the end of life, a sense of healing through forgiveness.

Then he shared this example:

> I remember Francisco, a 65-year-old Latino gentleman with an advanced cancer with poor prognosis, starting to process the reality that he would die from his illness. He identified his faith tradition as Catholic and had a wife and grandson living at home with him. During one of my visits to Francisco, he was sharing his life history with me, and as I was actively listening. I was intentionally seeking information to assess his spiritual pain, identify his inner resources, and assist him to articulate his wish going forward.
>
> Francisco told me his reason to live was the responsibility he had for raising his 13-year-old grandson, the joy in his life. He also said that was his reason to get more treatment and to endure all the side effects and procedures. He didn't understand at that point that all the treatments were palliative, symptom management and not curative.
>
> I asked, "What worries you the most?" He responded, "Dying and not being there for my grandson as he grows up." His response concerned the spiritual dimension, a source of spiritual pain caused by concern for a significant relationship. His response comprised the three most frequent fears of death, fear of physical suffering, and fear of what comes after life.
>
> After I shared the information with other members of the palliative care team, they talked to Francisco about palliative treatments that are not meant to cure, and goals of care that address quality of life rather than length of life. He decided to go to home to enjoy life with his grandson, and to finish a few projects with his grandson and his wife. Addressing his spiritual pain around the fear of not being there for his grandson, he created a plan with his wife and other relatives so that his grandson would continue receiving support when Francisco was not in this world for him. Francisco accepted support from the children's life specialist and from me to engage his grandson in an honest conversation breaking the bad news and creating memories with a picture book, a letter, and a handprint.
>
> Francisco was thankful for the support through honest conversations. He went home with good symptom management and an extra support group. He lived for several months.

He concluded:

> It is not about not dying. All of us will die. It is about where, how, and surrounded by whom. As a palliative care chaplain, I feel privileged to make a difference in people's lives by not

just listening but engaging those uncomfortable conversations that lead to healing, joy, peace, and reconciliation–that is, to a good death.

This chapter addresses the spiritual care needed for many people, like June and her family, and like Francisco and his family, who are choosing or in palliative care. It provides an overview of Catholic health care's pastoral care within palliative care, and will explain how spiritual needs are assessed, and what spiritual care services are provided. It will also address the chaplains or spiritual care specialists, like the ones mentioned above, who are needed to identify emotional and spiritual distress, and work within the interdisciplinary palliative care team to support of the patient and family.

## 7.3 The Spiritual Dimension of Pastoral Care

The World Health Organization (WHO) includes the identification and treatment of spiritual-based issues within its description of palliative care. It says that palliative care is "an approach that improves the quality of life of patients and their families facing the problem associated with life-threatening illness, through the prevention and relief of suffering by means of early identification and impeccable assessment and treatment of pain and other problems, physical, psychosocial and spiritual" (WHO 2014). In *Caring Even When We Cannot Cure*, the Catholic Health Association also explains, "In addition to assessing and treating physical and psychological symptoms of illness, palliative care can help relieve emotional and spiritual distress by providing resources and support for you and your family" (CHA 2016).

Part Two of the *Ethical and Religious Directives for Catholic Health Care Services* (ERDs), on "The Pastoral and Spiritual Responsibility of Catholic Health Care Services," uses both "pastoral" and "spiritual" as terms describing the spiritual dimension of palliative care (USCCB 2018). Often these terms are used interchangeably, which can be confusing. The professional chaplaincy associations show a growing consensus to use the language of spiritual care over pastoral care in deference to the interfaith acceptability of the term "spiritual," as "pastoral" connotes a more Christian theological underpinning. It is worthwhile, however, for this chapter to explore distinctive meanings that can be given to both these two terms within Catholic health care and its palliative care context.[1]

Emmanuel Lartey explains that the Christian view of "pastoral theology may be understood as a critical reflection on the nature and caring activity of the divine, and of human persons in relation to the divine, with the personal, social, communal, and cultural contexts of the world" (Lartey 2012). In light of this context, pastoral care assumes a very specific understanding in Catholic health care as "the nature and caring activity of the divine" and "of the human persons in relation to the divine"

---

[1] For example, in 2009 the Canadian Association for Pastoral Practice and Education (CAPPE/ACPEP) changed its name to Canadian Association for Spiritual Care (CASC/ACSS).

(Lartey 2012). We continue the healing ministry of Jesus, which is a pastoral ministry of the *Pastor Bonus* or Good Shepherd (John 10:1–18).

The Introduction of Part Two of the ERDs highlights the nature and caring activity of both God who created us and Christ who redeemed us, who are created in God's image and given a noble end:

> The dignity of human life flows from creation in the image of God (Gn 1:26), from redemption by Jesus Christ (Eph 1:10; 2 Tm 2:4-6), and from our common destiny to share a life with God beyond all corruption (1Cor 15:42–57).

The Introduction also explains how human activity should imitate the divine caring:

> Catholic health care has the responsibility to treat those in need in a way that respects the *human dignity* and *eternal destiny* of all. The Words of Christ have provided inspiration for Catholic health care: "I was ill and you cared for me" (Mt 25:36).

Along with honoring the human person's dignity (creation) and destiny (redemption), pastoral care further emphasizes the integrity or wholeness of the human person. This care "embraces the physical, psychological, social and spiritual dimensions of the person" (USCCB 2018).

The importance of pastoral care in Catholic health care in general also extends to its palliative care services. The palliative care team in Catholic health care emulates and participates in God's creative activity and Christ's redemptive healing action, and participates in a "community of healing and compassion" for the patient and the patient's family (USCCB 2018). The pastoral care focus of Catholic palliative care uniquely rests on the palliative care team being this community of healing and compassion, as it attends to the daily drama of patients and their families whose lives are impacted by their human (biological, psychological, and spiritual) experiences of illness, aging, and dying. Through those compassionate interactions, the palliative care team provides the support needed by the patient to confront her or his plight and make sense out of all of it. As the patient and family experience the palliative care team's holistic care and consistent, enduring, and endearing respect for their dignity and destiny, the patient and family are able to explore safely and face courageously what is driving their choice of treatment. Pastoral care asks what do they cherish, and how do they frame life's ultimate purpose? Choices are based on values and relationships, rather than purely functional or external realities. Whether one has an elective surgery might not be decided by the question "will I walk?" but rather by the question "who loves me and who cares whether I get better?"

The pastoral care mandate has everything to do with viewing and treating the patient and family as human beings deserving of love, respect, and care, and not as market share, objects for study, or a disease to cure. Within the pastoral care service where the palliative care team is experienced as a community of health and compassion, the human person encounters, lives with, finds meaning in, and stays connected to his or her life sources during illness, aging, and dying.

## 7.4 Spiritual Needs Within Palliative Care

The ERDs recognize that pastoral care is "directed to spiritual needs that are often appreciated more deeply during times of illness" (USCCB 2018). But how does one define and identify spiritual needs? "Have your emotional and spiritual needs been met?" is a Press Ganey patient satisfaction question used by many health systems. In this question, emotional needs are not separated from spiritual needs. This association is observed in Harold G. Koenig's enumerated ten emotional and spiritual needs which patients encounter during the vulnerable times of illness, disability, and dependency (Koenig 2003):

1. The need to make sense of the illness.
2. The need for purpose and meaning in the midst of illness. Patients need renewed purpose and meaning in order to continue to fight illness.
3. The need for spiritual beliefs to be acknowledged, respected, and supported
4. The need to transcend the illness and the self.
5. The need to feel in control and give up control.
6. The need to feel connected and cared for.
7. The need to acknowledge and cope with the notion of dying and death.
8. The need to forgive and be forgiven.
9. The need to be thankful in the midst of illness.
10. The need for hope.

The Press Ganey *White Paper* that accompanied Dr. Koenig's article argued that patients for the most part do not distinguish between an emotional and spiritual need. It contended that most needs seemed to fall into a cluster of "broad psychological concepts" that included such essentials as "search for meaning, hope, alleviation of fear, alleviation of loneliness, transcendence, desire to maintain religious practices, and presence of God" (Press Ganey 2003). Such a clustering of spiritual and emotional needs remains common practice.

Distinctions can be made, however, between *spiritual* and *religious* needs. About a decade ago, researchers observed the growing importance to distinguish spiritual needs from religious, and the need to "differentiate between the terms spiritual and religious" in health care research (Speck et al. 2004). However, a more recent review of research literature of spiritual needs in palliative care patients found that the terms spiritual and religious were often used together, sometimes differentiated, and sometimes understood jointly as spirituality/religiosity. For example, the research of Mark Cobb and his colleagues shows that "spiritual care responds to both religious and humanistic needs by meeting the requirements of faith and the desire for an accompanying person to 'be there', 'to listen', and 'to love'" (Cobb 2012).

Several authors have provided helpful explanations of spiritual and emotional needs in palliative care. A useful framework for understanding spiritual and emotional needs is expressed in a threefold model of spiritual health that addresses the need for meaning and direction, a sense of self-worth and belonging, and an ability to love and be loved – that often includes reconciliation of broken relationships

(McKnight 2016). Daniel Sulmasy identified three key needs of a dying patient as meaning, value and relationship (Sulmasy 2006). Ira Byock, in *The Four Things That Matter Most,* identified and explained the emotional/relational needs behind the four phrases: *Please forgive me; I forgive you; Thank you;* and *I love you* (Byock 2014). A qualitative hospice study collated spiritual needs under six themes: need for religion; want of companionship; need to be involved and in control; desire to finish business; need to experience nature; and the need of a positive outlook (Puchalski et al. 2016). One could argue that all of these aforementioned needs are profoundly spiritual, even religious, issues that are prominently addressed in palliative care.

Much research has also been done to show the value and impact of addressing or ignoring spiritual or religious needs. In a study on coping with cancer involving 230 cancer-diagnosed patients with a prognosis of less than 1 year, 68% identified religion as very important (Fitchett 2017). While some research studies note that the majority of patients do not have these issues addressed, the National Consensus Project for Quality Palliative Care's (NCP) *Clinical Practice Guidelines for Quality Palliative Care 4th Edition* emphasizes identifying, assessing, and meeting spiritual, emotional, and existential needs of palliative care (NCP 2018). In general, spiritual, religious, and existential needs are grouped under the term "spirituality," which is the common term used by the NCP to refer to these needs. In February 2009 in Pasadena, California, an NCP Consensus Conference focused on naming and improving the quality of spiritual care as an important dimension of palliative care. Under the leadership of Christina Puchalski, and Betty Ferrell, the Conference goal was to identify points of agreement about spirituality as it applies to health care and to make recommendations aimed at advancing the delivery of top quality spiritual care in palliative care (Puchalski et al. 2009).[2] The Consensus Conference defined typical spiritual/religious needs such as forgiveness, or sacramental/ritual needs, but also included "philosophical, religious, spiritual, and existential issues that arise in the clinical setting." The definition stated that "Spirituality is the aspect of humanity that refers to the way individuals seek and express meaning and purpose and the way they experience their connectedness to the moment, to self, to others, to nature, and to the significant or sacred" (Puchalski et al. 2009). This definition continues to be referenced in palliative care literature.[3]

---

[2] Much gratitude goes to Christina Puchalski, MD, and Betty Ferrell, RN, PhD, for their work with the Consensus Conference, February 17–18, 2009, sponsored by the Archstone Foundation of Long Beach, California.

[3] In UpToDate (2017) one can read: "Because of the need to standardize a definition for spirituality in palliative care, a group of interdisciplinary experts in palliative and spiritual care produced a consensus definition of spirituality as the aspect of humanity that refers to the way individuals seek and express meaning and purpose, and the way they experience their connectedness to the moment, to self, to others, to nature, and to the significant or sacred. It may include religion and other worldviews but encompasses far more general ways these experiences are expressed, including through the arts, relationships with nature and others, and for some, through the concept of secular humanism, the latter of which emphasizes reason, scientific inquiry, individual freedom and responsibility, human values, compassion, and the needs for tolerance and cooperation.

## 7.5 Spiritual Distress, Suffering, and Pain

Addressing these needs within palliative care is usually undertaken in a diagnostic process that uses diverse terms, such as "spiritual distress" or "spiritual or existential suffering" to determine "spiritual pain." *Spiritual distress* can be described as the "impaired ability to experience and integrate meaning and purpose in life through connectedness with self, others, art, music, literature, nature, and/or a power greater than oneself" (Burkhard 2014). *Spiritual suffering* is another term used to describe that experience in which the human spiritual needs of love, faith, hope, virtue and beauty are not being fulfilled. Spiritual care helps a person process the unfilled feelings, and come to place of resolve and peace (Bartel 2004).

Spiritual pain is one of the symptoms managed in palliative care. Attending to spiritual pain aids overall pain management in palliative care. Cecily Saunders, the founder of the modern hospice movement, contributed the concept of "total pain" – caring for the whole person and family with respect to spiritual, emotional, social, and physical pain. The Spiritual Health Assessment, for instance, designed with this total pain management in mind, identifies four dimensions of "existential suffering" as meaning, forgiveness, relatedness and hope, which need to be addressed in palliative care (Groves and Groves 2014).

Whether one names the common stressors as distress, suffering, or pain, much research now exists that evidences the toll taken on patients when these are not addressed. The importance of addressing common stressors is demonstrated in research that shows the negative impact on care and health outcomes when spiritual and religious resources are not provided to patients and families. If patients receive help to cope with their spiritual pain and suffering, then the caretakers' roles are a little easier for both family and the healthcare team (Fitchett 2017).

The research of Ken Pargament on coping with stress and spiritual distress in times of serious illness has been critically important for health care chaplaincy (2000, 2001, 2004). To chaplains, his research within palliative care has evidenced the many different ways that people will use religious and spiritual resources to cope with significant stressful life situations, such as serious chronic and acute illness. Religious/spiritual coping helps patients and families find meaning, experience more comfort, and promote life transformation (Pargament 1997). For example, among 170 patients with advanced cancer in one study, scores on a measure of positive religious coping (e.g., seeking God's love and care) were associated with better ratings of quality of life (Fitchett 2017). Pargament's work has informed chaplaincy practice with respect to spiritual assessments and more effective care overall. The body of evidence-based research continues to grow, supporting the convictions that spiritual care is a vital patient need, that spiritual distress is an integral part of overall distress, and that the spirituality of a patient and the patient's family affects health care decisions made, as well as health outcomes, such as quality of life, coping, and pain management (Balboni and Balboni 2016).

## 7.6 Spiritual Care Services Within Palliative Care

The foundation for spiritual care services is building a relationship and an ongoing dialogue of discovery with the patient and his/her family. Timothy Daalman and his colleagues explain that "spiritual care at the end of life as a series of highly fluid interpersonal processes in the context of mutually recognized human values and experiences, rather than a set of prescribed and proscribed roles" (Daalman et al. 2008). Consistent with all-encompassing whole- person care, all members of the palliative care team are in communication and coordination with one another to ensure that the patient's and family's desire for dignity and peace are respected and supported. Spiritual care services attend to the whole person with a focus on the person's spirituality, i.e., the person's need to express meaning and purpose, and to experience connectedness to the moment, self, others, nature and the Sacred.

NCP's Domain 5 provides a more developed definition of spirituality along with four guidelines that capture the core spiritual care services within palliative care. Its definition emphasizes that "Spirituality is recognized as a fundamental aspect of compassionate, patient and family-centered palliative care." It expands the 2009 definition by including the word "transcendence" with meaning and purpose to what individuals seek, and by adding community and society to the list of relationships that an individual experiences. Finally, it identifies how spirituality is expressed "through beliefs, values, traditions, and practices" (NCP 2018). NCP 5.1 guideline underscores the need to assess and respect the spiritual beliefs and practices of patients and families. The guideline cautions that there is a need to identify the possible diversity of beliefs among patient, family, and the Interdisciplinary Team (IDT), and to make sure they are "recognized and supported." It also emphasizes the professional board certified chaplain's role to assess and meet these needs, and that the IDT be educated in the "spiritual aspects of care" (NCP 2018). However, it also makes the very important point that "opportunities are provided to engage the staff in self-care and self-reflection regarding their own spirituality" (NCP 2018). This need for the entire palliative care team to be attentive to and nurture their own spirituality is greatly stressed in Catholic health care. The spiritual care specialist/board certified chaplain is to aid the palliative care team to care for themselves and to reflect on their own beliefs in the midst of care for seriously ill and dying patients. It is of utmost importance that the palliative care team members themselves also experience wellness as part of creating a community of healing and compassion.

NCP 5.2 delineates the spiritual assessment process that includes spiritual screening, a spiritual history, and an in-depth spiritual assessment. This guideline distinguishes these three methods from one another, and notes the diverse elements that could be part of this assessment process, including concerns about relationships, death and quality of life, and about sources of meaning, purpose and spiritual strength (NCP 2018).

The NCP 5.3 devotes itself to the importance of addressing the patient's and the family's religious, spiritual and cultural needs, "maximizing patient and family spiritual strengths" and incorporating the "symbols and language" of the patient and family's "cultural and spiritual practices." All needs to be documented in the plan of care, including "expected outcomes of care" (NCP 2018). This 4th edition added a fourth guideline, NCP 5.4, on ongoing care, recognizing the changing conditions of patients and care settings. It emphasizes being vigilant to these circumstances, watchful for "new or emerging issues" and, thus, the importance of an evolving care plan (NCP 2018).

## 7.7 Other Examples of Spiritual Care Services

Beyond these NCP Guidelines for spiritual care services, there are many ways patients and families are assisted in integrating their spirituality into their vulnerable life situation. These include: storytelling as part of meaning-making and legacy-naming; assisting with addressing unfinished business among family members or healing broken relationships; advance care planning; ensuring access to and utilization of meaningful religious resources; and connecting to appropriate religious leaders (O'Gorman 2006). Francisco, in the earlier story, was greatly helped as he spoke of his fears and regrets, and made choices about this remaining time that were empowering, satisfying, and endearing. His final weeks and months would have been very difficult without this support to bring closure and peace to his life story. The importance of being attentive to the patient's story is articulated in Atul Gawande's *Being Mortal*: "As people become aware of the finitude of their life, they do not ask for much, they do not seek riches. They ask only to be permitted, as far as possible, to keep shaping the story of their life in the world – to make choices and sustain connections to others according to their own priorities" (Gawande 2014, pp. 146–147).

Honoring and utilizing important religious, spiritual and cultural rituals and practices based on the patient's and family's needs and what might be most meaningful and relevant to them relies on the chaplain's professional awareness, skills, and creativity. Patients and families of all faiths, or no faith, might have rituals and practices that are meaningful to them. In the case of Catholic patients and their families, practices have included using scripture or a meditation on "the agony in the garden" for a patient struggling with what they are confronting (the chaplain is very selective as appropriate for the circumstances). The rosary might be a source of comfort, or the use of a ritual/prayer of healing to let go of past hurts. For some Catholics both the sacrament of Communion (or *Viaticum* – bread for the journey) and the Sacrament of Anointing can provide strength and comfort in end of life care. Recently, a palliative care chaplain shared the encounter with a patient who told the chaplain he was dying, and that he wanted to be anointed, even as the palliative care physician was assessing the possibility of the patient being discharged back to a skilled nursing.[4] The chaplain was able to get a priest to anoint the patient, and then later that day gave the patient Communion. The patient died the following morning. The chaplain recalled the patient's statements, "I am not scared of dying," "Death where is your sting," and "If I were honest, I am a bit scared." For this patient, the

[4] Author communication with an ACHPC chaplain.

Sacrament(s) of Anointing (with absolution) and Communion were a comfort and strength to help him fulfill his journey (O'Gorman 2006).

As noted above, the NCP 5.2 makes reference to three distinct types of interventions to identify a palliative care patient's spiritual health, distress, and quality of life: spiritual screening; spiritual history; and spiritual assessment. Each of these interventions is an important step in ensuring a holistic approach to identify the place of spirituality this patient's life – to understand what are the patient's spiritual stresses, concerns, strengths, and resources. To be done well, all three processes require sufficient education and training for each palliative care team member, since diagnosing such distress is complex. Sometimes spiritual suffering may show itself in psychological distress or in physical pain. As Puchalski and Farrell write, "Suffering is difficult to diagnose based on symptoms alone" (Puchalski and Farrell 2010). They note that identifying the type and source of suffering is not simple, since suffering includes physical, social, psychological and spiritual elements.

## 7.8 Spiritual Screening

The basic process of spiritual screening is conducted prior to a patient being referred to palliative care, since the screening is utilized most often during the admission process, or by a social worker or nurse during their first visit with a patient. One of the goals of spiritual screening is to ascertain the level of spiritual distress/pain. This screening is usually done through the use of one or more questions, such as: "Is religion or spirituality important to you as you cope with your illness? If yes, how much strength and comfort do you get from your religion/spirituality right now? If no, has there ever been a time when religion or spirituality been important to you?" Some health care entities employ different questions to determine how much or little a person is involved with his/her spiritual base (Fitchett and Risk 2009).

These spiritual screening questions either lead to a referral to a spiritual care specialist/board certified chaplain for a more in-depth spiritual assessment, or lead to a more general spiritual history taken by a social worker or nurse. In the case of palliative care, such screening might be done when the patient/family is discerning the palliative care philosophy.

Ongoing training of those who can provide this spiritual screening is vital for the improvement of palliative care. In a recent *MedScape* article on a training program to improve the spiritual care screening skills of the palliative care team and its importance to improving overall palliative care, Betty Ferrell made the following important point: "Expanding the generalist level of spiritual care is central to improving palliative care and addressing the very real workforce demands, optimal use of our chaplains as spiritual care specialists, and raising the bar for all clinicians" (Ferrell 2016). She noted that the training "demonstrated improvement in 15 spiritual generalist skills within three areas (spiritual screening and care plan, provision of spiritual care, and professional development)," and that "Fourteen of the 15 skills remained significantly higher at the 3-month follow-up, suggesting that these skills were retained."

## 7.9 Spiritual History and Assessment

Taking a spiritual history of the patient has served to provide clinicians a deeper background of a person's spiritual/religious life. The process of recording a patient's entire medical history also allows the patient to share spiritual, religious, cultural, existential beliefs and values that ground and guide that person's life, as well as to identify sources of nurture and strength that have provided support during vulnerable times. The clinician gains an understanding of the individual's key spiritual themes and struggles and their impact on the patient's current illness, treatment choices or plan of care, so that this knowledge can be called upon to empower the person, as needed, in the treatment process.

Several instruments, such as FICA, SPIRIT and HOPE, are utilized for this history-taking. The **FICA** tool organizes the inquiry around four letters: F – **F**aith and belief. I – the **I**mportance of spirituality; C – the role of spiritual or religious **C**ommunity; and A – how the palliative care team could **A**ddress issues. **SPIRIT** uses the interpretative filters of S – Spiritual belief system; P – Personal beliefs; I – **I**ntegration into a faith community; R – Rituals one might use or restrictions that are important to consider; I – **I**mplications of what is learned for care planning; and T – **T**erminal events planning. **HOPE** frames the discovery process with the following factors: H – seeking the sources of **H**ope; O – the role of **O**rganized religion; P – one's **P**ersonal spirituality and practices; and E – the **E**ffects of what is learned from the medical care and end-of-life decisions (Puchalski and Ferrell 2010). Validation and comparison of these tools in palliative settings are performed on a continuing basis (Borneman et al. 2010; Blaber et al. 2015).

At times during the spiritual screening, a more serious spiritual/religious distress or concern is identified that might potentially impact a patient's care. On these occasions, the patient is referred to a spiritual care specialist/board certified chaplain who will follow-up with a professional spiritual assessment. The assessment might include determining whether the spiritual/religious distress is caused by dissonance between the patient's beliefs and how the patient is interpreting the present illness, and issues such as a lack of meaning, alienation from God, or a lack of reconciliation in one's life. These issues could also be impacting psychological symptoms, such as anxiety, depression, anger, guilt, or physical symptoms, such chronic or acute pain.

While diverse instruments are employed to conduct these spiritual assessments, these more intensive assessments are less-scripted, more in-depth, extensive, and often entail an ongoing conversation in which the specialist listens to a patient's story to understand the patient's needs, and resources. Fitchett and Canada provide a helpful description of such assessments as "a more extensive (in-depth, on-going) process of active listening to a patient's story as it unfolds in a relationship with a professional chaplain and summarizing the needs and resources that emerges in that process. The summary includes a spiritual care plan with expected outcomes which should be communicated to the rest of the treatment team" (Fitchett and Canada 2010). The ongoing goal of the spiritual assessment is to assist the person in drawing on spiritual resources for coping with his/her life situation. The spiritual care specialist/chaplain utilizes the diagnosis to develop spiritual care interventions and plans to address these issues, and that ensure these goals are integrated into the patient's overall plan of care for the palliative care team (Shields et al. 2015).

## 7.10 Supportive Care Coalition's Spirituality in Goals of Care Project

The Spirituality in Goals of Care (SGOC) project conducted by the Supportive Care Coalition (SCC) was an excellent initiative to implement the NCP Domain 5 guidelines, as it aimed to "embed spirituality into palliative care practice, placing greater focus on identifying spiritual strengths and addressing sources of suffering, and emphasizing that spiritual care is the responsibility of every palliative care team member" (SCC 2017). The SGOC project sought to improve communication skills and collaboration among members of the palliative care team, and strengthen the palliative care team members' own self-assessment, with the aim of promoting "high-quality Goals of Care conversations for people living with serious illness" (SCC 2017). The Project recommended eight care team practices. They embody well the characteristics of a Catholic approach to spiritual care services in palliative care, as these practices honor the sacredness of the person being treated, the process of the goals of care conference, and foster the spirituality of the palliative care team members as they become a community of healing and compassion.

The first practice followed by the team asks that all the members of the palliative care team be "spiritually grounded and present." This resonates with the emphasis in the ERDs that the Catholic institution is a healing ministry. In the case of one patient, June, whose story is told above, prior to going on the home visit to see her, the palliative care team met in the office and reviewed her chart. Then the physician led a moment of intentional silence to help them enter into the present moment. Some palliative care teams might take turns reading a quote, a brief scripture passage, meditating on a word, or using an acronym that helps them acknowledge, affirm and detach or let go – all so that they can be present in the moment with the patient and family.

The second practice identified by the SGOC project poses the 'dignity' question: "What do we need to know about you as a person to give you the best care possible?" It may not be asked verbatim, and it is usually answered during the conference when talking with patients about who they are as persons, and hearing their and their families' stories. Their responses give the palliative care team a good indication how patients lived their lives and values, what quality of life means to them, and how this information influences their decision making. In the case of June, she mentioned that she might not be "all that with it" right now, but she is a well-educated person and wants to be told the truth. "Why do I need certain medications and what is it going to do to help me?"

The third SGOC practice asks about the patient's "spirituality, hopes and fears." During the conversation, there is an opportunity to identify what might be a road block for the patient or something that is keeping them from being more open and accepting of their care. One might ask, "What is your hope, now during this hospital stay? What is your hope at this time in your life?" To learn of their biggest concerns or fears can open an important dialogue. June's biggest fear was being abandoned. She loved her daughter, but she had the most tension living with her daughter. June's hope was to become more independent.

During this inquiry one will most likely need to engage in the fourth practice of honoring "silence that may facilitate deeper listening and sharing." When the palliative care team visited with June and her caregiver, Maggie, their moments of silence lasted 1–2 min. It took some time for June to respond to each question. The palliative care team needed to be patient with that length of time, as the temptation was to rush through the silence. The team realized that patients require additional time to process questions as many are sleep deprived, away from the comfort of their home, taking different medications, and spending the majority of their time lying flat in a hospital bed.

The fifth SGOC practice, assessing spiritual distress or suffering, aligns with the NCP 5.1, assessing and addressing the spiritual, religious and existential dimensions of care. June's spiritual distress was related to her loss of independence and autonomy. She was not able to do what she once did, and she missed being part of her church community. One of her goals was to be at home and there was grief and loss when this did not happen. Maggie was dealing with her own inadequacy as a caregiver and guilt in not being able to care for her mother. She was also dealing with caregiver burnout.

The spiritual distress assessment needs to be complemented by the sixth SGOC practice of discovering the patient's and family's spiritual strengths that include their faith, beliefs and values that also inform the goals of care (GOC) in general. June had two daughters and a son. They were all involved in her care. However, Maggie was the closest, and assumed the major role as caregiver. The children spoke on a weekly basis and were involved in every family care conference and GOC conversation. They supported each other. Although they no longer attended a Catholic Church, their faith was still important; it played a part in honoring their mother's wishes, and their decision making.

The final two SGOC practices focus on reverencing all the relationships that have been involved in this sacred discovery and directional process. In the seventh practice the palliative care team expresses their gratitude to the patient and family for the sacredness of this exchange with them. In the case of June, the team thanked June and Maggie as well as the other siblings for the privilege of being able to be with them during this difficult time in June's life, and assured them of their continued support as needed.

The eight practice of the palliative care team is a self-evaluation/reflection that is vital to ensuring their ongoing professional and spiritual growth as a palliative care team and a community of healing and compassion. June's care team met, after their GOC consult, to reflect on three questions. First, how had they felt about the conference? They reflected on how they experienced the conference as a little disjointed, and realized that it was due in part to the patient being in a manic state the past few weeks. They noted they had to move slowly with questions, and sit in a way to make eye contact with both June and Maggie, the caregiver. The second question was what they learned about the patient. They talked about the tension experienced between June and Maggie. They felt the Maggie wanted them there to "witness" how difficult it is for her to care for her mother. June and Maggie were open to a parish nurse coming to visit. The final question was what they learned about themselves while working together. The team noted how important it was to have more than one discipline present to observe the non-verbal communication between the patient and daughter. They spoke about their personal awkwardness when being uncertain as to when to initiate a different focus, such as addressing spiritual and emotional distress and needs, while still being sure to address the medical symptoms and needs.

These final practices illustrate well SCC's understanding that "spiritual care is the responsibility of every team member," while the chaplains were specialists, "uniquely prepared to be spiritual care mentors and models for their interprofessional colleagues." The chaplain played a significant role in helping develop "a team culture around spirituality and introduce specific team behaviors to provide whole-person care as well as build team resiliency and enhance spiritual well-being" (SCC 2017).

## 7.11 The Role of the Spiritual Care Professional/Chaplain in Palliative Care

The Association of Professional Chaplains (APC) outlines the *Standards of Practice for Professional Chaplains*. The first section of Standards, from 1 to 7, address the practices for patient and family care: assessment, delivery of care, access to information and documentation of care, teamwork and collaboration, ethical practice, confidentiality, respect for diversity. The second section of Standards, from 8 to 10, addresses the practices for the staff and the organization: care for staff, care for the organization, and the chaplain as leader. The third section of Standards, from 11 to 13, addresses maintaining competent chaplaincy care: continuous quality improvement, research, and knowledge and continuing education (APC 2016).

These Standards of Practice, especially Standards 1–7 directed to patient and family care, have been valuable in describing the chaplain's roles. However, more recently it has become helpful to describe the chaplain in palliative care as the spiritual care specialist on a team of specialists. Each member of the palliative care team is both a generalist and specialist. Each needs to be a generalist in the other specialties while providing the specialist knowledge and skill in one's own profession. The chaplain is the spiritual care specialist on the team, and must also be a competent generalist in the other palliative care team members' specialties (CAPC 2017).

In many places, the palliative care team is referred to as a transdisciplinary, versus interdisciplinary, team. This connotes the work and practice of each specialty being open to each other's specialty by contributing to a high-communication, mutual learning and mentoring culture that assists one another to become generalists in the other's field so that the most comprehensive, best care possible is provided the patient and family (Piotrowski 2012). The palliative care spiritual care specialist/chaplain is often viewed as the communication specialist in that the chaplain is involved in the SGOC conference to flesh out what is most important to the patient and family, and to ensure that what is said in the conversation is honored throughout the palliative care process or until death. This relationship building and the communication process among the palliative care team are vital to referencing spirituality within palliative care.

What is the spiritual care specialist/chaplain role? A study of chaplains in pediatric palliative care found that the medical directors in the study identified three critical issues where chaplains were needed and valued: addressing patient/ family spiritual suffering, improving the communication between family and staff, and offering appropriate rituals for the family and staff (Fitchett et al. 2011). These are helpful groupings describing core contributions for the spiritual care specialist. However, the spiritual care specialist's role also includes contributions to and care for the palliative care team.

As we saw in the SCC SGOC project above, the spiritual specialist/chaplain takes the lead for the palliative care team to address issues related to spiritual care. This includes assisting the other palliative care team members, through mentoring and coaching, to grow comfortable and confident in their abilities to perform spiritual screenings and histories. The chaplain also ensures all spiritual interventions are documented in the patient records, translating those concerns into understandable language of the palliative care team.

The CPG 5 provided the content for the development of specialty certification in hospice and palliative for board certified chaplains who are working in palliative care and hospice settings. Some chaplaincy organizations, such as HealthCare Chaplaincy Network (HCCN), teamed with a higher education institution to provide chaplains education in palliative care, and offer a certificate in palliative care chaplaincy (HCCN 2017). Other professional chaplaincy associations, such as the APC (APC 2016) and the National Association of Catholic Chaplains (NACC) (NACC 2016) developed specialty certifications in Hospice and Palliative Care.

These specialty certifications are vital for ensuring that those chaplains who are part of the palliative care teams are prepared and positioned to be clinical partners in the care of palliative care patients and families. In the case of the APC and the NACC, they both developed competencies for the Hospice and Palliative Care specialty certifications, grounding them in the NCP's CPG's, and organizing them under the four headings of the Common Qualification and Competencies for Certification for board certified chaplains. These four areas address the knowledge and theory of the (palliative care/hospice) practice, the identity and conduct of the chaplain, expected professional practices, and functioning effectively in the professional relationships and environment within which one's palliative care/hospice profession is provided. In the past two years, APC and NACC have partnered to revise their palliative care and hospice speciality certifications, joining together to create and implement one advanced certification process that resulted in a BCC-PCHAC (Board Certified Chaplain - Palliative Care and Hospice Advanced Certification) that signifies that the applicants have mastered the spiritual care competencies necessary for palliative care/hospice, including end-of-life care and care of those with life-limiting conditions (APC 2017, NACC 2016).

As mentioned above, to be an effective participant in the generalist/specialist model requires that palliative care chaplains be broadly trained in all dimensions (physical, social, psychological) of palliative care so that they can function competently in understanding the patient's situation, and explore with them implications, challenges, wishes, values and coping strategies, and communicate effectively with the other palliative care team members. Such training is obtained through programs, such as The Education on Palliative and End-of-Life Care (EPEC 2017) Project or the Center to Advance Palliative Care (CAPC 2017).

Beyond the above-mentioned competencies, the National Association of Catholic Chaplains (NACC) also expressed the importance of the spiritual and human development of the chaplain in this specialty certification given the chaplain's critical role of modelling/mentoring the other members of the palliative care team. The NACC Introduction to its first palliative care and hospice specialty certification explained that

> Palliative Medicine is a discipline that cultivates in each practitioner (nurse, physician, chaplain, social worker etc.) a servant's heart and a sage's mind... One's effectiveness in this specialty can never be adequately assessed within a set of competencies and/or prescribed guidelines but rather in the human crucible of an integrated, interdisciplinary practice that draws equally from each one's soulfulness as well as from each one's skill and life experiences. (NACC 2016)

This poetic language is grounded in the human and theological convictions that the patient and chaplain share a common humanity, and in the ministering of one to another they are "blest and broken" and become "sacrament for each other and a living sign of God's healing and redemptive love in our world." With these beliefs in mind, the certification interview process for this hospice and palliative care specialty certification would also include the candidate's ability to evidence how he/she has integrated major losses in life, and how he/she shows a profound sense of self-awareness and understanding (NACC 2016).

## 7.12 Final Remarks

In the context of Catholic health care, spiritual care in palliative care is grounded in the NCP's Clinical Practice Guidelines, and relies upon the nationally accepted process of spiritual screening, history taking, and assessment to identify a patient's spiritual needs, concerns, hopes, and beliefs and to ensure that in the GOC conferences the person's spirituality is fully identified and understood so as to be integrated into the total plan of care, and ensure the entire person and his/her family is respected and supported.

As noted in the opening remarks on the pastoral mission of the palliative care team in the setting of Catholic health care, the team understands their pastoral care as emulating and participating in God's creative activity and Christ's redemptive healing action. The palliative care team aims to become "a community of healing and compassion" for the patient and the patient's family. Thus, it is important that the palliative care team members reflect on their and support one another's their growing spirituality, as their palliative care practice is influenced by this patient and family engagement. The chaplain on a palliative care team leads this integration of spirituality, models and mentors the other palliative care team members in their specific roles, and helps to facilitate that deeper reflection on their own spirituality.

The mission statement of the Support Care Coalition expresses well this focus and the relationship of spiritual care within palliative care: *The Supportive Care Coalition fosters and strengthens the presence of spiritual care in palliative care team practice and promotes deeper spiritual engagement with patients and families, and within the interdisciplinary palliative care team* (SCC 2017).

## References


Association for Clinical Pastoral Education (ACPE). 2017. www.acpe.edu. Accessed 16 Nov 2016.

Association of Professional Chaplains. 2017. http://www.professionalchaplains.org/files/professional_standards/standards_of_practice/standards_of_practice_hospice_palliative_care.pdf. Accessed 12 Dec 2016.

Association of Professional Chaplains – BCCI (APC). 2017. http://bcci.professionalchaplains.org/content.asp?pl=45&sl=42&contentid=47. Accessed 29 Dec 2016. Also, http://bcci.professionalchaplains.org/content.asp?admin=Y&pl=42&sl=42&contentid=45. Accessed 11 Dec 2018.

Association of Professional Chaplains (APC). 2017. www.professionalchaplains.org. Accessed 16 Nov 2016.

Balboni, Michael, and Tracy Balboni. 2016. Influence of spirituality and religiousness on outcomes in palliative care patients. *UpToDate* 2017. https://www.uptodate.com/contents/influence-of-spirituality-and-religiousness-on-outcomes-in-palliative-care-patients?source=search-result&search=influence-of-spirituality-outcomes-in-palliative-care&selectedTitle=1~150. Accessed 6 Dec 2016.

Bartel, M. 2004. What is spiritual? What is spiritual suffering? *Journal of Pastoral Care and Counseling* 58(3):187–201. https://www.ncbi.nlm.nih.gov/pubmed/15478953. Accessed 29 Dec 2016.

Blaber, Michael, June Jones, and Derek Willis. 2015. Spiritual care: Which is the best assessment tool for palliative settings? *International Journal of Palliative Nursing* 21 (9): 430–439.

Borneman, Tami, Betty Ferrell, and Christina Puchalski. 2010. Evaluation of the FICA tool for spiritual assessment. *Journal of Pain Symptom Management* 40 (2): 163–173. https://doi.org/10.1016/j.jpainsymman.2009.12.019 Epub 2010 Jul 8.

Burkhart, Lisa. 2014. Nursing outcomes classification. In *Nursing diagnosis handbook*, ed. Betty J. Ackley and Gale B. Ladwig, vol. 762, 10th ed. Maryland Heights: Mosbey Elsevier.

Byock, Ira. 2014. *The four things that matter most: A book about living – 10th anniversary edition.* New York: Simon and Schuster.

Catholic Health Care Association of the United States. 2016. Palliative and hospice care: Caring even when we cannot cure. https://www.chausa.org/docs/default-source/ethics/3441_cha_end-of-life-guide_pc_english-lowres.pdf?sfvrsn=0. Accessed 28 Dec 2016.

Center to Advance Palliative Care (CAPC). 2017. https://www.capc.org/providers/palliative-care-videos-podcasts/defining-the-role-of-the-palliative-care-chaplian-rao85uGBb10/. Accessed 28 Dec 2016.

Cobb, Mark, Christopher Dorwick, and Mari Lloyd-Williams. 2012. What can we learn about the spiritual needs of palliative care patients from the research literature? *Journal of Pain and Symptom Management* 43 (6): 1105–1119.

Daalman, Timothy, Barbara Usher, Sharon Williams, Jim Rawlings, and Laura Hanson. 2008. An exploratory study of spiritual care at the end of life. *Annuals of Family Medicine* 6(5): 406–411. https://www.ncbi.nlm.nih.gov/pmc/articles/PMC2532767/. Accessed 29 Dec 2016.

Ferrell, Betty. 2016. Training in spiritual care: On the right course. Medscape. http://www.medscape.com/viewarticle/873263. Accessed 28 Dec 2016.

Fitchett, George. 2014. *Assessing spiritual needs in a clinical setting*. http://www.ecrsh.eu/mm/Fitchett_-_Keynote_ECRSH14.pdf. Accessed 15 Dec 2016.

———. 2017. *Recent progress in chaplaincy-related research*. Unpublished Manuscript.

Fitchett, George, and A.L. Canada. 2010. The role of religion/spirituality in coping with cancer: Evidence, assessment, and intervention. In *Psycho-oncology*, ed. J.C. Holland, 3rd ed. New York: Oxford University Press 2015.

Fitchett, George, and James Risk. 2009. The role of religion/spirituality in coping with cancer: Evidence, assessment, and intervention. *Journal of Pastoral Care and Counseling* 63(4): 1–12. http://www.thehastingscenter.org/wp-content/uploads/Fitchett-et-al-2011-%E2%80%93-Role-of-Professional-Chaplains-on-Pediatric-Palliative-Care-Teams.pdf. Accessed 29 Dec 2016.

Fitchett, George, Kathryn Lyndes, Wendy Cadge, Nancy Berlinger, Erin Flanagan, and Jennifer Misasi. 2011. The role of professional chaplains on pediatric palliative care teams: Perspectives from physicians and chaplains. *Journal of Palliative Medicine* 14 (6): 704–707.

Gawanda, Atul. 2014. *Being mortal: Medicine and what matters in the end.* New York: Metropolitan Books.

Groves, Richard, and Mary Groves. 2014. *Spiritual health assessment form*. https://sacredartofliving.org/index.php?option=com_content&view=category&id=84&Itemid=517. Accessed 29 Dec 2016.

HealthCare Chaplaincy Network. 2017. https://www.healthcarechaplaincy.org/professional-continuing-education/online-certificate-courses.html; and https://csupalliativecare.org/programs/chaplaincy/. Accessed 29 Dec 2016.

Koenig, Harold. 2003. Meeting the emotional and spiritual needs of patients. *Satisfaction Snapshot*. http://www.pressganey.com.au/snapshots/2013/Meeting%20the%20Emotional%20and%20Spiritual%20Needs%20of%20Patients.pdf. Accessed 12 Dec 2016.

Lartey, Emmanuel. 2012. Pastoral theology in healthcare stings: Blessed irritant for holistic human care. In *Oxford textbook of spirituality in healthcare*, ed. Mark Cobb, Christina M. Puchalski, and Bruce Rumbold, 293–297. Oxford: Oxford University Press.

McKnight, Whitney. 2016. Chaplains play important part of integrated palliative care. *Clinical Psychiatry*. http://www.mdedge.com/clinicalpsychiatrynews/article/107563/geriatrics/chaplains-play-important-part-integrated-palliative. Accessed 20 Dec 2016.

National Association of Catholic Chaplains (NACC). 2016. http://www.nacc.org/certification/applying-for-certification/. Accessed 28 Dec 2016. November. Also, https://www.ncc.org/certification/palliative-care-and-hospice-advanced-certification/. Accessed 11 Dec 2018.

National Consensus Project (NCP) for Quality Palliative Care 2018. *Clinical practice guidelines for quality palliative care*. 4th ed. Richmond, VA: National Coalition for Hospice and Palliative Care.

Neshama: Association of Jewish Chaplains (NAJC). 2017. www.najc.org. Accessed 16 Nov 2016.
O'Gorman, Mary Lou. 2006. Spirituality in end-of-life care from a catholic perspective. In *A time for listening and caring: Spirituality and the care for the chronically ill and dying*, ed. Chistiana M. Puchalski, 139–154. Oxford: Oxford University Press.
Pargament, Kenneth. 1997. *The psychology of religion and coping: Theory, research, practice*. New York: The Guilford Press.
Pargament, Kenneth, Harold Koenig, and Lisa Perez. 2000. The many methods of religious coping: Development and initial validation of the RCOPE. *Journal of Clinical Psychology* 56 (4): 519–543.
Pargament, Kenneth, Harold Koenig, N. Tarakeshwar, and J. Hahn. 2001. Religious struggle as a predictor of mortality among medically ill elderly patients: A two-year longitudinal study. *Archives of Internal Medicine* 161: 1881–1885.
———. 2004. Religious coping methods as predictors of psychological, physical and spiritual outcomes among medically ill elderly patients: A two-year longitudinal study. *Journal of Health Psychology* 9 (6): 713–730.
Piotrowski, Linda. 2012. Transdisciplinary relationships. In *Professional & spiritual care*, ed. Rabbi Stephen B. Roberts, 240–248. Woodstock: Skylight Paths Publishing.
Press Ganey. 2003. *Patient satisfaction with emotional and spiritual care*. http://www.pressganey.com.au/snapshots/2013/Meeting%20the%20Emotional%20and%20Spiritual%20Needs%20of%20Patients.pdf. Accessed 12 Dec 2016.
Puchalski, Christina, and Betty Ferrell. 2010. *Making health care whole: Integrating spirituality into patient care*, 6–7. West Conshohocken: Templeton Press.
Puchalski, Christina, Betty Ferrell, and Rose Virani. 2009. Improving the quality of spiritual care as a dimension of palliative care: The report of the consensus conference. *Journal of Palliative Care Medicine* 12 (10): 885–904.
Puchalski, Christina, Betty Ferrell, Shirley Otis-Green, and George Handzo. 2016. Overview of spirituality in palliative care. *UpToDate*. https://www.uptodate.com/contents/overview-of-spirituality-in-palliative-care/print?source=search_result&search=palliative-care-spirituality&selectedTitle=1~150. Accessed 6 Dec 2016.
Shields, M., A. Kestenbaum, and L.B. Dunn. 2015. Spiritual AIM and the work of the chaplain: A model for assessing spiritual needs and outcomes in a relationship. *Palliative Supportive Care* 13 (1): 75–89.
Speck, Peter, Irene Higginson, and Julie Addison-Hall. 2004. Spiritual needs in health care. *British Medical Journal* 329: 123–124. http://eprints.soton.ac.uk/17604/1/Addington-Hall,_J._BMJ_-_Spiritual_needs_-_Jul_2004.pdf. Accessed 12 Dec 2016.
Spiritual & Existential Care – Nurses Learning Network. https://nurseslearning.com/courses/.../CPFSpiritualExistential/QS14ECPFSpiritualCare. Accessed 12 Dec 2016.
Sulmasy, Daniel. 2006. Spiritual issues in the care of the dying. *Journal of the American Medical Association (JAMA)* 296 (11): 1385–1392.
Supportive Care Coalition. 2017. http://supportivecarecoalition.org/wp-content/uploads/2016/06/SCC-Stages-Tools-Goals-of-Care-2014.pdf. Accessed 7 Jan 2017.
The Education on Palliative and End-of-Life Care (EPEC) Project. 2017. http://www.nhpco.org/link/education-palliative-and-end-life-care-epec-project. Accessed 29 Dec 2016.
United States Conference of Catholic Bishops. 2018. *Ethical and religious directives for catholic health care services*. 6th ed. Washington, DC: United States Conference of Catholic Bishops.
UpToDate. 2017. https://www.uptodate.com/contents/overview-of-spirituality-in-palliative-care. Accessed 28 Aug 2017.
World Health Organization (WHO). 2014. *67th world health assembly*. http://www.who.int/cancer/palliative/definition/en/. Accessed 14 Nov 2016.

# Chapter 8
# Psychological Issues in Catholic Palliative Care: The Challenge of Requests to Hasten Death

**Daniel P. Dwyer**

## 8.1 Introduction

"Will you help me hasten my death?" It is a question caregivers and family members are increasingly likely to hear as more states consider and approve legislation authorizing physician assisted suicide and "aid in dying." Physician assisted suicide is now legal in Oregon, Washington, Vermont, California, Colorado, and Montana. Each year, critics and proponents are waging new battles over aid-in-dying bills and citizen initiatives. While these political debates are necessary and often impassioned, they may also distract from the deep human distress that often underlies patient requests to hasten death, distress that in many cases contributes to and is exacerbated by undiagnosed or undertreated psychological and psychiatric illnesses.

For those of us who serve in Catholic health care, the question "Will you help me hasten my death?" is an invitation to participate in a sacred encounter with dying patients and their families. A "sacred encounter" recognizes that the dying person, like all human beings, is made in the image and likeness of God and is therefore worthy to have this inherent dignity recognized and to be treated with respect and reverence. Caring for persons who are approaching the end of their lives, who may be depressed and overwhelmed with feelings of hopelessness, requires a deeper form of connection and an approach to health care that is compassionate, wholistic, and open-hearted. Palliative care and hospice services, supported by Catholic social and moral teachings, can address the myriad medical, psychological, and spiritual needs of men and women at the end of life, including those persons experiencing emotional distress so overwhelming that a hastened death appears to them to be their only option.

---

D. P. Dwyer (✉)
Providence St. Joseph Health, Renton, WA, USA
e-mail: daniel.dwyer@stjoe.org



P. J. Cataldo, D. O'Brien (eds.), *Palliative Care and Catholic Health Care*, Philosophy and Medicine 130, https://doi.org/10.1007/978-3-030-05005-4_8

This chapter will focus on the psychological domain of palliative care and explore several issues critical to enhancing our understanding of palliative care as a wholistic approach to end-of-life care, particularly in response to the significant psychological challenges often faced by dying persons. How can we deal compassionately with the realities of grief, hopelessness, and the depression they may engender? What is the relationship between hopelessness and depression and the patient's request to hasten death? How effectively does Catholic-inspired palliative care, in particular, support the dying person's journey, even in the midst of those aspects that appear to contradict Catholic moral teaching? Finally, how can palliative care programs "act in communion with the Church" by assisting parishes and dioceses in ministering to their members who may have questions about end-of-life care, including physician assisted suicide?

## 8.2 The Catholic Moral Tradition

In the Catholic tradition, the dignity of the human person is a foundational principle. It affirms the nature of the person as a unity and integration of the physical body, the mind, the emotions, and the spirit from the moment of conception until the last breath. The Catholic view affirms that life is a gift from God; we are stewards over that life but not its owners. We have a moral duty to protect and sustain life but, significantly, we are not required to prolong it with non-beneficial or excessively burdensome treatments or conditions (Ethical and Religious Directives 2018). While Catholic moral tradition recognizes the redemptive power of suffering, the Church does not demand passive submission to suffering if there are positive, life-affirming means of alleviating it. Consider Catholic teaching on this subject as recapitulated by St. John Paul II:

> In modern medicine, increased attention is being given to what are called 'methods of palliative care', which seek to make suffering more bearable in the final stages of illness and to ensure that the patient is supported and accompanied in his or her ordeal. Among the questions which arise in this context is that of the licitness of using various types of pain-killers and sedatives for relieving the patient's pain when this involves the risk of shortening life. While praise may be due to the person who voluntarily accepts suffering by forgoing treatment with pain-killers in order to remain fully lucid and, if a believer, to share consciously in the Lord's Passion, such "heroic" behavior cannot be considered the duty of everyone. (John Paul II 1995)

Catholic-sponsored palliative care proactively addresses physical, emotional, and spiritual pain, reflecting a wholistic vision of compassionate care at the end of life. Palliative caregivers, working in concert with doctors and other medical providers, can also address the fears of many patients about pain and the unnecessary prolongation of the dying process through aggressive but ultimately futile interventions. Fear of pain (as opposed to actual pain) can be a determinative factor in patient inquiries about physician assisted suicide and other forms of aid in dying. Within the Catholic moral tradition, palliative caregivers are not required to ignore such inquiries, cognizant as they are of the existential suffering that prompted them. Instead, palliative care and hospice workers have an ethical obligation to acknowledge moral distress and then provide effective alternatives to physician assisted suicide, reassuring patients and their families that a "good death" does not require – and in fact, discourages – the premature ending of a life.

## 8.3 The Psycho-Social Dimensions of Palliative Care

The palliative care movement has long recognized the significant psychological and emotional needs of the person facing an impending death. Palliative caregivers with psycho-social expertise (primarily psychologists and social workers) are valued members of palliative care teams, and professional organizations and funding sources require appropriate levels of mental health credentialing in exchange for accreditation and reimbursement of end-of-life services. In addition to the treatment of physical and psychological pain, care of the whole patient requires expertise in social systems theory and practice, including appreciating the role of the family and social factors such as religious, ethnic, and socio-cultural influences (NQF 2006; CAPC 2006; National Consensus Project 2018).

While mental health professionals play an important role in palliative care, most models assume that all members of a palliative care team (physicians, nurses, social workers, and chaplains) are capable of and willing to address psychological issues as they arise (Weissman and Meier 2008). Physical pain and other uncomfortable treatment-related symptoms are rarely isolated phenomena. They frequently co-present, or give rise to, deeply distressing feelings of helplessness and hopelessness, which in turn may evoke clinical conditions such as anxiety and depression. The palliative care team serves as a multi-dimensional resource for early-stage detection, assessment, and treatment of psychological distress, whether rooted in physical pain, spiritual uncertainty, or worries about family well-being.

Another significant source of psycho-social distress for the dying person may be the context and content of end-of-life health care itself. In principle, all caregiver-patient encounters within a Catholic ministry are sacred, but in practice many patients feel lost and frightened in a modern hospital, a complicated medical-technical universe with its own fast-paced culture and unfamiliar language. Many dying patients have experienced significant, often painful medical treatment including surgery, radiotherapy, chemotherapy, and dialysis; hundreds of lab tests, x-rays, and scans; and disorienting stays in high-tech intensive care units. The accumulation of physically invasive experiences and physical and emotional suffering can trigger or exacerbate clinical conditions like anxiety and depression. By providing a gentler alternative to aggressive, potentially painful, and ultimately futile medical intervention, palliative care programs help dying patients manage the physical trauma and fear that too often prompt questions about physician assisted suicide (Byock 2017).

## 8.4 Palliative Approaches to Grief, Hopelessness, and Depression

One of the most challenging and complicated emotions faced by dying persons and their families is grief, a multifaceted and very normal response to loss and anticipated loss. Preparatory grief is exhibited when a person begins to mourn future losses – loss of agency and control, loss of simple pleasures, loss of beloved family, friends, and pets, loss of self-identity, loss of future possibilities, and a perceived loss of meaning and purpose (Widera and Block 2012). Grief over these losses can be expressed in numerous ways, including anger, denial, withdrawal, and deep sadness. Grief can be physically painful and emotionally wrenching, both for the dying person and their loved ones. As painful and difficult as the grieving process may be – and it differs from person to person – it is an entirely rational and often necessary response to loss. Skilled palliative caregivers trained in grief work can acknowledge and truly hear the dying person's anguish without turning away, becoming lost in it, or treating it as a pathology in immediate need of pharmacotherapy (Periyacoil and Hallenbeck 2002). By allowing grief to exist, and by working through the grieving process as slowly or quickly as the patient desires, palliative caregivers have an opportunity to ease the emotional distress that, unaddressed, can deepen into hopelessness and depression.

In the medical literature, there is no "bright line" that determines where normal grieving ends and clinical depression begins. Behavior that may make sense and even be beneficial in a grieving context – withdrawing temporarily from social interaction to take stock and assess one's life, for example – can quickly spiral, in the depressed individual, into isolation and extended withdrawal from human interaction (Hughes 2011). Studies have also found that unaddressed and untreated depression can manifest in the kind of despair that leads dying individuals to consider physician assisted suicide (Block and Billings 1994). In fact, "a persistent, active desire for an early death in a patient whose symptomatic and social needs have been reasonably met is suggestive of clinical depression" (Periyacoil and Hallenbeck 2002). Although physical suffering can be a source of deep distress and a strong motivator for hastening death, the perception that physical suffering alone generates requests for physician assisted suicide overlooks the significance of psychological distress. Numerous studies completed over the past 20 years find there is a strong correlation between depressive symptoms and hopelessness and a desire for hastening death. For instance, Max Cochinov noted this in the mid-1990s:

> Concerns about euthanasia and assisted suicide have been expressed by mental health professionals who have cautioned that the patients most likely to request assisted suicide would be 'the elderly, those frightened by illness, and the depressed of all ages.' This concern is based on the view that a desire for death even among the terminally ill, may be indicative of a potentially treatable psychiatric disorder. Although leading proponents of the right to die sometimes dismiss this argument out of hand, in fact the limited psychiatric data that are available provide evidence consistent with it. (Cochinov et al. 1995)

Ezekiel Emanuel, writing 10 years later, observed: "The extent to which potentially treatable psychiatric disorders may influence patients' decisions for aid in dying has been debated. For people at the end of life, depression, hopelessness, and psychological distress are among the strongest correlates of desire to hasten death" (Emanuel 2005). In 2014, Barry Rosenfeld found that "Elevated levels of depression and hopelessness, and low levels of spiritual well-being, have emerged as the strongest predictors [of requests to hasten death] regardless of illness or disease status" (Rosenfeld et al. 2014). Since depression correlates with a higher rate of requests for assistance in hastening death, Catholic palliative care providers opposed to assisted suicide on moral grounds must be vigilant and proactive in caring for the psychological and psychiatric needs of their patients. Consequently, a request for physician assisted suicide should trigger a psychological evaluation and a conversation about the degree of hopelessness being experienced by the individual.

Addressing hopelessness in patients calls for the expertise of the chaplain on the palliative care team, working collaboratively with other medical, nursing, and social work professionals. This is an extremely important skill, because palliative care programs that take care to address the whole person notice signs of depression and hopelessness sooner and respond more aggressively and effectively with psychological assessment. The Hospital Anxiety and Depression Scale, for example, is a 14-item, self-administered assessment that identifies concerns of patients with a life-threatening illness (Widera and Block 2012). Such assessments can help caregivers determine the need for a referral to a mental health professional. Susan Block's study shows that individuals served by palliative care professionals are more likely to receive services to address their depression. Patients who get early referral to palliative care generally have more time for staff to identify and address these psychological symptoms and provide mental health counseling and pharmacological treatment, if appropriate. This is important because some anti-depressant medications take several weeks before they become effective. With shortened lengths of stay in hospice, these pharmaceuticals may not have the runway necessary for effective symptom control (Block 2000).

Finding meaning in one's death is the final challenge for the human person. This is unlikely to happen in a room with the shades pulled, self-exiled from human contact and dialogue. Patients struggling with depression and hopelessness are often inclined to isolate themselves. They require "responsible closeness" from their professional caregivers, family, and community. The co-existence of feelings of depression and a sense of hopelessness is familiar to palliative caregivers. Perhaps unique to the Catholic health care ministry is a commitment to addressing the predicament of hopelessness at the end of life. Cardinal Bernardin, the late Archbishop of Chicago, was suffering from end-stage pancreatic cancer when he wrote "A Sign of Hope": "Our distinctive vocation in Christian healthcare is not so much to heal better or more efficiently than anyone else; it is to bring comfort to people by giving them an experience that will strengthen their confidence in life. The ultimate goal of our care is to give to those who are ill, through our care, a reason to hope" (1995).

## 8.5 The Catholic Response to Aid-in-Dying

Catholic moral teaching is unambiguous about the sanctity of life and the dignity of the whole person and united in its opposition to aid in dying legislation. That opposition has been vocal and fierce, and the source of some misunderstanding about the Catholic approach to end-of-life care. In 2016, Catholic health care systems and the Catholic Church in California (12 dioceses and 4 system health care ministries, including St. Joseph Health, Providence Health, Dignity Health, and Scripps Mercy) began a coordinated and determined opposition to the California End of Life Option Act. While this coalition was unsuccessful in its efforts to stop passage of the law, it did have an unintended consequence. Many advocates of aid-in-dying legislation erroneously surmise that Catholic opposition to physician assisted suicide means that Catholic health care organizations are unsympathetic to the needs of the dying person and hold a disproportionately negative position toward end of life care in general. Speaking for many of the Catholic leaders who engaged in soul searching after the law's passage, Ira Byock observed, "The public knows what we are against but can be forgiven for wondering what we are for" (Byock 2017).

The Whole Person Care Initiative (WPCI) is an organized response to this uncertainty. Created by Church and health care ministry leaders, the WPCI encourages collaboration at the grass roots level between local parishes and local palliative care and hospice programs to create communities of concern and support for members of congregations suffering from end of life challenges. Diocesan staff and leaders from the Alliance of Catholic Health Care and the California Catholic Conference crafted a proposal to build or strengthen already existing bridges between health care ministries and parishes for the purpose of supporting caregivers and patients (WPCI 2017). Embedded in this initiative is the realization that palliative and hospice care takes place in the homes and neighborhoods of communities as well as in local community hospitals. The initiative endorses a seamless continuum of care and support for dying persons and promotes a Catholic vision of end-of-life care to treat the whole person while acting in communion with the Church (CHA 2014).

During the first year following the passage of the End of Life Option Act in California, a Catholic hospice in Northern California served over 1000 patients. Of that number, 70 requested assistance in dying under the law, and 30 progressed to the point of obtaining the necessary drugs to complete the process. Of the 30, 13 acted to hasten their deaths (Johanson 2018). While the Catholic hospice did not assist in or encourage the use of life ending drugs, it did not abandon those patients who expressed an interest in or even obtained them. As a result, the hospice was able to support 930 patients who did not pursue aid in dying, as well as dozens of patients who pursued it to the point of obtaining the drugs. Each patient who requested aid in dying received increased attention to symptoms of depression and hopelessness. They were not told, "We don't do this here," and discharged from the service. On the contrary, their cries for help were a stimulus for a comprehensive and multi-disciplinary effort to assess and address the underlying distress they were experiencing.[1]

Fortunately, requests for aid in hastening one's death are still rare, particularly from patients already being cared for in a Catholic palliative care or hospice program. Promoting and supporting palliative care and hospice programs should be a priority for Catholic health care systems and hospitals as continued efforts to promote physician assisted suicide and aid in dying dominate the attention of persons facing their future demise. Palliative care and hospice should be the first responders to cries for assistance in hastening one's death.

Medicine tends to compartmentalize the physical and non-physical, considering pain as primarily a physical issue. This downplays the formidable role of a person in making rational sense of physical pain and may relegate spiritual pain to a status outside the purview of scientific medicine. Whole person care, a foundational model for palliative care, rejects the compartmentalization of illness and responses to it. The palliative caregiver agrees to accompany the patient and family on a journey that crosses a terrain full of challenges and threats to the whole person.

Psychologically, the palliative care team offers a safe space for patient and family to look at the reality of a life-threatening illness and an impending death without judgment or pressure. Partnering with family and friends, the palliative care team practices what Pope Francis recently described as "responsible closeness" by showing love and solidarity during a final illness (2017). As the *Ethical and Religious Directives for Catholic Health Care Services* state: "Above all, as a witness to its faith, a Catholic health care institution will be a community of respect, love, and support to patients or residents and their families as they face the reality of death" (Ethical and Religious Directives 2018).

## 8.6 What We Stand for

Sister Karin Dufault, S.P. has also written about hope as an essential challenge and opportunity for service to those facing death:

> When an object of hope is abandoned, health professionals can strengthen the person's ability to mourn the loss, to identify what is possible, and eventually to identify new hopes. When the grief of abandoned hope is not resolved, and investing further in the hoping process seems impossible, health professionals can share their own beliefs regarding what is possible and what is not. We can carry the flickering candle of hope for them until they feel ready to carry it themselves. (2008)

Catholic palliative care and hospice programs are like bright candles hidden under a basket too often obscured by justified but fervent opposition to euthanasia and

[1] Personal communication with Gary Johanson in preparation for California select committee on End of Life Act hearing. January 3, 2018.

physician assisted suicide. There is a false impression among some groups that patients requesting hastening of their dying will be abandoned by Catholic providers. Nothing should be further from the truth. From a Catholic view, the question is an invitation from another human person to be treated with dignity and compassion as a whole person. The invitation demands that one clears the mind of preconceived assumptions and judgments. When sitting in a patient's room in a Catholic hospital at the moment of the "ask," one's personal opinion of the act of physician assisted suicide itself is irrelevant. Catholic social and moral teaching about the dignity of the human person and the sanctity of human life demand that we resist the urge to say, "We don't do that here." These principles do not allow for the abandoning of the person, even though Catholic health care does not participate in physician assisted suicide.

In the Catholic moral theological tradition, the whole person must be encountered at these times when the end of life is near and the person is thinking about and requesting assistance in hastening the dying process. Simply giving them information about the process for hastening their death misses an opportunity to address their real needs in a more comprehensive and compassionate way. Palliative care services are specifically designed and organized to address the suffering of people reaching out for help in hastening their dying. Physical pain and distressing physical symptoms, psychological and psychiatric problems, and spiritual distresses are addressed by the team and referrals are made when special expertise is demanded.

Parker Palmer has written eloquently about standing in the "tragic gap" with another person. He defines this as "the gap between what is and what could and should be, the gap between the reality of a given situation and an alternative reality we know to be possible because we have experienced it." A patient motivated to ask for help in hastening their death is standing in such a tragic gap. Palmer suggests that this is a moment when a patient might assume a "broken open heart." In a gesture of compassionate mutuality, we as caregivers are also invited to assume a broken open heart. As he says, "There is no way to be human without having our heart broken… We know that heartbreak can become a source of compassion and grace because we have seen it happen with our own eyes as people enlarge their capacity for empathy and their ability to attend to the suffering of others" (Palmer 2009). This means accompanying the other person, being fully present with them, empathizing with their pain and suffering, and helping them to understand its meaning and implications. This means helping the person acknowledge and name their suffering – without judgment or coercion, helping them to move directly to the heart of their suffering and feel the pain fully, finally allowing the palliative care process to create a space of safety where the turmoil suffered can settle down and be experienced, with the patient surrounded by the love and support of family, friends and caregivers.

Palliative care in the context of a Catholic health care ministry must provide all required services of an accredited program. The professional palliative caregiver provides assessment and treatment for pain and psychological distress. But as a ministry in the Catholic tradition it calls itself to an even higher standard by meeting a suffering human person in a sacred encounter with a broken open heart.

## References

Bernardin, Joseph. 1995. *A sign of hope: Pastoral letter on healthcare*, 4–5. Chicago: Archdiocese of Chicago.
Block, Susan. 2000. Assessing and managing depression in the terminally ill patient. *Archives of Internal Medicine* 132 (3): 209.
Block, Susan, and Andrew Billings. 1994. Patient requests to hasten death. *Archives of Internal Medicine* 154 (18): 2039.
Byock, I. 2017, November/December. We must earn confidence in end-of-life comfort care. *Health Progress* 98(6): 19–24.
Catholic Health Association. 2014. *A shared statement of identity*. https://www.chausa.org/mission/a-shared-statement-of-identity. Accessed 3 Feb 2018.
Center to Advance Palliative Care. 2006. *Policies and tools for hospital palliative care programs*. https://www.capc.org/topics/hospice-and-palliative-care/. Accessed 3 Feb 2018.
Chochinov, Max, Keith Wilson, Murray Enns, Neil Mowchin, Sheila Lander, Martin Levitt, and Jennifer Clinch. 1995. Desire for death in the terminally ill. *American Journal of Psychiatry* 152: 8.
Dufault, Karin, S.P. 2008, May/June. Offering a flickering light. *Health Progress* 89 (3): 59.
Emanuel, Ezekiel. 2005. Depression, euthanasia, and improving end of life care. *Journal of Clinical Oncology* 23: 6456.
Hughes, Virginia. 2011. *When does mourning become a mental illness?*https://www.scientificamerican.com/article/shades-of-grief/. Accessed 3 Feb 2018.
Johanson, Gary. 2018. *Personal communication in preparation for California select committee on End of Life Act hearing*. January 3, 2018.
John Paul II. 1995. *The gospel of life*. http://w2.vatican.va/content/john-paul-ii/en/encyclicals/documents/hf_jp-ii_enc_25031995_evangelium-vitae.html. Accessed 3 Feb 2018.
National Consensus Project. 2018. *Clinical practice guidelines for quality palliative care*. https://www.nationalcoalitionhpc.org/ncp-guidelines-2013/. Accessed 3 Feb 2018.
National Quality Forum. 2006. *Framework and preferred Practice for palliative and hospice care quality*. www.qualityforum.org/publications/reports/palliativecare. Accessed 3 Feb 2018.
Palmer, Parker. 2009. The broken open heart: Living with faith and hope in the tragic gap. *Weavings*. March/April 24: 2.
Periyacoil, Vyjeyanthi, and James Hallenbeck. 2002, March. Identifying and managing preparatory grief and depression at the end of life. *American Family Physician* 65(1): 883–891.
Pope Frances. 2017. *Message of his holiness Pope Francis to the participants in the European regional meeting of the world medical association*. https://w2.vatican.va/content/francesco/en/messages/pont-messages/2017/documents/papa-francesco_20171107_messaggio-monspaglia.html. Accessed 3 Feb 2018.
Rosenfeld, Barry, H. Pessin, A. Marziliano, C. Jacobson, B. Sorger, and J. Abbey. 2014, June. Does desire for hastened death change in terminally ill cancer patients? *Social Science & Medicine* 111:35–40.
United States Conference of Catholic Bishops. 2018. *Ethical and religious directives for catholic health care services*. 6th ed. Washington, DC: United States Conference of Catholic Bishops.
Weissman, David, and Diane Meier. 2008. Operational features for hospital palliative care programs. *Journal of Palliative Medicine* 11: 9.
Whole Person Care Initiative. 2017. *A collaborative project of the California bishops and catholic health care*. Draft proposal. October 3, 2017.
Widera, Eric, and Susan Block. 2012. Managing grief and depression at the end of life. *American Family Physician* 1 (3): 259–264.

# Chapter 9
# Care of the Family and Social Aspects of Care

**Sarah E. Hetue Hill**

In this chapter, I will explore the social aspects of caregiving and palliative care, examining and analyzing research on the impact that caregiving has on family members who deliver various levels of care and support to ailing loved ones. I will also examine the need for perinatal palliative care services for families facing a life-limiting fetal diagnosis, and explore all these issues in light of Catholic teaching on the family and palliative care. I will conclude with some recommendations for public policy to improve the social and family aspects of care.

I begin with a case study that illustrates the powerful impact on families and persons that a palliative care intervention in the social aspects of care can have in the trajectory of serious life-threatening illness.

> Mr. Danny,[1] veteran of the Vietnam War and retired fireman and paramedic, became a frequent flyer to one of our hospitals and nearby nursing and rehab facilities. According to staff, calling him cantankerous and obstreperous would have been fitting and diplomatic. The staff simply resorted to considering him a "grumpy old man." Mr. Danny suffered with multiple comorbidities including lung cancer and COPD with a matching attitude that not even the most attentive nurse could improve. Without known family or friends, he spent his days in his bed alone and seemed to have nothing for which to desire to live. Several of the physicians, including his attending physician, decided perhaps it was time to consider making Mr. Danny a DNR, as each exacerbation and hospitalization seemed to get worse, and his quality of life seemed to diminish. As with any time physicians consider a DNR for patients at this hospital, the attending called the palliative care team to talk with Mr. Danny about his needs, goals of care, and so on. The core of providing palliative care is the interdisciplinary team who seek to provide holistic, person-centered care to patients, which means focusing on spiritual and psychosocial needs along with physical needs. Upon screening Mr. Danny for spiritual/existential needs, the team asked him whether he had any regrets and Mr. Danny shared that he had an unfulfilled "bucket list" wish. Years before, Mr.

[1] Name has been changed to protect identity.

S. E. Hetue Hill (✉)
Ascension, St. Louis, MO, USA
e-mail: SEHill@ascension.org



P. J. Cataldo, D. O'Brien (eds.), *Palliative Care and Catholic Health Care*, Philosophy and Medicine 130, https://doi.org/10.1007/978-3-030-05005-4_9

Danny had a falling out with his family, moved to a very remote area, and had not been in contact with them since, essentially living in isolation. His wish was to reconnect and reconcile with his children and grandchildren, whom he had not seen for years. The problem was that his family had all moved far away and his physicians felt he was not healthy enough to travel to see them. The palliative care social worker set out to find the family and succeeded in connecting with them. To say they were elated is an understatement. They wanted nothing more than to reconnect with and care for their father. The team devised a plan that would enable him to get strong enough and well enough to travel across the country to his family. Providentially, one of his grown grandchildren was now a firefighter paramedic himself. He felt confident in his skills and was willing to drive his grandfather down with plenty of oxygen and needed medications for his journey to his new assisted living apartment that his children secured for him close to their homes. Danny's children in the interim flew to his home, packed up his limited belongings, and drove them to his new space to make it feel like "home" even before he arrived. Danny's drive was difficult, but he made it and was reunited, reconciled, and living happily near his family for the last days of his life. Danny is still flourishing, and his children recently sent a letter to the palliative care team with pictures of him smiling, laughing, and filled with love—quite a different picture of the man that presented as a "grumpy old man." Now, his once lonely world is filled with love, laughter, and care.

Above all, this narrative emphasizes the importance of caring for patients as persons and of understanding that at times, the deepest ailment that needs healing is not something a medication can fix. Any healthcare professional can share a story about a patient whose healing did not occur through a medical intervention alone. The story highlights how important it is for us to care for more than patients' physical needs and to include their family and caregivers in the care we offer. Ira Byock, in his book *The Best Care Possible* (2012), says that "to provide the best possible care and support for people during the waning phases of life—effectively and affordably—we need to be *person-* and *family*-centered." To greatly improve health care in the United States, he states, "we must get beyond seeing people solely as individuals (a.k.a. patients) and begin seeing people as individuals *within* families and communities" (Byock 2012, p. 254).

Due to the necessity of caring for families and social aspects of care, palliative care experts included Social Aspects of Care as the fourth domain of the eight domains of palliative care in the *Clinical Practice Guidelines for Quality Palliative Care* (NCP 2004). Both the first and second editions of the *Guidelines*, published in 2004 and 2009, had the following guideline under the fourth domain, which is the Social Aspects of Care: "Comprehensive interdisciplinary assessment identifies the social needs of patients and their families, and a care plan is developed to respond to these needs as effectively as possible" (NCP 2004, 2009). Authors of the *Guidelines* presented two preferred practices that aligned with the guideline for the fourth domain. Preferred practice n. 18 required interdisciplinary teams to facilitate patient/family care conferences during which goals of care discussions occurred. Preferred practice n. 19 required the creation and implementation of comprehensive social care plans.

The third edition of the *Clinical Practice Guidelines for Quality Palliative Care* (2013) shares the same preferred practices but split the single guideline into two, which aligns more closely with the preferred practices. Guideline 4.1 under this domain states that "the interdisciplinary team assesses and addresses the social aspects of care to meet patient-family needs, promote patient-family goals, and maximize patient-family strengths and well-being" (NCP 2013, p. 24). Guideline 4.2 directs the palliative care team to ensure that "a comprehensive, person-centered interdisciplinary assessment identifies the social strengths, needs, and goals of each patient and family" (NCP 2013, p. 24).

The revisions made to the fourth edition of the *Clinical Practice Guidelines for Quality Palliative Care* (2018) recognized the importance of family caregiver assessment, support, and education and thus this edition references these items in numerous domains. Additionally, the fourth edition now contains four guidelines under the fourth domain. Guideline 4.1 states "the palliative care IDT has the skills and resources to identify and address, either directly or in collaboration with other service providers, the social factors that affect patient and family quality of life and well-being" (NCP 2018, p. 26). Guideline 4.2 requires that the IDT "screens for and assesses patient and family social supports, social relationships, resources, and care environment based on the best available evidence to maximize coping and quality of life" (NCP 2018, p. 27). Guideline 4.3 states that "in partnership with the patient, family, and other providers, the IDT develops a care plan for social services and supports in alignment with the patient's condition, goals, social environment, culture, and setting to maximize patient and family coping and quality of life across all care settings" (NCP 2018, p. 28). Finally, Guideline 4.4 requires that "a palliative care plan addresses the ongoing social aspects of patient and family care, in alignment with their goals and provides recommendations to all clinicians involved in ongoing care" (NCP 2018, p. 28).

The palliative care experts who identified this domain among the eight domains of palliative care had many reasons for doing so. The fourth domain is vitally important because it annually affects over 43 million family caregivers. These caregivers risk ill physical health, mental health, and even higher risks of death from the stress of caregiving for a loved one. Spouses in particular who feel they experience the most strain (both physical and emotional) from caregiving have a much higher risk of death. Caregiving also often takes a financial toll on families due to lost time at work, lost productivity, and lost wages. Caregivers are often adult daughters with their own families. When health care systems support families well, families have better outcomes, less risk of negative outcomes, and even report feeling good about caregiving for their loved ones (Byock 2012; Meier et al. 2010). To be adequately supported, families need information and communication, advanced care planning, psychosocial and spiritual support, continuity of care, access to support services, and bereavement support following the death of their loved one (Lynn et al. 2007). Many of these services can be provided by a palliative care team.

Unfortunately, the health care system does not currently do enough to assist patients and families, especially given that most families do not understand health care and disease management yet are expected to steer through the courses of a complex and fragmented system that often leaves them to manage for themselves. It is difficult enough for people who work in healthcare to navigate the system as family caregivers, much less for those without healthcare experience. Take Carol Levine for example. Although she was a highly skilled and competent healthcare professional, even Carol had great difficulty navigating the system when forced to become a caregiver herself (Levine 1999, 2004). Levine's resume includes working on health policy and health care ethics for groups such as the United Hospital Fund, Hastings Center, and who was a MacArther Foundation Fellow for her work on AIDS policy (2004). She became a longtime caregiver for her husband after he was in a car accident that left him hospitalized and eventually home bound. Despite her adept knowledge about healthcare, she described her frustration with her experience: "I feel that I am challenging Goliath with a tiny pebble. More often than not, Goliath just puts me on hold" (Levine 1999, p. 1587). She also shared how dealing with this problem changed her family dynamics and strained relationships, noting that "I stopped being a wife and became a family care giver" (Meier et al. 2010, p. 407). This issue of inadequate care and support for the seriously ill and dying is so important that the National Academies of Science and Medicine (NASEM), formerly called the Institute of Medicine (IOM), convened an ad hoc workgroup to develop recommendations for policies to support family caregivers and to minimize the barriers that caregivers face in attempting to care for their loved ones. The culmination of their research and meetings was captured in the publication, *Families Caring for an Aging America* (2016).

Too often researchers in this area focus only on caregivers of the elderly or caregivers of adult or pediatric disabled. Not enough attention is given to research on the unmet needs of parents and families facing a life-limiting fetal diagnosis. Similar to the caregivers of the elderly patient, parents facing life-limiting illness of their unborn or newborn baby also need spiritual and psychosocial support, anticipatory guidance and information, and social supports in the home if the baby lives long enough to live at home for a period of time. Unfortunately, in most cases, the current health care system does not adequately train or prepare medical professionals in how to care for these parents and family members in their needs. Offering perinatal palliative care to parents facing a life-limiting fetal diagnosis is one significant way to meet their needs. I will say more about this later.

## 9.1 Background

Without family and other informal caregivers, the United States healthcare system simply could not exist. According to the National Alliance for Caregiving and the AARP, in 2015 alone, 43.5 million caregivers provided unpaid care to family members, friends, or other loved ones, and 34.2 million provided care for someone 50 or older (NAC and AARP 2015). Authors suggest several definitions for family caregivers. For example: "one or more family members giving aid or assistance to other family members beyond that required as part of normal everyday life, which is a result of the need by the elder on another individual for activity essential to daily living" (Walker et al. 1995, pp. 402–403). Bruhn and Rebach, in *The Sociology of Caregiving* (2014), define family caregivers as those who care for loved ones with chronic illness or disability. They also utilize Carol Levine's definition, which notes the importance of distinguishing between family and caregiver:

> 'Family' denotes a special personal relationship with the care recipient; one based on birth, adoption, marriage, or declared commitment. 'Caregiver' is the job description, which may include providing personal care, carrying out medical procedures, managing a household, and interacting with the formal health care and social service systems on another's behalf. Caregivers are more than the sum of their responsibilities; they are real people with complex and often conflicted responses to the situations they face. (Levine 2004)

Most family caregivers care for only one other adult (most often an adult parent), but 15% of caregivers care for two adults and 3% provide care to three or more adults. Most caregivers are female, although the amount of male caregivers has been steadily increasing. The average age of caregivers is 69.4 (NAC and AARP 2015). The increasing number of the baby boomers joining those aged 65 and older—termed the "silver tsunami" by health researchers and economists—is projected to double, from 43.1 million in 2012 to 83.7 million by 2050, when the surviving baby boomers will be over the age of 85 (Ortman et al. 2014). Given advances in technology and medications that enable people to live with chronic illnesses, the longer people are living, the more complex their care may be, due to their multiple comorbidities (Xu et al. 2016). The need for palliative care will increase proportionately with such developments.

Caregivers provide both direct care, which includes assisting patients with activities of daily living such as bathing, grooming, toileting, or emotional support along with indirect care, which includes coordinating schedules of medical providers, assistance with handling medical information and medication management, or patient advocacy. Caregivers also often provide complex care—which includes management of symptoms and providing medications—and they are often trained to operate high tech medical equipment needed by their loved one. Despite having to offer support in all of these ways, they often have little understanding about the disease or about the care needed, because they also often become caregivers quickly after their loved one has an exacerbation of symptoms that requires a hospital stay and discharge to home with little to no support.

The economic value and the toll of this caregiving is astounding. The economic value of services provided by nonpaid caregivers is steadily increasing, and at last calculation in 2013 reached $470 billion. Conversely, families take out loans, mortgages, and even go into bankruptcy to meet the needs of their ailing loved ones (Emanuel et al. 2000). Many caregivers also maintain full time jobs and their duties as caregivers affect work productivity, given they often spend 20 h or more on caregiving. Studies show decreased productivity by 18.5% and a higher incidence of leaving work altogether (Pavalko et al. 2016). The toll on caregivers is not only economic. Caregivers also report that their productivity in their non-work life is reduced by 27.2% and the burden of caregiving has three times the effect on personal life than it does on work life (McKinlay et al. 1995).

The health of family caregivers is also negatively affected. Two-thirds of caregivers report emotional or psychological strain which is related to higher morbidity and mortality (Schulz and Sherwood 2008). Caregivers who state that they "didn't have a choice" to become a caregiver generally report the highest amount of strain. Those with the highest amount of strain have a 63% increase in risk of death (Schulz and

Beach 1999). Additionally, researchers conducted a study of 50,000 nurses who also were caregivers at home for more than 9 h per week for a seriously ill or disabled spouse, and found they had an 80% increased risk of cardiac death than noncaregivers (Lee et al. 2003).

To prevent all of this, Joanne Lynn and colleagues (2007) recommend that needs of caregivers include "information on disease progression, training in patient care, support for medical-decision making, financial help, respite care, and emotional and spiritual support" (p. 96). Palliative care by its nature offers many of the resources families need to be able to successfully navigate the healthcare system and to provide their loved one with the holistic care they need. Without palliative care, patients often experience a stark difference between the care they desire and the care they receive. This result is due in part to a lack of communication and missed opportunities around areas of person-centered care, such as the timely management of difficult symptoms and basing patient care recommendations on a prior understanding of patient goals and values (Curtis et al. 2005; Desbiens et al. 1996; Lynn et al. 1997; Schenker et al. 2012).

A study of 51 family care conferences in the ICU sought to identify missed opportunities for physicians to provide information and support to families. Frequent areas missed were: opportunities to acknowledge and address emotions; opportunities to listen and respond to the family; and opportunities to address key principles of medical ethics and palliative care, including exploration of surrogate decision-making, patient preferences and affirmation of non-abandonment (Curtis et al. 2005, p. 844). Misunderstanding about diagnosis or prognosis adds additional stress and anxiety to family caregivers attempting to make treatment decisions. Physicians who provide clear communication about diagnosis and prognosi assist patients and families to overcome their stress and anxiety and give families reasons to feel relieved about the decisions (Hebert et al. 2009). Pediatric parents have been shown to find peace and relief in decision-making if they understand the diagnosis, prognosis, and the true nature of their child's disease (Kars et al. 2010). Additionally, family meetings with clear goals of care conversations decrease post-traumatic stress and depression and even the grief of surviving family members of patients who die in the ICU (Azoulay et al. 2005; Daly et al. 2010; Lilly et al. 2003). Palliative care teams are required to deliver clear communication to patients and families. They often provide insights to "get everyone on the same page" regarding all of the specialists who provide sometimes siloed care. As noted earlier, Domain 4 also requires palliative care teams to host interdisciplinary family meetings where goals of care are discussed.

All of the above research centers on patients who have family caregivers to support them. Unfortunately, there are many elderly who have no one to support them and deal with their illness in loneliness and desperation. These patients are at high risk of readmission. However, even a small amount of support—for example something as simple as meals on wheels—could assist them. A Health Affairs study showed that if 1% more elderly received meals delivered to their homes, 1722 Medicaid patients would not need nursing home care. Twenty-six states would save money by doing so, because costs saved on nursing care would offset the price of the food visits. Furthermore, those who received daily meals also benefitted from decreased loneliness and higher quality of life, which may contribute to their better outcomes (Thomas et al. 2016; Thomas and Mor 2013). Again, palliative care teams can assist in analyzing the social aspects of these patients' care and provide recommendations for meeting their needs. Unfortunately, studies continue to show those with low socioeconomic status have less access to and barriers to palliative care services and thus worse outcomes (Byers et al. 2008; Krakowsky et al. 2013; Walshe et al. 2009). Increasing access to palliative care for these patients is vital to their improved care. Another group that could benefit from great palliative care involvement is parents facing a life-limiting fetal diagnosis.

## 9.2 Perinatal Palliative Care

A special but often overlooked topic is care of the family experiencing the diagnosis of a baby with a congenital abnormality incompatible with life. This diagnosis affects many parents each year. Unfortunately, as I noted above, this group is often neglected because most researchers in the social aspects of care focus on caregivers of the elderly or disabled. However, just like with caregivers of adults, these parents often do not need medical interventions. Rather, they need emotional support and anticipatory guidance to assist them with options, goals of care, and other social, emotional, and spiritual support (Bastek et al. 2005; Berth et al. 2009). Psychosocial and spiritual needs often go unmet, yet their needs are inadequately explored in current research, and physicians are inadequately trained in meeting their needs (Berth et al. 2009; Boss et al. 2009; Cote-Arsenault and Denney-Koelsch 2011; Leuthner et al. 2003). Below, I articulate reasons why I believe it is imperative we offer assistance and good care to these parents and babies, particularly in Catholic healthcare.

The leading cause of death in babies each year in the United States is congenital abnormalities. Along with the other top two reasons for infant death—low birthweight and sudden infant death syndrome—congenital abnormalities are responsible for 23,000–30,000 deaths each year (Mathews and MacDorman 2014). Parents whose babies are diagnosed with congenital abnormalities often feel like they must select between two grave choices—either to terminate or to carry the baby to term with little assistance or information on what to expect (Hoeldtke and Calhoun 2001). One way health care systems can support them in their decision-making and in their decision to carry to term is through the delivery of perinatal palliative care.

Perinatal palliative care focuses on providing palliative care to both the prenatally diagnosed imperiled baby and to the parents and family members (Munson and Leuthner 2007). The care focuses on both supporting the parents with their needs and on enhancing the quality of life of the fetus and/or baby (CBCHC 2000). Perinatal palliative care as a specialty has existed since at least 1982, when a group of physicians advanced the tenets of Cicely Saunders for adults and pediatrics into the neonatal realm (Whitfield et al. 1982a, b). The fact that parents have benefited from perinatal palliative care programs that deliver care beginning at the point of diagnosis, has resulted in a steady increase of programs since around the first published articles on the topic in early 2000 (Calhoun et al. 2003; Hoeldtke and Calhoun 2001; Howard 2006; Ramer-Chrastek and Thygeson 2005).

Data regarding perinatal programs is not widely published. The Perinatal Hospice and Palliative Care website (2017) lists programs in the U.S. and abroad. It is the only regularly updated listing of programs. The website currently lists 160 programs in the U.S. However, 20 of these are limited to providing bereavement for parents only after their baby with a life-limiting diagnosis has died. There are only 15 programs that offer both palliative care services while the baby is still living with a life-limiting diagnosis and bereavement services after death. None of these 35 programs offer any support to parents upstream or at the time of diagnosis. Of the remaining 125 programs posted on the site, data about programmatic elements are sparse. For example, there is inadequate information about whether the programs are staffed by interdisciplinary teams, who are devoted full time to the team or whether the support is limited to a part-time person who cannot by themselves offer robust and meaningful support. Within Catholic healthcare institutions themselves, 30 programs are listed, 4 of which are diocesan sponsored. Despite the growth of these programs, researchers still conclude that more studies need to be completed to elucidate the needs of parents and programmatic elements needed that would most assist them in continuing their pregnancy (Cote-Arsenault and Denney-Koelsch 2011).

Researchers have found that most parents facing a life-limiting fetal diagnosis are satisfied with their medical care but that their spiritual and psychosocial needs were unmet. Tellingly, physicians themselves report that not enough time is spent on training them to meet the needs of parents, particularly around spiritual and psychosocial issues. Often parents carry the baby to term against the recommendations of their own obstetric gynecologist, high-risk obstetric gynecologist, or genetic counselor, to terminate at the point of diagnosis or early in the pregnancy. These parents feel abandoned and unsupported in their decision to carry to term. Nancy English (2009), who has her PhD in nursing, experienced this feeling of abandonment after her own unborn baby was diagnosed with Trisomy 18. She reports that termination was recommended over the phone, as soon as possible, "as he was 18 gestational weeks old." She said that "no other options were provided or discussed" (English and Hessler 2013). A labor and delivery nurse herself, English understood the difficulties of diagnosis, but she still sought to carry the baby to term and desired additional assistance from her care providers which she felt was not adequately provided.

Parents of imperiled newborns have many needs throughout their diagnosis and delivery. Research shows that parents who received normal results of genetic testing have lower levels of anxiety about pregnancy, increased maternal bonding, and decreased fear of the unknown (Gudex et al. 2006) in contrast to parents who receive abnormal results, who develop severe anxiety, grief, and depression (Boyd et al. 1998; Leuthner et al. 2003). Additionally, parents who lose a baby due to a fetal anomaly often endure depression, anxiety, and even posttraumatic stress symptoms (Berth et al. 2009).

When medical teams offer psychological support to parents, along with education and full but easy to understand information about their child's diagnosis, parents experience better outcomes including reduced anxiety and fear of the unknown (Fioravanti 2002; Kemp et al. 1998; Statham et al. 2000). The American Academy of Pediatrics recommends the inclusion of clinical information such as long term outcomes, quality of life, and parental preferences in prenatal consultations for imperiled newborns due to the increased burden on parents in these situations (AAP 2012). Despite this, researchers found in a 2005 regional survey of neonatologists that most providers frequently discussed clinical realities, but did not touch upon ethical, social, or religious issues that could have influenced parents' choices. Only 25% of the neonatologists discussed religious or spiritual beliefs in any capacity (Bastek et al. 2005). In another more recent web-based national study, researchers surveyed graduating neonatal fellows in their final months of fellowship and found similar results. The fellows felt very competent and well trained in medical management of the infants but over 40% reported no training on communication in role-play, didactic sessions, or simulated patient scenarios. Furthermore, they expressed lack of training regarding supervision of and feedback about family meetings led by them, and they indicated that they had little training to discuss "palliative care, families' religious and spiritual needs," and to manage "conflicts of opinion between families and staff or among staff." Fellows perceived that communication skills seemed less of a priority for faculty than for the fellows, and 93% believed that training should be improved (Boss et al. 2009).

Another study focused on determining whether obstetric-gynecologists and neonatologists spoke with parents about how they would be supported if they chose to carry the baby to term (Tosello et al. 2016). In this study, 24.7% of parents decided to terminate, 35.4% chose to carry to term, and the rest miscarried, or the physicians were unsure of their decision because they did not return to their office. Only 52.9% of the physicians had any conversation at all about how the parents would be supported if they chose to carry to term.

In contrast, physicians caring for parents carrying a healthy baby normally offer anticipatory guidance on what to expect. Multitudes of books and other resources also assist parents in knowing the unknown. When parents face a life-limiting fetal diagnosis, knowledge of what to expect also assists them in planning and lessens their emotional and psychosocial fears of the unknown (Aite et al. 2011; Kuebelbeck 2011). Furthermore, as with any parent, these parents have hopes and dreams for their unborn child. Unfortunately for these parents, most of those hopes and dreams will be unable to come true, so these parents must be assisted through coping with anticipatory grief and bereavement which for them begins even at the point of diagnosis and often lasts long after the baby dies (Bennett et al. 2011; Sandelowski and Barroso 2005). One major tool that can assist with these needs offered through perinatal palliative care is the "birth plan," sometimes called an advanced care plan (Limbo et al. 2009).

Birth plans can vary but usually include items such as: (a) what types of clinical interventions the parents would want once the baby is born; (b) who they might want in the delivery room or there immediately after, including family, photographer, familiar spiritual care provider (to complete a baptism or other blessing ritual); (c) what type of remembrances the family may want from the baby (a lock of hair, foot or hand prints in clay or inked, pictures); and (d) other items relevant to what the parents want for their child during and after the time of delivery (Calhoun et al. 2003; Leuthner and Jones 2007; Limbo et al. 2009).

Wool and colleagues (2016) found in one study that the satisfaction rate of 405 parents who faced a life-limiting fetal diagnosis who decided to carry their babies to term was 76%. More importantly, however, those who were satisfied with their care were 1.9 times more likely to report that their care was consistent; 1.8 times more likely to report their care was compassionate; and 1.8 times more likely to have been assisted with coping with their emotions. These all seem like potential reasons that parents who hear and learn more about perinatal palliative care are less likely to choose abortion, as perinatal palliative care offers all these things.

For many of the above reasons, parents whose physicians spoke with them about perinatal palliative care and its benefits were less likely to choose termination and more likely to choose to carry the baby to term with perinatal palliative care support (Breeze et al. 2007; Calhoun et al. 2003; Chitty et al. 1996). Especially within Catholic healthcare, these findings are important because of the Catholic view of the fetus as being a member of the human family from the moment of conception and who has the right to be protected and to be carried to term if possible, and to be loved and cared for.

Within the context of perinatal palliative care, a duty or obligation to provide care for the parents *and* the baby from the point of conception forward exists for the Catholic health ministry. This duty is underscored by research showing that parents who feel alone and not supported in their decision to carry the baby to term, or who believe their baby will suffer, may choose abortion because they feel their needs are unmet. By offering exceptional support to parents facing a life-limiting fetal diagnosis—including care that focuses on their social, emotional, and spiritual needs—they may decide against abortion by virtue of our offering them supportive care, which is a good in and of itself. Catholic teaching on the family and the social aspects of care and palliative care support the fourth domain of palliative care. It will now be helpful to examine the Church's teaching in these areas.

## 9.3 Catholic Teaching on Family and Social Aspects of Care

The Catholic Church has long cared for persons dealing with serious illness and their families and loved ones. The Church's history of caring for the sick is rooted in the Gospel narrative of Jesus' ministry as loving healer. The Gospels highlight numerous times when Jesus healed the sick, whether by curing the blind (Mark 10:46–52; Matthew 20:29–34), the mute (Luke 11:14), or the paralytic (Mark 2:1–12). Jesus' healing was not only physical but simultaneously spiritual and restorative because through his actions he returned many of those he healed back into the community. Many suffered from afflictions which made them "unclean" under

Jewish law and therefore unable to participate in the community until they were healed. Examples of this include the woman with the hemorrhage (Matthew 9:20–22; Mark 5:25–34) or the many lepers he healed (Luke 17:11–19, Mark 1:40–45).

Jesus' great understanding of familial love and concern for family caregivers was evident in his raising people from the dead—including the widow's son (Luke 7:11–17), Jairus' daughter (Matthew 9:18–26; Mark 5:22–43; Luke 8:41–56) and his friend Lazarus (John 11:39). Jesus' close relationship with Lazarus and his sisters Martha and Mary, and his compassion for them, is evident in the Gospel when it tells us that "Jesus wept" upon learning that Lazarus died (John 11:35). As some have suggested, he did not weep because his friend was dead—as he knew he could restore him to life and also knew that he would have eternal life with him—but he wept with Martha and Mary in compassion for their sorrow (Pope Benedict XVI 2011). The call to care for one another, even the stranger who may have no family, is echoed in Jesus' call to see him in those for are ill–"I was sick and you looked after me" (Matthew 25:36)—and in his call for us to imitate the Good Samaritan (Luke 10:37).

Beyond the Gospels, the Catholic moral tradition also provides direction on caring for those facing serious illness in many forms of Church teaching including magisterial teaching (which are included in the appendix at the end of this book). The *Catechism of the Catholic Church* tells us that "those whose lives are diminished or weakened deserve special respect. Sick or handicapped persons should be helped to lead lives as normal as possible" (USCCB 1997, n. 2776). The call to care is not limited to caring for family members or neighbors but for communities of people. Those in Catholic healthcare are compelled to provide compassionate care to those who need it by the belief that Jesus works through us and is present in those we care for, including the least among us. Pope Francis's *Address to the International Federation of Catholic Medical Associations* (2013) challenges us to this call when he states:

> In a frail human being, each one of us is invited to recognize the face of the Lord … [a]nd every elderly person, even if he is ill or at the end of his days, bears the face of Christ. They cannot be discarded, as the 'culture of waste' suggests! They cannot be thrown away! (Pope Francis 2013)

He also calls for support of families and caregivers who care for their loved ones: "The provision of adequate assistance and services which respect the dignity, identity and needs of patients is important, but the support of those who assist them, whether family members or healthcare professionals, is also important" (Pope Francis 2013; see also Pope Francis 2015). Pope Francis reiterated the importance of caring for the elderly in his Exhortation *Amoris Laetitia* (2016):

> Most families have great respect for the elderly, surrounding them with affection and considering them a blessing. A special word of appreciation is due to those associations and family movements committed to serving the elderly, both spiritually and socially… In highly industrialized societies, where the number of elderly persons is growing even as the birth rate declines, they can be regarded as a burden. On the other hand, the care that they require often puts a strain on their loved ones. (Pope Francis 2016, n. 17)

Saint Pope John Paul II prophetically proclaimed in his Apostolic Exhortation *Famliaris Consortio* (1981) his fears of a culture where the elderly are pushed aside as a burden and articulates the need for us to recognize our commitment to care for them: "Other cultures, however, especially in the wake of disordered industrial and urban development, have both in the past and in the present set the elderly aside in unacceptable ways. This causes acute suffering to them and spiritually impoverishes many families" (Pope John Paul II 1981, n. 27).

John Paul also prophetically witnessed the increase of abandonment of the elderly, sick, and disabled and how this leads to loneliness and other psychosocial ills. Again, in *Familiaris Consortio*, he recognizes the need for us to care for our families and make sure the elderly or sick do not feel they are burdens to their families, but an integral part of the family unit:

> Today, even more preoccupying than child abandonment is the phenomenon of social and cultural exclusion, which seriously affects the elderly, the sick, the disabled, drug addicts, ex-prisoners, etc.... But there is also the burden of loneliness, more often psychological and emotional rather than physical, which results from abandonment or neglect on the part of children and relations. (nn. 44 and 77)

He also reflected on the need to care for children dealing with illness, stating that "special attention must be devoted to the children by developing a profound esteem for their personal dignity, and a great respect and generous concern for their rights. This is true for every child, but it becomes all the more urgent the smaller the child is and the more it is in need of everything, when it is sick, suffering or handicapped" (John Paul II, n. 26).

Both our commitment to caring for the least among us and our commitment to holistic, person-and family-centered care can be enhanced by utilizing palliative care services, so it is not surprising that palliative care is seen as a hallmark of and mission imperative of Catholic health care. Dan O'Brien expressed this same point at the 2013 Supportive Care Congress to highlight how palliative care is a hallmark of Catholic healthcare: "This belief has inspired Christians, individually and collectively, for *centuries*, to reach out to the sick, the hungry, the lame, the imprisoned, in order to heal, to comfort, to care, to console – and so touch the face of God" (O'Brien 2013, p. 4).[2]

Those in Catholic healthcare believe we are called to deliver holistic care that focuses as much on psychosocial and spiritual support as it does on medical interventions. Within this understanding, all human beings, from the unborn baby to the very sick elderly have an inherent dignity and corresponding rights, including the right to person-centered health care and palliative care itself as a model of person-centered care. In an address for the World Day of the Sick, Pope Benedict XVI said, "It is necessary to stress once again the need for more palliative care [services] which provide integral care, offering the sick the human assistance and spiritual accompaniment they need. This is a right belonging to every human being, one which we must all be committed to defend" (Pope Benedict XVI 2006).

[2] See also O'Brien's chapter in this volume, which, in large part, is based on that same talk.

The primary motivation for providing palliative care services in Catholic healthcare is distinctive but aligns well with the domains of palliative care articulated by the *Clinical Practice Guidelines* cited earlier (NCP 2018). In fact, the mission of Catholic healthcare aligns with the tenets of palliative care as palliative care actually finds its origins in the Catholic tradition of caring for those who could not be cured and who others would not care for, as exemplified by the work of the hospitaller orders (Order of the Knights of Malta 2016). Palliative care resonates with people's sense of justice and compassion, and it is also strategically important and assists health systems to accomplish the quadruple aim (better patient experience, better patient outcomes, and better provider experience, all at lower costs). But the *primary* motivation for Catholic health care systems to offer palliative is mission based: palliative care promotes human dignity and a holistic view of persons and families. Yet, despite the belief that caring for those dealing with serious illness and their families is intrinsically important, even within Catholic healthcare, social needs of patients and families still remain unmet. Therefore, the final portion of this chapter will reflect on improving the social aspects of care and how we can better support family caregivers in their vitally important role.

## 9.4 Improving Care

Improvements in the social aspects of palliative care need to be made at both the system and facility level and at the larger public policy level in order to better support patients and family caregivers. Obviously, a first step is that health systems should seek to increase the delivery of palliative care throughout the continuum of care to patients facing serious illness and their families, given that palliative care itself addresses a vast number of the identified patient and family needs and increases positive outcomes and decreases possible negative side effects of living through serious illness.

One pragmatic recommendation that quality experts make is that hospitals and other care facilities need to utilize social assessments. As I shared earlier, the *Clinical Practice Guidelines for Quality Palliative Care* require interdisciplinary teams to assess social aspects of care for the patient and family to identify needs, goals, and strengths of each family and patient unit, and to connect the patient and family to resources that can assist them with their needs and accomplishing their goals (NCP 2018, Domain 4). Several examples of social assessments and tools are available in the Tools and Resources section of the Guidelines (NCP 2018, pp. 75–76).

Joanne Lynn and colleagues also offer many practical recommendations for improving palliative care in general in *The Common Sense Guide to Improving Palliative Care* and specifically include a full chapter on caring for caregivers (Lynn et al. 2007). They recommend that any quality improvement efforts in this realm need to be based on the needs and settings of both patients and caregivers. When done well, this can lead to reduced readmission rates, caregivers who feel adequately assisted, and quality of life improvement for both patients and their caregiver/loved ones. For health system improvement, they recommend baseline data analysis of current and former caregivers of patients. Even a small number of phone calls—for example 10–20—of recently discharged patients and surveying staff can lead to understanding what problems exist to begin action plans to improve them.

Lynn and colleagues (2007) provide the example of Fairlawn Hospital's quality improvement project. Health care leaders at Fairlawn noticed problems such as lack of advanced care planning, lack of support for discharged patients, families feeling abandoned when there was disagreement between clinicians and families around goals of care, and so on. They surveyed staff and interviewed patients/caregivers which assisted them to learn much about what could be improved even though in general families felt their family was good. They created an overall aim to develop a Caregiver Support Services Program with several sub-aims with process and outcome measures attached. One such measure was the goal of 100% of patient caregivers participating in interdisciplinary family meetings within 72 h of admission. At baseline, this occurred only 25% of the time. In response, Fairlawn leaders developed standardized family meeting protocols and after implementing their use at the end of 6 months these meetings were occurring within 72 h 100% of the time. Similarly, leaders created a goal that 80% of caregivers would find family meetings useful or very helpful. At baseline only 30% of families felt these meetings were useful, so leaders developed a satisfaction survey and utilized other follow-up techniques, and within 6 months reached 100% of families reporting that they found the meetings useful or very helpful. These quality improvement techniques can be replicated in other health systems. Health systems can also employ many other simple and cost effective solutions including, but not limited to: quiet spaces for families; comfort carts for patients and families; counseling and respite care for families; educational materials and one on one education for family caregivers to a level that meets their needs; 24/7 availability of phone support and counseling for families who have urgent questions or needs; peer support groups for families; increased utilization of home health nurses; and the creation of No One Dies Alone programs or similar (Byock 2012; Lynn et al. 2007). Increased referral to hospice is also noted as something that benefits both patients and families (Byock 2012).

Another important recommendation would be to increase social services such as Meals on Wheels for seniors without family caregivers. As mentioned earlier, this relatively simple and low cost solution has been shown to decrease loneliness and readmission rates and allows the chronically ill elderly to age in the place where they prefer. Other innovative strategies should also be tested, analyzed and spread for their effectiveness in enhancing care for those who live alone without families or for those with families but who are still lonely because they are cared for in an elder care facility without family. For example, programs that have incorporated child care facilities into or next door to elder care communities, or "adopt-a-senior programs" for those living alone (Byock 2012).

At the national policy and advocacy level, many changes should be encouraged by health care leaders that could improve outcomes. Policymakers should encourage legislation that assists in protecting caregivers from the financial and health consequences of the role. For example, legislation could help enhance workplace policies and programs such as flexible work schedules or the option to work from home that allow for caregiving individuals to successfully work while also caring for their loved ones. Other recommendations include building infrastructure for accessible respite care, increasing the amount of trained direct-care workforce for assisting caregivers with patients at home, and coverage of assessments and reassessments of social needs of patients and families, including in the home.

The NASEM Committee on Families Caring for an Aging America also made four major recommendations. First, that the Administration that took office in 2017 should address the issue of the health, economic, and social issues facing family caregivers. Specifically, their recommendation was that:

> The Secretary of Health and Human Services, in collaboration with the Secretaries of Labor and Veterans Affairs, other federal agencies, and private-sector organizations with expertise in family caregiving, develop and execute a National Family Caregiver Strategy that, administratively or through new federal legislation, explicitly and systematically addresses and supports the essential role of family caregivers to older adults. This strategy should include specific measures to adapt the nation's health care and long-term services and supports (LTSS) systems and workplaces to effectively and respectfully engage family caregivers and to support their health, values, and social and economic well-being, and to address the needs of our increasingly culturally and ethnically diverse caregiver population. (NASEM 2016)

Their first recommendation also included numerous sub-recommendations, such as: finding mechanisms within Veterans Affairs, Medicare, and Medicaid to both identify and assess needs of family caregivers; creating CMS payment reforms to stimulate providers to include families in all aspects of care; strengthening training of providers to encourage engagement of family caregivers and to connect them to needed services; increasing funding for programs that explicitly support family caregivers; exploring policies that provide caregivers with economic supports; and expanding on research on family caregivers. Their second recommendation was that state governments lacking in addressing caregiver needs should learn from those states with successful programs and implement those programs. Their third recommendation was that "The Secretaries of Health and Human Services, Labor, and Veterans Affairs should work with leaders in health care and long-term services and supports delivery, technology, and philanthropy to establish a public-private, multi-stakeholder innovation fund for research and innovation to accelerate the pace of change in addressing the needs of caregiving families." The final recommendation was that in all previous recommendations, diversity be addressed (NASEM 2016).

Clearly, health care systems and policy makers need to work together to improve the social aspects of care, given their wide-reaching implications for tens of millions of Americans. Given the teaching on care of the family and palliative care, leaders in Catholic health care have an opportunity to lead by example in creating sustainable change and improvements in the way we care for both the person and their caregivers.

## References

Aite, L., A. Zaccara, N. Mirante, A. Nahom, A. Trucchi, I. Capolupo, and P. Bagolan. 2011. Antenatal diagnosis of congenital anomaly: A really traumatic experience? *Journal of Perinatology* 31 (12): 760–763. jp201122.

American Academy of Pediatrics (AAP). 2012. *Guidelines for perinatal care*. 7th ed.

Azoulay, E., F. Pochard, N. Kentish-Barnes, S. Chevret, J. Aboab, C. Adrie, and Group, Famirea Study. 2005. Risk of post-traumatic stress symptoms in family members of intensive care unit patients. *American Journal of Respiratory and Critical Care Medicine* 171 (9): 987–994. https://doi.org/10.1164/rccm.200409-1295OC.

Bastek, T.K., D.K. Richardson, J.A.F. Zupancic, and J.P. Burns. 2005. Prenatal consultation practices at the border of viability: A regional survey. *Pediatrics* 116 (2): 407–413.

Bennett, Joann, Janet Dutcher, and Michele Snyders. 2011. Embrace: Addressing anticipatory grief and bereavement in the perinatal population: A palliative care case study. *Journal of Perinatal and Neonatal Nursing* 25 (1): 72–76.

Berth, H., A.K. Puschmann, A. Dinkel, and F. Balck. 2009. The trauma of miscarriage – factors influencing the experience of anxiety after early pregnancy loss. *Psychotherapie Psychosomatik Medizinische Psychologie* 59 (8): 314–320. https://doi.org/10.1055/s-2008-1067540.

Boss, R.D., N. Hutton, P.K. Donohue, and R.M. Arnold. 2009. Neonatologist training to guide family decision making for critically ill infants. *Archives of Pediatrics and Adolescent Medicine* 163 (9): 783–788.

Boyd, P.A., P. Chamberlain, and N.R. Hicks. 1998. 6-year experience of prenatal diagnosis in an unselected population in Oxford, UK. *Lancet* 352 (9140): 1577–1581.

Breeze, Andrew C.G., Christoph C. Lees, Arvind Kumar, Hannoah H. Missfelder-Lobos, and Edile M. Murdoch. 2007. Palliative care for prenatally diagnosed lethal fetal abnormality. *Archives of Disease in Childhood-Fetal and Neonatal Edition* 92 (1): F56–F58.

Bruhn, John G., and Howard M. Rebach. 2014. *The sociology of caregiving*. Berlin: Springer.

Byers, T.E., H.J. Wolf, K.R. Bauer, S. Bolick-Aldrich, V.W. Chen, J.L. Finch, and Patterns of Care Study, Group. 2008. The impact of socioeconomic status on survival after cancer in the United States: Findings from the National Program of Cancer Registries Patterns of Care Study. *Cancer* 113 (3): 582–591. https://doi.org/10.1002/cncr.23567.

Byock, Ira. 2012. *The best care possible: A physician's quest to transform care through the end of life*. New York: Avery.

Calhoun, Byron C., Peter Napolitano, Melisa Terry, Carie Bussy, and Nathan J. Hoeldke. 2003. Perinatal hospice: Comprehensive care for the family of the fetus with a lethal condition. *Obstetrical & Gynecological Survey* 58 (11): 718–719.

Chitty, Lyn S., Chris A. Barnes, and Caroline Berry. 1996. For debate: Continuing with pregnancy after a diagnosis of lethal abnormality: Experience of five couples and recommendations for management. *BMJ* 313 (7055): 478–480. https://doi.org/10.1136/bmj.313.7055.478.

Committee on Bioethics and Committee on Hospital Care (CBCHC). 2000. Palliative care for children. *Pediatrics* 106: 351.

Cote-Arsenault, D., and E. Denney-Koelsch. 2011. "My baby is a person": Parents' experiences with life-threatening fetal diagnosis. *Journal of Palliative Medicine* 14 (12): 1302–1308. https://doi.org/10.1089/jpm.2011.0165.

Curtis, J.R., R.A. Engelberg, M.D. Wenrich, S.E. Shannon, P.D. Treece, and G.D. Rubenfeld. 2005. Missed opportunities during family conferences about end-of-life care in the intensive care unit. *American Journal of Respiratory and Critical Care Medicine* 171 (8): 844–849. https://doi.org/10.1164/rccm.200409-1267OC.

Daly, B.J., S.L. Douglas, E. O'Toole, N.H. Gordon, R. Hejal, J. Peerless, J. Rowbottom, A. Garland, C. Lilly, C. Wiencek, and R. Hickman. 2010. Effectiveness trial of an intensive communication structure for families of long-stay ICU patients. *Chest* 138 (6): 1340–1348. https://doi.org/10.1378/chest.10-0292.

Desbiens, N.A., A.W. Wu, S.K. Broste, N.S. Wenger, A.F. Connors Jr., J. Lynn, Y. Yasui, R.S. Philllips, and W. Fulkerson. 1996. Pain and satisfaction with pain control in seriously ill hospitalized adults: Findings from the SUPPORT research investigations. For the SUPPORT investigators. Study to Understand Prognoses and Preferences for Outcomes and Risks of Treatmentm. *Critical Care Medicine* 24 (12): 1953–1961.

Emanuel, E.J., D.L. Fairclough, J. Slutsman, and L.L. Emanuel. 2000. Understanding economic and other burdens of terminal illness: The experience of patients and their caregivers. *Annals of Internal Medicine* 132 (6): 451–459.

English, N.K., and K.L. Hessler. 2013. Prenatal birth planning for families of the imperiled newborn. *Journal of Obstetric, Gynecologic, and Neonatal Nursing* 42 (3): 390–399. https://doi.org/10.1111/1552-6909.12031.

Fioravanti, J. 2002. Issues related to prenatal diagnosis of congenital heart disease. *Neonatal Network – Journal of Neonatal Nursing* 21 (6): 23–29.

Gudex, C., B.L. Nielsen, and M. Madsen. 2006. Why women want prenatal ultrasound in normal pregnancy. *Ultrasound in Obstetrics and Gynecology* 27 (2): 145–150.

Hebert, R.S., R. Schulz, V.C. Copeland, and R.M. Arnold. 2009. Preparing family caregivers for death and bereavement. Insights from caregivers of terminally ill patients. *Journal of Pain and Symptom Management* 37 (1): 3–12. https://doi.org/10.1016/j.jpainsymman.2007.12.010.

Hoeldtke, Nathan J., and Byron C. Calhoun. 2001. Perinatal hospice. *American Journal of Obstetrics and Gynecology* 185 (3): 525–529.

Howard, E.D. 2006. Family-centered care in the context of fetal abnormality. *Journal of Perinatal and Neonatal Nursing* 20 (3): 237–242.

Kars, M.C., M.H. Grypdonck, A. Beishuizen, E.M. Meijer-van den Bergh, and J.J. van Delden. 2010. Factors influencing parental readiness to let their child with cancer die. *Pediatric Blood & Cancer* 54 (7): 1000–1008. https://doi.org/10.1002/pbc.22532.

Kemp, Jennifer, Mark Davenport, and Amanda Pernet. 1998. Antenatally diagnosed surgical anomalies: The psychological effect of parental antenatal counseling. *Journal of Pediatric Surgery* 33 (9): 1376–1379.

Krakowsky, Y., M. Gofine, P. Brown, J. Danziger, and H. Knowles. 2013. Increasing access – A qualitative study of homelessness and palliative care in a major urban center. *The American Journal of Hospice & Palliative Care* 30 (3): 268–270. https://doi.org/10.1177/1049909112448925.

Kuebelbeck, A. 2011. Waiting with Gabriel. *Current Problems in Pediatric and Adolescent Health Care* 41 (4): 113–114. S1538-5442(10)00210-5.10.1016/j.cppeds.2010.10.013.

Lee, S., G.A. Colditz, L.F. Berkman, and I. Kawachi. 2003. Caregiving and risk of coronary heart disease in U.S. women: A prospective study. *American Journal of Preventive Medicine* 24 (2): 113–119.

Leuthner, S., and E.L. Jones. 2007. Fetal concerns program: A model for perinatal palliative care. *MCN, American Journal of Maternal Child Nursing* 32 (5): 272–278.

Leuthner, S.R., M. Bolger, M. Frommelt, and R. Nelson. 2003. The impact of abnormal fetal echocardiography on expectant parents' experience of pregnancy: A pilot study. *Journal of Psychosomatic Obstetrics and Gynaecology* 24 (2): 121–129.

Levine, Carol. 1999. The loneliness of the long-term care giver. *The New England Journal of Medicine* 340 (20): 1587–1590. https://doi.org/10.1056/NEJM199905203402013.

Levine, Carol, and United Hospital Fund of New York (Levine). 2004. *Always on call: When illness turns families into caregivers*. Updated and expanded ed. Nashville: Vanderbilt University Press.

Lilly, C.M., L.A. Sonna, K.J. Haley, and A.F. Massaro. 2003. Intensive communication: Four-year follow-up from a clinical practice study. *Critical Care Medicine* 31 (5 Suppl): S394–S399. https://doi.org/10.1097/01.CCM.0000065279.77449.B4.

Limbo, R., R. Kobler, S. Toce, and T. Peck. 2009. *Resolve through sharing: Position paper on perinatal palliative care* (Ed. I. Gunderson Lutheran Medical Foundation). La Crosse.

Lynn, Joanne, J.M. Teno, R.S. Phillips, A.W. Wu, N. Desbiens, J. Harrold, M.T. Claessens, N. Wenger, B. Kreling, and A.F. Connors Jr. 1997. Perceptions by family members of the dying experience of older and seriously ill patients. SUPPORT Investigators. Study to Understand Prognoses and Preferences for Outcomes and Risks of Treatments. *Annals of Internal Medicine* 126 (2): 97–106.

Lynn, Joanne, Ekta Chaudhry, Lin Noyes Simon, Ann M. Wilkinson, and Janice Lynch Schuster. 2007. *The common sense guide to improving palliative care*. Oxford: Oxford University Press.

Mathews, T.J., and M.F. MacDorman. 2014. Infant mortality statistics from the 2010 period linked birth/infant death data set. *National Vital Statistics Reports* 62 (8): 1–26.

McKinlay, J.B., S.L. Crawford, and S.L. Tennstedt. 1995. The everyday impacts of providing informal care to dependent elders and their consequences for the care recipients. *Journal of Aging and Health* 7 (4): 497–528. https://doi.org/10.1177/089826439500700403.

Meier, Diane, Stephen Isaacs, and Robert Hughes. 2010. *Palliative care: Transforming the care of serious illness*. 1st ed. San Francisco: Jossey-Bass.

Munson, D., and S.R. Leuthner. 2007. Palliative care for the family carrying a fetus with a life-limiting diagnosis. *Pediatric Clinics of North America* 54 (5): 787–798, xii. doi: S0031-3955(07)00096-X [pii] 10.1016/j.pcl.2007.06.006.

National Academies of Sciences, Engineering, and Medicine (NASEM). 2016. *Families caring for an aging America*. Washington, DC: The National Academies Press. https://doi.org/10.17226/23606.

National Alliance for Caregivers (NAC) and AARP. 2015. *Caregiving in the U.S.* http://www.caregiving.org/wp-content/uploads/2015/05/2015_CaregivingintheUS_Final-Report-June-4_WEB.pdf. Accessed 1 Sept 2017.

National Consensus Project for Quality Palliative Care (NCP). 2018. *Clinical practice guidelines for quality palliative care*. 4th ed. Pittsburgh: National Consensus Project.

National Consensus Project for Quality Palliative Care (NCP). 2004. *Clinical practice guidelines for quality palliative care*. 1st ed. Pittsburgh: National Consensus Project.

———. 2009. *Clinical practice guidelines for quality palliative care*. 2nd ed. Pittsburgh: National Consensus Project.

———. 2013. *Clinical practice guidelines for quality palliative care*. 3rd ed. Pittsburgh: National Consensus Project. http://www.nationalcoalitionhpc.org/wp-content/uploads/2017/04/NCP_Clinical_Practice_Guidelines_3rd_Edition.pdf.

O'Brien, Daniel L. 2013. *Palliative care: A hallmark of catholic healthcare*. Paper presented at the fifth national palliative care congress of the supportive care coalition. https://www.chausa.org/docs/default-source/health-progress/palliative-care-the-biblical-roots.pdf?sfvrsn=2. Accessed 22 Sept 2017.

Order of the Knights of Malta. 2016. *History of the knights of Malta*. http://www.knightsofmalta.com/history/history.html. Retrieved 22 June 2016.

Ortman, Jennifer, Victoria Velkoff, and Howard Hogan. 2014. An aging nation: The older population in the United States, pp. 25–1140. Current Population Reports.

Pavalko, Eliza K., and Kathryn A. Henderson. 2016. Combining care work and paid work. *Research on Aging* 28 (3): 359–374.

Perinatal Hospice and Palliative Care. 2017. http://www.perinatalhospice.org/. Accessed 20 Sept 2017.

Pope Benedict XVI. 2006. *Message of his holiness Benedict XVI for 14th World Day of the Sick*.

———. 2011. General audience. The prayer of Jesus linked to his miraculous healing action. http://w2.vatican.va/content/benedict-xvi/en/audiences/2011/documents/hf_ben-xvi_aud_20111214.html. Accessed 10 Aug 2017.

Pope Francis. 2013. *Address of Holy Father Francis to participants in the meeting organized by the International Fedration of Catholic Medical Assocations*. https://w2.vatican.va/content/francesco/en/speeches/2013/september/documents/papa-francesco_20130920_associazioni-medici-cattolici.html. Accessed 10 Aug 2017.

Pope Francis. 2015. *Assisting the elderly and palliative care*. Address of His Holiness Pope Francis to participants in the plenary of the 21st general assembly of the Pontifical Academy for Life.

———. 2016. *Amoris Laetitia*.

Pope John Paul II. 1981. *Familiaris Consortio*. http://w2.vatican.va/content/john-paulii/en/apost_exhortations/documents/hf_jp-ii_exh_19811122_familiaris-consortio.html.

Ramer-Chrastek, J., and M.V. Thygeson. 2005. A perinatal hospice for an unborn child with a life-limiting condition. *International Journal of Palliative Nursing* 11 (6): 274–276.

Sandelowski, M., and J. Barroso. 2005. The travesty of choosing after positive prenatal diagnosis. *Journal of Obstetric, Gynecologic, and Neonatal Nursing* 34 (3): 307–318. 34/3/307 [pii] 10.1177/0884217505276291.

Schenker, Y., G.A. Tiver, S.Y. Hong, and D.B. White. 2012. Association between physicians' beliefs and the option of comfort care for critically ill patients. *Intensive Care Medicine* 38 (10): 1607–1615. https://doi.org/10.1007/s00134-012-2671-4.

Schulz, R., and S.R. Beach. 1999. Caregiving as a risk factor for mortality: The Caregiver Health Effects Study. *JAMA* 282 (23): 2215–2219.

Schulz, Richard, and Paula R. Sherwood. 2008. Physical and mental health effects of family caregiving. *The American Journal of Nursing* 108 (9 Suppl): 23–27. https://doi.org/10.1097/01.NAJ.0000336406.45248.4c. quiz 27.

Statham, H., W. Solomou, and L. Chitty. 2000. Prenatal diagnosis of fetal abnormality: Psychological effects on women in low-risk pregnancies. *Baillière's Best Practice & Research. Clinical Obstetrics & Gynaecology* 14 (4): 731–747. https://doi.org/10.1053/beog.2000.0108.

Thomas, Kali S., and Vincent Mor. 2013. Providing more home-delivered meals is one way to keep older adults with low care needs out of nursing homes. *Health Affairs* 32 (10): 1796–1802. https://doi.org/10.1377/hlthaff.2013.0390.

Thomas, K.S., U. Akobundu, and D. Dosa. 2016. More than a meal? A randomized control trial comparing the effects of home-delivered meals programs on participants' feelings of loneliness. *The Journals of Gerontology. Series B, Psychological Sciences and Social Sciences* 71 (6): 1049–1058. https://doi.org/10.1093/geronb/gbv111.

Tosello, B., G. Haddad, C. Gire, and M.A. Einaudi. 2016. Lethal fetal abnormalities: How to approach perinatal palliative care? *The Journal of Maternal-Fetal & Neonatal Medicine* 1–4. https://doi.org/10.1080/14767058.2016.1186633.

United States Conference of Catholic Bishops (USCCB). 1997. *Catechism of the Catholic Church: Revised in accordance with the official Latin text promulgated by Pope John Paul II*, United States Catholic Conference. 2nd ed. Vatican City: Libreria Editrice Vaticana.

Walker, A.J., C.C. Pratt, and L. Eddy. 1995. Informal caregiving to aging family members: A critical review. *Family Relations* 44 (4): 402–411.

Walshe, C., C. Todd, A. Caress, and C. Chew-Graham. 2009. Patterns of access to community palliative care services: A literature review. *Journal of Pain and Symptom Management* 37 (5): 884–912. https://doi.org/10.1016/j.jpainsymman.2008.05.004.

Whitfield, J.M., A. Glicken, R. Harmon, R. Siegel, and L. Butterfield. 1982a. Neonatal hospice program. *Pediatrics* 70 (3): 502–503.

Whitfield, J.M., R.E. Siegel, A.D. Glicken, R.J. Harmon, L.K. Powers, and E.J. Goldson. 1982b. The application of hospice concepts to neonatal care. *American Journal of Diseases of Children* 136 (5): 421–424.

Wool, Charlotte, John T. Repke, and Anne B. Woods. 2016. Parent reported outcomes of quality care and satisfaction in the context of a life-limiting fetal diagnosis. *The Journal of Maternal-Fetal & Neonatal Medicine* 30 (8): 894–899.

Xu, J., S.L. Murphy, K.D. Kochanek, and B.A. Bastian. 2016. Deaths: Final data for 2013. *National Vital Statistics Reports* 64 (2): 1–119.

# Chapter 10
# Integrating Palliative Care into the Treatment of Serious Illness

**Christopher W. Lawton and Diane E. Meier**

## 10.1 Introduction

In March 2015, while addressing the Pontifical Academy for Life, Pope Francis offered reflections on the role of palliative care in modern society.

> Palliative care is an expression of the truly human attitude of taking care of one another, especially of those who suffer. It is a testimony that the human person is always precious, even if marked by illness and old age. Indeed, the person, under any circumstances, is an asset to him/herself and to others and is loved by God. That is why, when life becomes very fragile and the end of their earthly existence approaches, we feel the responsibility to assist and accompany them in the best way. (Pope Francis 2015)

In these words, Pope Francis underscores the central responsibility we have as human beings to honor the dignity of every human person by walking with them in the midst of the suffering that comes with aging and serious illness.

As we will discuss in this chapter, accompanying patients and their loved ones in the face of serious illness is the central mission of palliative care. Healthcare providers practicing palliative medicine are called upon to tend to the physical, emotional and spiritual components of suffering. A physician adjusts a pain medication dose when a tumor compresses nearby nerves. A hospice nurse meets with a patient and his wife to discuss what medical care will look like when a patient returns home. A chaplain sits with a patient who seeks to find meaning in her physical suffering. The challenges found in Pope Francis's "taking care of one another" are many and diverse.

Among the greatest challenges in this journey with patients is that of integrating palliative measures and, when the time comes, appropriate end-of-life care includ-

C. W. Lawton
Palliative Care Physician, Milwaukee, WI, USA

D. E. Meier (✉)
Center to Advance Palliative Care, NY, USA
e-mail: diane.meier@mssm.edu



P. J. Cataldo, D. O'Brien (eds.), *Palliative Care and Catholic Health Care*, Philosophy and Medicine 130, https://doi.org/10.1007/978-3-030-05005-4_10

ing hospice. The challenge here stems from a host of factors. These include a historical shift in how we as a society respond to human suffering and the reality of death, inadequate access to palliative care services, a need for more education for patients, families and providers, and the emotional barriers patients and providers experience in incorporating palliative care into the treatment plan. In this chapter, we will explore these factors in detail and examine the ways in which modern medicine is beginning to find solutions to make this integration easier for all involved. Ultimately, palliative care is becoming an important vehicle by which providers journey with patients, respond to their unique needs, and, as Pope Francis writes, “assist and accompany them in the best way.”

## 10.2 Pressures to Readdress Care for Seriously Ill

Understanding the challenges of integrating palliative and hospice care in the treatment of the seriously ill requires an understanding of palliative care’s history. Palliative care’s beginning and subsequent rapid expansion into a field of medicine come at a time when a confluence of pressures including epidemiologic, medical and financial demanded a reevaluation of how we care for patients with serious illness.

The American population is aging. The average American today has a life expectancy of nearly 79 years, up from 47 years in the year 1900 (Xu et al. 2016; CDC 2011). By the year 2050, the population of adults over the age of 65 is expected to reach more than 83 million, nearly double what it was in 2012 (Ortman et al. 2014). With this aging comes an increase in the number of individuals suffering from serious illness. Alzheimer’s disease, the most common form of dementia and the sixth leading cause of death among US adults, is a prime example of a serious illness that many patients live with for years and even decades. While an estimated 5 million Americans over the age of 65 were living with Alzheimer’s disease in 2013, this number is expected to triple to nearly 14 million by the year 2050 (Hebert et al. 2013).

As we consider this large and expanding number of individuals living with serious illness, we are faced with the reality that historically there has been and continues to be widespread inadequacy and dissatisfaction with the care provided to seriously ill patients. In 1995, the SUPPORT trial provided new and staggering data regarding the negative experiences of patients and families facing serious illness. Among its findings was the fact that more than half of doctors did not know their patient’s preferences regarding CPR and that significant pain was common even among those who were actively dying (SUPPORT 1995). More recently, in a 2004 study of family members of Medicare decedents who were asked to evaluate the care their loved-one received at the end of their life, approximately one quarter felt that pain or difficulty breathing was inadequately treated. Approximately the same number felt that physician communication was inadequate (Teno et al. 2004).

Facing the reality of an aging population in a country where there have been serious deficiencies in how we care for those with advanced illness, financial pressures have further propelled the effort to reevaluate the care we provide. The increase in the number of individuals living with serious illness has led to an enormous increase in current and projected health care spending. Again looking at dementia, in 2010 the total cost of this illness for the US economy was estimated to be between $159 billion and $215 billion. By the year 2040, it is expected that this cost will more than double (Hurd et al. 2013). Medicare spending offers a broader lens through which we can see the rise of US healthcare costs. Between 2010 and 2015, Medicare expenditures rose from approximately $523 billion to more than $647 billion (CMS 2011, 2016). Reevaluation of the way we care for the seriously ill and those at the end of life has also been required in the setting of these significant current and projected financial burdens.

## 10.3 The Establishment of Palliative Care

Palliative care's emergence as a field of medicine comes in response to the need for the American healthcare system to respond to the complex interplay of these pressures: an aging population, gross inadequacies in care of the seriously ill, and mounting financial pressures. A key component of this response was the Institute of Medicine's 1997 report *Approaching Death: Improving Care at the End of Life*. The report made recommendations aimed at restructuring the healthcare system to better meet the needs of patients living with serious and life-limiting illness. Among its recommendations were the establishment of "palliative medicine" as a new medical specialty that delivers palliative care to patients and their families. The report also called for research and educational reform to support the expansion of palliative care (IOM 1997). Since the time of this report, this expansion has included the National Consensus Project for Quality Palliative Care's establishment of guidelines for the delivery of high-quality palliative care. The project defines the field's work as follows:

> Palliative care is both a philosophy of care and an organized, highly structured system for delivering care. The goal of palliative care is to prevent and relieve suffering and to support the best possible quality of life for patients and their families, regardless of the stage of the disease or the need for other therapies. Palliative care expands traditional disease-model medical treatments to include the goals of enhancing quality of life for patients and family members, helping with decision-making, and providing opportunities for personal growth. Palliative care can be rendered along with life-prolonging treatment or as the main focus of care. (NCP 2004)

The Project's guidelines call for patient assessment by an interdisciplinary team, care planning based on the patient and family's preferences, values and goals, and thorough attention to a patient's physical, psychological, social and spiritual needs. Today, the delivery of this model of palliative care occurs most commonly through inpatient hospital consultation teams. These interdisciplinary teams include physicians, nurses, social workers and chaplains and in many cases psychologists, massage therapists and other experts. Teams address physical symptoms such as pain, nausea and fatigue as well as other dimensions of suffering such as psychological and spiritual distress. They assist in clarifying patient's goals, recommending treatment plans based on those goals and support the coordination of care among various providers involved in a patient's care. Because of the passage of the Affordable Care Act, and the increasing attention to value (improving quality and reducing costs), palliative care teams are being expanded to the outpatient setting as well.

## 10.4 Palliative Care and Hospice

As we consider the rise of palliative care, it is important to draw a distinction between palliative care and hospice care. While palliative care is provided at any stage of an illness and in tandem with life-prolonging and curative therapies, hospice is an interdisciplinary system of care that provides palliative care as the focus of care for patients in the last months of life. In the United States, federal programs and insurance providers including Medicare (the government-run insurance program for people over age 65 and some people with disabilities) and Veterans Affairs are largely responsible for establishing criteria for hospice care. Under such programs, patients are eligible for hospice services if they have a prognosis of less than 6 months. Hospice services are a benefit of Medicare Part A. In electing the hospice benefit, patients choose to focus on treatments that promote comfort and quality of life and give up coverage for any therapy aimed at disease modification or cure. Hospice services are provided in a variety of settings. Most commonly, hospice services are provided in a patient's home, but can also be provided in a nursing home, assisted living facility, residential or inpatient hospice facility or in a hospital that has a contract with a hospice agency. Ultimately, hospice is a program of care limited to people who are clearly and predictably dying that falls under the larger umbrella of palliative care (Kelley and Morrison 2015).

## 10.5 Cure Versus Care

The development of the Medicare hospice benefit in the US provides an important lens through which we can better understand the challenges of integrating palliative care into the treatment of serious illness. The creation of the Medicare hospice benefit in 1982 was followed in 1986 by the establishment of hospice eligibility criteria. One of the purposes of these criteria was to restrict access to hospice services in order to control costs. Through these criteria, and specifically the provision that in order to elect the hospice benefit a patient has to forego curative or disease-modifying therapy, hospice became in the eyes of many in the medical community and the public at large a form of "giving up" the fight to live. By giving the impression that at some point in their disease trajectory, a seriously ill patient must choose to either "fight" or "give up," the Medicare hospice benefit has contributed to the widely accepted but incorrect assumption that modern medicine has two mutually exclusive endpoints: to cure disease and extend life or to provide comfort-focused care. As a result, it remains all too common that the introduction of any treatments focused on comfort and quality of life occurs only after disease-modifying and life-prolonging therapies are no longer an option. This dichotomy and the corresponding cultural misconception of how we should approach serious illness is known as the "care versus cure" phenomenon. An important and problematic result of this dichotomy is that for many patients, the introduction of not just hospice, but any palliative care therapies often comes very late in the course of an illness when patients have little time left to reap their benefits (Meier 2010).

## 10.6 The Simultaneous Care Model

The expansion of non-hospice palliative care services has become an important way that some in the medical community have attempted to respond to "care versus cure." If palliative care is defined, as we have discussed, as care that is provided "regardless of the stage of the disease," then when should patients start to receive palliative care and what role does it play as illness evolves? The answers start with an understanding of the "simultaneous care model." This model of care argues that suffering whether physical, psychological, social or spiritual, is not just experienced by patients near the end of life, but rather throughout the trajectory of an illness, starting at the time of diagnosis. If palliative care's work is to alleviate suffering in all its forms, then it follows that palliative care is best provided starting at the time of diagnosis and then continued *simultaneously* with other disease-modifying or curative treatments (Meyers and Linder 2003). The figure below shows where palliative care fits in the course of an illness (NQF 2006).

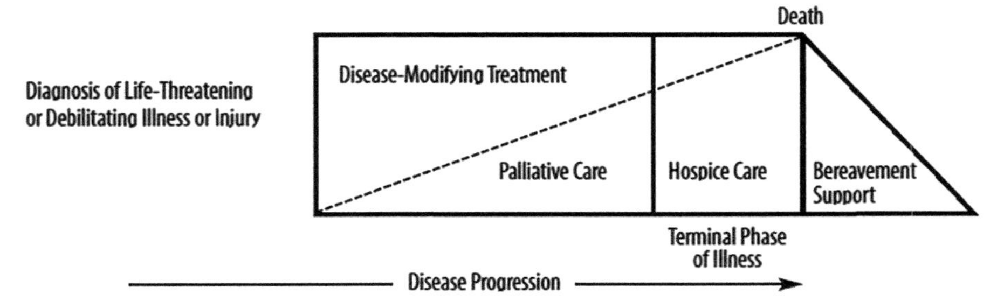


**Reprinted with permission from** *A National Framework and Preferred Practices for Palliative and Hospice Care Quality: A Consensus Report.* **by the National Quality Forum, Washington, D.C.**

In keeping with this model, a major part of the effort to expand access to palliative care has focused on incorporation of palliative care services earlier in the course of illness, in tandem with life-prolonging therapies.

## 10.7 The Benefits of Palliative Care

There is a substantial and increasing body of evidence supporting the benefits of palliative care for patients and their families. It has been repeatedly found that the involvement of a palliative care leads to improvement in patient's pain and other symptom control (Higginson et al. 2003; Higginson and Evans 2010). A study of nearly 600 patients seen by an inpatient palliative care consultation team found that pain control improved in 87% of patients (O'Mahony et al. 2005). Another study of over 400 cancer patients seen in an outpatient palliative care clinic found that patients experienced significant improvement in pain as well as a host of other symptoms including fatigue, nausea, shortness of breath, anxiety and depression (Yennurajalingam et al. 2011). Patients seen by palliative care teams have been found to receive improved information sharing and communication from medical teams about serious illness, increased emotional and spiritual support, and an improved sense of well-being and dignity (Casarett 2008). Palliative care programs have been shown to lead to improved patient and family satisfaction with care as well as to help patients avoid hospitalization and remain adequately cared for at home (Brumley et al. 2007).

In addition to these benefits, we now know that early integration of palliative care leads to the potential to prolong life. A recent study of lung cancer patients found that, as compared to patients receiving usual oncologic care, patients who started receiving palliative care shortly after diagnosis lived nearly 3 months longer. These patients were also found to have improved overall quality of life and decreased depressive symptoms (Temel et al. 2010). While the reason for this survival benefit is still being investigated, it has been posited that avoidance of both the hazards of hospitalization and of toxic treatments and improved symptom management and quality of life may improve patient's function and ability to complete beneficial treatments. Another theory is that the support palliative care provides patients may result in patients foregoing disease-modifying therapies at a point in their disease course when such therapies may be more harmful than beneficial, again resulting in a survival benefit (Bruera and Yennurajalingam 2012). These promising results have led to an expert consensus opinion of the American Society of Clinical Oncology that palliative care services be regularly integrated into standard oncologic care for patients diagnosed with metastatic or advanced cancer (Smith et al. 2012). Caregivers are also supported by early palliative care involvement. One recent study of caregivers of advanced cancer patients receiving early palliative care involvement found that they were more satisfied with the care their loved ones received than caregivers of patients receiving standard cancer care (McDonald et al. 2016).

## 10.8 Barriers to Palliative Care Integration

Despite this clear and expanding evidence base for palliative care's important role in the care of the seriously ill, integrating palliative care into the treatment of the seriously ill, and specifically in tandem with curative therapies, is fraught with several barriers. These include the cultural shift that palliative care represents in the context of modern healthcare, a lack access to palliative care services, widespread misconceptions and inadequate education about the role of palliative care, and the emotional difficulty inherent to facing serious illness and mortality.

### *10.8.1 Cultural Shift in the Understanding of Death*

As we have discussed, the field of palliative medicine was born out of societal pressures that demanded a shift in the way we care for the seriously ill. Today, more and more patients and families, healthcare professionals and policy makers appreciate the need for such a shift and the benefits of palliative care's work. But this shift requires reframing of our cultural conception of mortality and our expectations of the healthcare system. As a society, we have become less accustomed with mortality and death. Historically, death commonly occurred in one's home, an expected and accepted part of the life cycle. Even though the majority of Americans continue to desire to die at home, today most die in institutions where the end-stages of disease and death itself are simply less visible and are in many ways hidden from society (Grunier et al. 2007). Simultaneously, we have become a society that is often more focused on the extension of life than on the assurance of quality of life and alleviation of suffering. Today we have therapies for illnesses such as advanced cancer and heart failure that allow patients to live longer with serious illness than ever before. When a person becomes acutely ill, we have amazing and ever-expanding technology to extend life, including ICU level care where the function of vital organs like the heart, lungs and kidneys can be sustained by machines even when these organs have failed. As we continue to focus our research and clinical efforts on doing all we can to sustain life, often at the cost of quality of life and suffering, we as a society operate as though death itself were a beatable condition. As Americans, we're asking ourselves, even if unconsciously, "No matter what the cost, how can we stave off death, even for just a little longer?" Death comes to represent a failure; of not working hard enough to find the right answer to an impossible question. This implicit assumption, that death is somehow avoidable, is a major reason why for many, palliative care represents such a paradigm shift. Our societal pendulum has swung far toward the sustaining of life and cure of disease, away from alleviation of suffering and acceptance and mortality. The idea that we can bring that pendulum back toward the middle, pursuing quality of life and quantity of life simultaneously, requires an ongoing shift in our societal understanding of how we best face illness, treatment and ultimately death.

### 10.8.2 Inadequate Access

Ensuring access to palliative services remains an ongoing challenge to the effort to better integrate palliative care in our healthcare system's approach to serious illness. As we have discussed, the diverse benefits of palliative care have been well evidenced and when palliative care is introduced late in a disease course, patients get minimal access to its therapies. One study which surveyed providers at cancer centers across the US found that the average time between referral to palliative care and death was only 3 months for patients being seen in an outpatient palliative care clinic and just 1 week for hospitalized patients who received a palliative care inpatient consultation (Hui et al. 2010).

The sheer number of palliative care programs and providers also contributes heavily to the access barrier. While the growth of hospital palliative care program has been well documented in recent decades, many people are still not being reached. As of 2015, one third of US hospitals had no form of palliative care services. In community settings such as home, nursing home and assisted living facilities, palliative care services are largely focused on those who are actively dying (i.e., hospice-eligible), leaving many individuals suffering with serious illness without adequate care. Geographic disparity also leads to segments of our population being particularly underserved. Patients in the south-central US and specifically Arkansas, Mississippi and Alabama are most deprived. In these three states, less than one third of hospitals reported having an inpatient consultation team (Morrison et al. 2015). Another significant component of the access barrier is an inadequate palliative care workforce. According to 2010 data, there were approximately 4400 hospice and palliative medicine (HPM) physicians who were board certified or members of the American Academy of Hospice and Palliative Medicine (the professional organization for physician specialists). Yet as many as four times that number of providers would be needed to overcome the work force shortage facing the field (Lupu 2010).

### 10.8.3 Educational and Evolutionary Barriers

Beyond this broader paradigm shift, integration of palliative care services is further challenged by lack of awareness of what palliative care is and misconceptions about its work. According to a recent public opinion survey, 70% of participants reported having no knowledge of palliative care. For many patients and even some providers, the term "palliative care" is still misunderstood as equivalent to "hospice" or "only for the dying" (CAPC 2011). Without proper education, patients and families may resist palliative care involvement because of the misconception that such care requires that disease-modifying or curative therapies be discontinued. Some patients or providers may resist involvement because they mistakenly believe that palliative therapies are not effective, or worse, will decrease survival.

For many providers, difficulty in knowing when a palliative care referral is appropriate presents another challenge. As medical science continues to offer new and varied treatments options across all fields of medicine, providers must fit palliative care within increasingly complex treatment plans. And even with the expanding evidence base for palliative care's benefit and role, many research topics require ongoing investigation. These include ongoing study of optimal pain and symptom management therapies, the role of palliative care in non-cancer care (such as heart and kidney disease) and the mechanism by which palliative care obtains its host of benefits including improved survival (Kelley and Sean Morrison 2015).

### *10.8.4 Emotional Barriers*

Even with adequate understanding of palliative care's role, many patients and families as well as physicians find it emotionally challenging to consider palliative care involvement. This may occur for a host of reasons. For some patients and families, accepting palliative care support may be hindered by the fact that the term "palliative care" stirs an increased awareness of mortality and the real possibility of disease progression. A family member may resist palliative care involvement for an ill loved one so that their loved one doesn't "lose hope" for the best possible outcome. These barriers are particularly troublesome since an important part of palliative care's work is to tend to the emotional components of suffering.

Physicians and other providers may face their own unique emotional barriers. Building on the above, some physicians may fear that introducing palliative care will take away a patient's "hope" for cure or better control of disease progression, even if the chance of control or cure is exceedingly small. Some oncologists may even continue active cancer treatment in an effort to preserve hope while knowing that such treatments are highly unlikely to change the course of the disease. Other physicians may feel that palliative care involvement represents an abandonment of their patient, as they are not trained in how to care for patients once disease-directed treatments are no longer helpful (Meier 2014). In our culture where for so long medical care and technological advancement has focused on cure and prolongation of life, physicians may feel that involving palliative care represents a "failure" of their effort to bring about cure. When this happens, palliative care consultation may be sought only when there no further disease-modifying therapies available and providers feel, "there is nothing more I can do." Again, waiting until this point affords less opportunity for palliative interventions to benefit patients and families.

## 10.9 Race and Ethnicity in Palliative Care

A discussion of the challenges inherent to integrating palliative care in the treatment of seriously ill patients would not be complete without addressing the significant racial and ethnic disparities found in the use of palliative care services. Minority patients have been found to receive lower quality palliative care with studies showing decreased satisfaction with communication and pain management. The same populations access palliative care services less often. For example, among Medicare beneficiaries who died in 2010, nearly 46% of whites were enrolled in hospice compared to only 34% of African Americans and 37% of Hispanics. To date, the reason for these disparities is not well understood. It has been posited that a combination of racial and ethnic differences in understanding of disease, treatment preferences and cultural beliefs are likely at play. There is a clear need for ongoing research in this area to help ensure improvement in meeting the needs of these specific populations (Johnson 2013).

## 10.10 Improving Integration of Palliative Care

Amidst the barriers described above, there is much work underway seeking to overcome these obstacles to incorporating palliative and, when appropriate, end-of-life care in our treatment of the seriously ill. This work has led to solutions which address the array of challenges we described above: improvements in access to palliative care, education to promote enhance awareness and understanding, and greater support to clinicians, patients and families so that emotional barriers can be addressed. Only by taking these steps can we begin to change the culture of American medicine and more fully embrace a palliative approach to care.

### *10.10.1 Overcoming Access and Staffing Barriers*

If we are going to effectively integrate palliative and end-of-life care, families and providers must have adequate access. Improving access is multifaceted and involves ongoing expansion of clinical palliative care programs, innovative models of care, the clinician workforce and efforts aimed at healthcare policy.

Within US hospitals today, more patients and families have access to palliative care teams than ever before. From 2008 to 2015, the number of US hospitals with a palliative care program on-site increased by more than 13%. Among hospitals with greater than 300 beds, more than 90% of hospitals now report having a palliative care program (Morrison et al. 2015). Today many hospital-based palliative care programs have rapidly expanding outpatient practices, allowing palliative care teams to be involved in a patient's care during periods of greater clinical stability. In the outpatient setting, symptom management and discussions of goals of care can be undertaken outside of the stress of a hospitalization. Here too teams can build relationships with patients and loved ones, easing the introduction of palliative care services over time instead of only at moments of crisis. An important way in which palliative care clinicians are working with other specialists to smooth this transition is by offering outpatient palliative care services in an integrated fashion, embedded within other specialty clinics. Palliative care providers may work within an outpatient oncology clinic where a patient can be seen by their oncologist and their palliative medicine specialist on the same day. A more recent trend is for a similar embedded model within advanced heart failure clinics. And we have evidence that these outpatient models are working. Patients seen by outpatient palliative care providers have been shown to have improved symptom control as well as higher satisfaction with their medical care. Moreover, many patients who received outpatient palliative care services wished that they had received them earlier in their disease course (Rabow et al. 2003, 2004).

The expansion of access to hospice services is similarly critical in ensuring that patients can make smooth transitions in care during the final months of life. Since the first hospice program opened in the US in 1974, thousands of hospice programs have been established. And the number of programs is on the rise, today reaching 6100 programs, up from just over 5000 5 years ago. In 2014, more than 1.6 million Americans received hospice services, an increase by nearly 20% compared to 2010 (NHPCO 2015).

Beyond these more traditional care delivery models, an important component in improving access to palliative care has been the development of alternative models of care that seek to ease access to services. These models focus on providing access to palliative care services earlier than allowed by hospice alone and in settings other than the acute care hospital. Home-based palliative care programs are an important example of such model. In these programs, providers deliver traditional palliative services at a patient's home, often in conjunction with an interdisciplinary team including a social worker, chaplain and other therapists. Here, palliative interventions including pain and other symptom management as well as emotional and spiritual support are provided often in tandem with disease-modifying treatments such as ongoing radiation or chemotherapy. These programs may be run by providers affiliated with a home health care agency or a nearby medical center. Teams providing such services are able to address the needs of patients and families earlier in the course of disease, becoming involved before the estimated final 6 months of life required for hospice services. By developing longer relationships with patients and loved ones, these providers can also later help ease transitions to hospice when appropriate. Patients served by such programs have been shown to have higher satisfaction with their medical care and higher likelihood of dying at home, an unmet wish of many patients with serious illness (Brumley et al. 2003; The Dartmouth Atlas 2016).

For patients who are eligible for hospice, models of care have been developed to help patients obtain the benefits of hospice while not necessarily forgoing therapies usually precluded by hospice, including disease-modifying therapy. This model is known as "concurrent care" or "open access." In such models, a hospice agency may offer beneficiaries coverage for treatments including chemotherapy, radiation, blood transfusions, hemodialysis and intravenous medications in tandem with traditional hospice services. Such programs are often provided by larger hospice agencies that, in caring for more patients, can spread out the higher costs of concurrent treatments. These models are important as they work to overcome a major psychological barrier to hospice, namely, the need to forgo disease-modifying therapy. For certain populations, including children and veterans, such models are used commonly. Under the Affordable Care Act, Section 2302 entitled, "Concurrent Care for Children," state Medicaid programs have been required to pay for disease-modifying and curative treatments concurrently with hospice services for eligible patients less than 21 years of age. While important barriers remain as we work to better incorporate palliative services in pediatrics, this law is an important step forward in improving the care provided to children at the end of life (NHPCO 2012). Similarly, the Veterans Health Administration (VA) healthcare system offers an example of the great potential of the concurrent care model. The VA does not require, as Medicare does, that patients choose between curative therapy and hospice, instead allowing patients who desire it to receive concurrent care in the last 6 months of life. One recent study from within the VA system that looked at healthcare utilization over a period of years showed that while the use of disease-directed therapy including chemotherapy and radiation remained stable, the use of concurrent care with hospice services increased (Mor et al. 2016). Such trends suggest that not forcing patients to choose between disease-modifying therapy and hospice may allow more patients to reap the benefits of hospice. Concurrent care becomes a clear example of how we can more effectively incorporate palliative care in our healthcare system's approach to serious illness.

If these and other innovative models are to be enacted to better integrate palliative care, there will have to be a sufficient work force to staff them. To date, important steps have been made to promote the expansion of the hospice and palliative medicine work force. As we have alluded to, the Institute of Medicine's 1997 report was an important step forward as it called for the designation of palliative medicine as a new medical specialty in hopes that it might improve patient care, expand research and enhance education for healthcare providers around palliative and end-of-life care. By 2007, the American Board of Medical Specialties approved "Hospice and Palliative Medicine" as a new subspecialty field of medicine. A year later, The Accreditation Council for Graduate Medical Education approved fellowship training in HPM and the first board certification exam was offered (Meier 2010). And while the growth of the physician work force has allowed for significant expansion of palliative care and hospice programs in the US, addressing the workforce shortage and educating more palliative care specialists will be an important step in meeting the needs of patients and families facing serious illness.

Ongoing healthcare policy efforts are similarly critical in ensuring expanded access to palliative care. At both the state and federal level, lawmakers will continue to play a key role in ensuring this expansion. According to recent recommendations from the Center to Advance Palliative Care and the National Palliative Care Research Center, expanded access requires Congress working in conjunction with organizations including Centers for Medicare & Medicaid Services, the National Institutes of Health and the Agency of Healthcare Research and Quality. Policy changes must support innovative models of palliative care delivery, appropriate payment schemes and resource allocation to help expand the workforce of palliative care providers (Morrison et al. 2015).

### *10.10.2 Overcoming Educational Barriers*

Improved incorporation of palliative care will continue to depend on ongoing efforts to overcome educational barriers. This means improving awareness of the role and benefits of palliative care for both patients and providers. It also means educating ourselves as a medical community by expanding research efforts in the field such that we provide the highest-quality treatments possible.

Efforts to expand awareness of the role and benefits of palliative care for patients and families exist in a number of forms. Of course, a critical mechanism is through the expansion of inpatient and outpatient palliative care programs. In such programs, physicians, nurses and other team members can provide education for patients referred to palliative care to ensure understanding of its services and benefits as well as address any misconceptions of its work. Various members of a program's interdisciplinary team may be uniquely positioned to address educational barriers or misconceptions. This might include a nurse practitioner on the palliative care team clarifying the distinction between palliative care and hospice, or a social worker alleviating a family's anxiety by explaining that palliative care and hospice services can be provided in a variety of care settings. As we overcome common educational barriers for patients and families, we ease the incorporation of palliative care and ensure those in need reap its benefits. Further below (and in Chap. 13 of this volume) we discuss how the Catholic Church can have an important role in the education of patients and families regarding palliative care and advance care planning.

There are also various national organizations working to educate the public about the work of palliative care. The National Hospice and Palliative Care Organization (NHPCO) provides ongoing educational resource materials to the public to promote awareness and education about hospice and palliative care (NHPCO 2016). A host of organizations work to educate patients and families to prepare them for the kinds of decisions that might need to be made in the setting of serious illness or at the end of life. Aging with Dignity is a non-profit organization which works to promote quality care for patients facing serious illness by educating patients and families about advanced care planning (in which a patient decides how he or she would want to be cared for in the event that he or she were not able to participate in decision-making). Their work includes the dissemination of the "Five Wishes" tool, which helps patients document choices about how they want to be cared for as they near the end of their life (Aging With Dignity 2015).[1]

In addition to addressing patient and family educational barriers, provider education is also critically important. All providers should have an understanding of palliative care and its work and there is an expanding body of resources to support providers towards this end. Today the Liaison Committee for Medical Education (the accreditation body for medical schools in the U.S.) requires that all accredited schools pro-

[1] See Chap. 14 of this volume on Catholic teaching about advance care planning.

vide students education in palliative care. The American Medical Association has created a curriculum around palliative care, the Education for Physicians on End-of-Life Care, which has reached more than 90,000 providers. Its counterpart, the End-of-Life Nursing Education Consortium, is a project that aims to educate nurses-in-training in the basic tenants of the field (Von Gunten and Ferrell 2014).

For physicians already in practice, there is a need for education in what is widely known as, "primary palliative care." These are palliative care skills that all physicians should have, including skills in basic pain and symptom management and in discussing topics such as prognosis, goals of care and code status (in contrast to "specialty palliative care" in which board-certified palliative care specialists manage refractory or more complex symptoms as well as issues of conflict resolution around goals of care). Today specific programs seek to educate physicians in primary palliative care. One example is OncoTalk, a communication training program for oncologists-in-training which teaches basic skills in palliative care. Organizations including the Center to Advance Palliative Care, the American Academy of Hospice and Palliative Medicine and the Agency for Healthcare Research and Quality are also promoting the development of new curricula to address palliative care educational needs for all physicians (Quill and Abernethy 2013). Palliative care providers in programs across the country must continue to work not only with the patients they serve, but also with other medical providers and administrative leaders to ensure adequate education about the field and its benefits, as such relationships are critical to any program's maintenance and success.

Finally, in addition to enhancing education for patients, families and providers, a key component of overcoming educational barriers will be continuing to educate ourselves as a medical community through research to build the palliative care evidence base. An improved evidence base allows us to better understand the mechanisms that underlie the success of palliative care interventions and improve our knowledge of the highest quality therapies and models of care.

### *10.10.3 Overcoming Emotional Barriers*

Creating a healthcare system that more fully integrates palliative care requires overcoming not only the access and educational barriers we have discussed, but also addressing the emotional hurdles that patients, families and physicians face. For the individual patient or family member having difficulty with the introduction or acceptance of palliative care, there is much we can do. When patients or families fear loss of hope or "giving up," a primary care doctor or oncologist may reassure them with a message of non-abandonment, reminding patients and families that we can always accompany and support them no matter how their disease evolves. In this setting, the incorporation of palliative treatments may happen more gradually, at first with a simple explanation of its aims, with more information and a referral happening over time. Palliative care teams themselves can often be helpful in addressing these emotional barriers. Such teams work routinely to build trusting relationships with patients and loved ones, even over a short period of time. In a hospital ICU, for example, a palliative care team may start by simply introducing themselves to a family struggling with decisions about whether to continue aggressive, life-prolonging therapies. A first visit may be spent building rapport by listening and getting to know a patient and family: exploring their understanding of an illness, who they are as people and the values they hold dear. The palliative care team, in conjunction with the referring providers, seeks to earn the patient and family's trust that the work of the entire team is always to honor the patient's unique goals and wishes.

Overcoming the emotional barriers providers face in integrating palliative care can happen in a number of ways. With improvements in access and provider education, palliative care programs help forge relationships between palliative care specialists and providers in all specialties. Through such relationships, palliative interventions can come to be seen are part of the best practice and multidisciplinary medical care we offer patients, rather than a sign of "abandonment." As such relationships strengthen, individual providers can know they have not only a resource to support their patients, but also a place to come to share their own fears, sadness and feelings of vulnerability when it comes to caring for patients with serious illness. Ultimately, a healthcare system best prepared to meet needs of seriously ill patients is one in which everyone on the treatment team feels that they have a community where they are supported emotionally as they engage this challenging work.

## 10.11 Looking Forward

Helping patients and families navigate the journey through serious illness is a complex task. It requires the work of many hands across our healthcare system to ensure that palliative services are available and that patients, families and providers are not only well educated about palliative care's work, but also have their emotional needs met. The Institute of Medicine's 2015 report, *Dying in America: Improving Quality and Honoring Individual Preferences Near the End of Life*, offers an important example of the ongoing effort to advance the mission and work of palliative care as we look toward the future. In the report, leaders in medicine define key components of quality care for patients facing serious illness and offer recommendations for ongoing national improvement in end-of-life care (IOM 2015).

**Recommendations of the Institute of Medicine's 2015 Report: Dying In America.***

1. *Government health insurers and care delivery programs as well as private health insurers should cover the provision of comprehensive care for individuals with advanced serious illness who are nearing the end of life.*
2. *Professional societies and other organizations that establish quality standards should develop standards for clinician patient communication and advance care planning that are measurable, actionable, and evidence-based. These standards should change as needed to reflect the evolving population and health system needs and be consistent with emerging evidence, methods, and technologies. Payers and health care delivery organizations should adopt these standards and their supporting processes, and integrate them into assessments, care plans, and the reporting of health care quality. Payers should tie such standards to reimbursement, and professional societies should adopt policies that facilitate tying the standards to reimbursement, licensing, and credentialing to encourage.*

- all individuals, including children with the capacity to do so, to have the opportunity to participate actively in their health care decision making throughout their lives and as they approach death, and receive medical and related social services consistent with their values, goals, and informed preferences.
- clinicians to initiate high-quality conversations about advance care planning, integrate the results of these conversations into the ongoing care plans of patients, and communicate with other clinicians as requested by the patient; and.
- clinicians to continue to revisit advance care planning discussions with their patients because individuals' preferences and circum- stances may change over time.

3. *Educational institutions, credentialing bodies, accrediting boards, state regulatory agencies, and health care delivery organizations should establish the appropriate training, certification, and/or licensure requirements to strengthen the palliative care knowledge and skills of all clinicians who care for individuals with advanced serious illness who are nearing the end of life.*
4. *Federal, state, and private insurance and health care delivery programs should integrate the financing of medical and social services to support the provision of quality care consistent with the values, goals, and informed preferences of people with advanced serious illness nearing the end of life. To the extent that additional legislation is necessary to implement this recommendation, the administration should seek and Congress should enact such legislation. In addition, the federal government should require public*

*reporting on quality measures, outcomes, and costs regarding care near the end of life (e.g., in the last year of life) for programs it funds or administers (e.g., Medicare, Medicaid, the U.S. Department of Veterans Affairs). The federal government should encourage all other payment and health care delivery systems to do the same.*

5. *Civic leaders, public health and other governmental agencies, community-based organizations, faith-based organizations, consumer groups, health care delivery organizations, payers, employers, and professional societies should engage their constituents and provide fact-based information about care of people with advanced serious illness to encourage advance care planning and informed choice based on the needs and values of individuals.*

*Reprinted with permission from *Dying in America: Improving Quality and Honoring Individual Preferences Near the End of Life, 2015* by the National Academy of Sciences, Courtesy of the National Academies Press, Washington, DC.

As emphasized in this report, the goal of the ongoing integration and expansion of palliative care becomes the provision of the highest quality care throughout the course of illness and to the very end of life. Incorporating palliative care into the treatment of our patients requires not only addressing the barriers we have discussed above, but also progress on a societal level in how we conceptualize serious illness and death. If we remain a society focused so heavily on the extension of life with little focus on, or worse, at the cost of quality of life and alleviation of suffering, then helping patients receive much needed palliative care becomes only more challenging. If instead we embrace the philosophy of palliative medicine and more fully walk with patients and families in the face of illness, we take part in a paradigm shift in American medicine. It is a shift which embraces quality of life and alleviation of suffering concurrently with life-prolonging interventions, seeks to understand and meet the unique goals of each patient, and supports patients and families in all ways to the end of life. Making this shift will require the ongoing work of many hands working in healthcare and health policy. It is work in which we all must be invested as we seek the best possible care for all those affected by serious illness and those they love.

## 10.12 The Integration of Palliative Care and the Catholic Church

We overcome these barriers to integration of palliative care more easily when advocates stand in solidarity with the mission of palliative care. The Catholic Church has been and continues to be an important advocate in this regard, standing as an ally in the mission to expand the field of palliative medicine. The Church has done this uniquely in its call to non-abandonment of those who are seriously ill and its commitment to the inherent dignity of the human person.

As we said at the outset of this chapter, Pope Francis has called for greater attention to the accompaniment of all those experiencing illness, suffering and eventually death. To bring this call alive, Church leaders have offered scriptural reference points. Among these is Jesus' teaching regarding the Good Samaritan. In The Gospel of Luke, Jesus teaches a man how he can obtain salvation. To make his point Jesus tells the story of the Good Samaritan, a man who comes to the aid of a man who has been robbed and left half dead, even as others pass the victim by. The Samaritan man tends to the wounds of his neighbor, brings him to a place of shelter and pays for all the expenses of the man's care. Jesus concludes by explaining that the mercy shown by the Samaritan is a model for all of us (NRSV 2010). Speaking in 2004 to the Pontifical Council for Health Pastoral Care, Saint John Paul II discussed the Good Samaritan as he described our call to lovingly accompany patients even when illness brings about great burden (Saint John Paul II 2004).

The introduction and incorporation of palliative treatments is an important part of this journey. Often the burden here is not only the disease, but also the treatments themselves with the suffering they can cause. We know too that even as patients come to the end of their lives, the experience of suffering and of caring for those who suffer can be extraordinarily burdensome. In keeping with the Pope's message, the mission of palliative care is to stay with the suffering, attempting to ease this burden to the very end of life. Pope Benedict noted that such accompaniment is "a right belonging to every human being, one which must be committed to defend" (Pope Benedict XVI 2009). And Pope Francis in his 2015 Address to the Pontifical Academy for Life makes this call to accompaniment even more practical by calling people to train in palliative care (Pope Francis 2015). Here he effectively advocates for attention to the challenges described above, helping ensure adequate access to palliative care services through expansion of the provider workforce.

Why is it so critical to the Church that we accompany patients and families on the journey of illness? It is because, as the Church tells us, we must honor the dignity of every human person. As Saint John Paul II notes, "Suffering, old age, a comatose state or the imminence of death in no way diminish the intrinsic dignity of the person created in God's image" (Saint John Paul II 2004). And this desire to honor the dignity of each person could not be more in line with the work of palliative medicine; it is truly complicit with the mission of the field. Walking with patients from early on in the course of a disease, working to alleviate suffering, easing points of transition and accompanying patients even until death, are all at the heart of palliative care. The field stands united with the Catholic Church in her commitment to advancing this work. Through this shared vision, there is more hope than ever that palliative care will become ever more integrated in our efforts to provide the highest quality care to the seriously ill.

## References

Aging With Dignity. 2015. *Five wishes*. https://www.agingwithdignity.org/five-wishes/about-five-wishes. Accessed 9 Nov 2016.

Bruera, Eduardo, and Sriram Yennurajalingam. 2012. Palliative care in advanced cancer patients: How and when? *The Oncologist* 17: 267–273.

Brumley, Richard, Susan Enguidanos, and David A. Cherin. 2003. Effectiveness of a home-based palliative care program for end-of-life. *Journal of Palliative Medicine* 6 (5): 715–724.

Brumley, Richard, et al. 2007. Increased satisfaction with care and lower costs: Results of a randomized trial of in-home palliative care. *Journal of the American Geriatrics Society* 55 (7): 993–1000.

CAPC. Center to Advance Palliative Care. 2011. *Public opinion research on palliative care: A report based on research by public opinion strategies*. http://www.capc.org/tools-for-palliative-care-programs/marketing/public-opinion-rese arch/2011-public-opinion-research-on-palliative-care.pdf. Accessed 9 Nov 2016.

Casarett, David. 2008. Do palliative care consultations improve patient outcomes? *Journal of the American Geriatrics Society* 56 (4): 593–599.

CDC. Centers for Disease Control. 2011. *National center for health statistics*. http://www.cdc.gov/nchs/data/hus/2011/022.pdf. Accessed 12 Sept 2016.

CMS. Centers for Medicare & Medicaid Services. 2011. *Medicare trustees report*. https://www.cms.gov/Research-Statistics-Data-and-Systems/Statistics-Trends-and-Reports/ReportsTrustFunds/downloads/tr2011.pdf. Accessed 9 Sept 2016.

———. 2016. *Medicare trustees report*. https://www.cms.gov/Research-Statistics-Data-and-Systems/Statistics-Trends-and-Reports/ReportsTrustFunds/downloads/tr2016.pdf. Accessed 9 Sept 2016.

Grunier, Andrea, et al. 2007. Where people die: A multilevel approach to understanding influences on site of death in America. *Medical Care Research and Review* 64 (4): 351–378.

Hebert, Liesi E., et al. 2013. Alzheimer disease in the United States (2010–2050) estimated using the 2010 census. *Neurology* 80 (19): 1778–1783.

Higginson, Irene, and Catherine Evans. 2010. What is the evidence that palliative care teams improve outcomes for cancer patients and their families? *Cancer Journal* 16 (5): 423–435.

Higginson, Irene, et al. 2003. Is there evidence that palliative care teams alter end-of-life experiences of patients and their caregivers? *Journal of Pain and Symptom Management* 25 (2): 150–168.

Hui, David, et al. 2010. Availability and integration of palliative care at US cancer centers. *Journal of the American Medical Association* 303 (11): 1054–1061.

Hurd, Michael D., et al. 2013. Monetary costs of dementia in the United States. *NEJM* 368 (14): 1326–1334.

Institute of Medicine (IOM). 1997. *Approaching death: Improving care at the end of life*. Washington, DC: National Academies Press.

———. 2015. *Dying in America: Improving quality and honoring individual preferences near the end of life*. Washington, DC: National Academies Press.

Johnson, Kimberly. 2013. Racial and ethnic disparities in palliative care. *Journal of Palliative Medicine* 16 (11): 1329–1334.

Kelley, Amy S., and R. Sean Morrison. 2015. Palliative care for the seriously ill. *NEJM* 373 (8): 747–755.

Lupu, Dale. 2010. Estimate of the current hospice and palliative medicine physician workforce shortage. *Journal of Pain and Symptom Management* 40 (6): 899–911.

McDonald, Julie, et al. 2016. Impact of early palliative care on caregivers with advanced cancer: Cluster randomized trial. *Annals of Oncology*. https://doi.org/10.1093/annonc/mdw438.

Meier, Diane. 2010. The development, status and future of palliative care. In *Palliative care: Transforming the care of serious illness*, ed. Diane Meier, 4–37. San Francisco: Jossey-Bass.
———. 2014. 'I don't want Jenny to think I'm abandoning her': Views on overtreatment. *Health Affairs* 33 (5): 895–898.
Meyers, Frederick J., and John Linder. 2003. Simultaneous care: Disease treatment and palliative care throughout illness. *Journal of Clinical Oncology* 12 (7): 1412–1415.
Mor, Vincent, et al. 2016. The rise of concurrent care for veterans with advanced cancer at the end of life. *Cancer* 122 (5): 782–790.
Morrison, R. Sean, Diane Meier, Tamara Dumanovsky, and Maggie Rogers. 2015. *America's care of serious illness*. https://reportcard.capc.org/wp-content/uploads/2015/08/CAPC-Report-Card-2015.pdf. Accessed 11 Oct 2016.
National Consensus Project for Quality Palliative Care. 2004. http://www.nationalconsensuspro-ject.org/DisplayPage.aspx?Title=What%20Is%20Palliative%20Care?. Accessed 14 Sept 2016.
NHPCO. National Hospice and Palliative Care Organization. 2012. *Pediatric palliative care*. http://www.nhpco.org/sites/default/files/public/ChiPPS/Continuum_Briefing.pdf. Accessed 8 Nov 2016.
———. 2015. *NHPCO's facts and figures hospice care in America 2015* Edition. http://www.nhpco.org/sites/default/files/public/Statistics_Research/2015_Facts_Figures.pdf. Accessed 26 Oct 2016.
———. 2016. *Hospice & palliative care*. http://www.nhpco.org/about/hospice-care. Accessed 9 Nov 2016.
NRSV. 2010. *The catholic prayer bible: Lectio Divina Edition*. Mahwah: Paulist Press.
NQF. National Quality Forum. 2006. *A national framework and preferred practices for palliative and hospice care quality: A consensus report*. Washington, DC: National Quality Forum.
Ortman, Jennifer M., Victoria A. Velkoff, and Howard Hogan. 2014. *An aging nation: The older population. U.S. department of commerce economics and statistics administration*. https://www.census.gov/prod/2014pubs/p25-1140.pdf. Accessed 12 Sept 2016.
Pope Benedict XVI. 2009. *Address of his holiness Pope Benedict XVI, visit to hospice sacro coure*. https://w2.vatican.va/content/benedict-xvi/en/speeches/2009/december/documents/hf_ben-xvi_spe_20091213_hospice.html. Accessed 10 Nov 2016.
Pope Francis. 2015. *Address of his holiness Pope Francis to participants in the plenary of the pontifical academy for life*. Vatican. https://w2.vatican.va/content/francesco/en/speeches/2015/march/documents/papa-francesco_20150305_pontificia-accademia-vita.html. Accessed 7 Sept 2016.
Saint John Paul II. 2004. *Address of Pope John Paul II to the participants in the 19th international conference of the pontifical council for health pastoral care*. http://www.fjp2.com/us/john-paul-ii/online-library/speeches/15921-to-the-participants-in-the-international-conference-sponsored-by-the-pontifical-council-for-health-pastoral-care-november-12-2004. Accessed 10 Nov 2016.
Quill, Timothy E., and Amy P. Abernethy. 2013. Generalist plus specialist palliative care – creating a more sustainable model. *NEJM* 368 (13): 1173–1174.
Rabow, Michael, et al. 2003. Patient perceptions of an outpatient palliative care intervention: "It had been on my mind before, but I did not know how to start talking about death...". *Journal of Pain and Symptom Management* 26 (5): 1010–1015.
———. 2004. The comprehensive care team: A controlled trial of outpatient palliative medicine consultation. *Archives of Internal Medicine* 164 (1): 83–91.
O'Mahony, Sean, et al. 2005. The benefits of a hospital-based inpatient palliative care consultation service: Preliminary outcome data. *Journal of Palliative Medicine* 8 (5): 1033–1039.
Smith, Thomas J., et al. 2012. American society of clinical oncology provisional clinical opinion: The integration of palliative care. Into standard oncology care. *Journal of Clinical Oncology* 30 (8): 880–887.
Temel, Jennifer, et al. 2010. Early palliative care for patients with metastatic non-small cell lung cancer. *NEJM* 363: 733–742.

Teno, Joan M., et al. 2004. Family perspectives on end-of-life care at the last place of care. *Journal of the American Medical Association* 291 (1): 88–93.

The Dartmouth Atlas. 2016. *End of life care*. http://www.dartmouthatlas.org/keyissues/issue.aspx?con=2944. Accessed 31 Oct 2016.

The SUPPORT Principal Investigators. 1995. A controlled trial to improve care for seriously ill hospitalized patients: The study to understand prognoses and preferences for outcomes and risks of treatments (SUPPORT). *JAMA* 274: 1591–1598.

Von Gunten, C., and B. Ferrell. 2014. Palliative care: A new direction for education and training. *Health affairs blog*. http://healthaffairs.org/blog/2014/05/28/palliative-care-a-new-direction-for-education-and-training/. Accessed 9 Nov 2016.

Xu, Jiaquan, et al. 2016. Deaths: Final data for 2013. *National Vital Statistics Reports* 64 (2): 1–14.

Yennurajalingam, Sriram, et al. 2011. Impact of a palliative care consultation team on cancer-related symptoms in advanced cancer patients referred to an outpatient supportive care clinic. *Journal of Pain and Symptom Management* 41 (1): 49–56.

# Part III
# Catholic Social Teaching and Institutional and Societal Issues Relating to Palliative Care

# Chapter 11
# Palliative Care and the Common Good

**James P. Bailey**

> The Church wishes to support the incurably and terminally ill by calling for just social policies which can help to eliminate the causes of many diseases and by urging improved care for the dying and those for whom no medical remedy is available. There is a need to promote policies which create conditions where human beings can bear even incurable illnesses and death in a dignified manner. Here it is necessary to stress once again the need for more palliative care centres which provide integral care, offering the sick the human assistance and spiritual accompaniment they need. This is a right belonging to every human being, one which we must all be committed to defend. Pope Benedict XVI (2006)

## 11.1 Introduction

This chapter aims to connect a relatively concrete and specific form of medical intervention—palliative care—to the relatively abstract and expansive concept of the common good. Each of these terms is connected to long histories and varied interpretations. So it may be helpful therefore to begin with some basic definitions, recognizing that doing so will inevitably oversimplify. Palliative care is understood here not as care "limited to end-of-life care, or care for those who are actively dying" but, rather, as "a form of care needed over a human life span, especially for persons living with serious chronic illnesses" (Farley 2011). The goal of palliative care will be understood in this chapter to be "the alleviation of suffering, the optimization of quality of life…and [in the case of terminal illness] the provision of comfort in death" (Foley 2002). Implicit in this understanding of palliative care is not only a deep and abiding concern for the well-being and dignity of persons but also for the sanctity of human life itself.

J. P. Bailey (✉)
Duquesne University, Pittsburgh, PA, USA
e-mail: baileyj@duq.edu



P. J. Cataldo, D. O'Brien (eds.), *Palliative Care and Catholic Health Care*,
Philosophy and Medicine 130, https://doi.org/10.1007/978-3-030-05005-4_11

Understood in this broad sense, palliative care is a form of care that can, arguably, be traced back to the beginnings of medicine.[1] Of course, it is true that, until relatively recently, medicine lacked powerful pharmaceuticals capable of attenuating or even suppressing agonizing chronic pain and the pain that often accompanies life-threatening illness. But palliative care has never been limited to the administration of medicines or other forms of bodily interventions, nor has human pain and suffering been limited to the body. Even when physicians and others who care for the sick have been impotent in the face of failing bodies, and lacking in effective pharmaceutical analgesics or other ways to reduce pain, they have nevertheless sought to comfort patients, if only by being present to them in their suffering. This latter kind of care was represented by Luke Fildes' painting titled simply "The Doctor," a painting said to be inspired by the devotion given by the doctor to Fildes' own son as he lay dying. Commenting on the painting, Frances Wells captures why it is that many find the painting so captivating: "The study of emotion and caring expressed in the pose and the face of the doctor encapsulates all that is good in medicine. The face exposes a deeply thoughtful and analytical mind. The body language, with the head projected forwards and resting gently on the left hand, reveals one human being caring deeply for another: emotion and intellect entwined for the betterment of the child." (Wells 2006). Properly understood, palliative care is consistent with the spirit of care — one human being caring deeply for another — portrayed in this dramatic painting.

[1] Randall and Downie (2006) suggest the administration of palliative care can be traced back to Greek practices that pre-date Hippocrates.

Like palliative care, the principle of the common good has a long history. The principle of the common good was central to Greek and Roman philosophy, elaborated upon by influential theologians in the tradition of Christianity, including Augustine and Aquinas, and features prominently in modern Catholic social teaching (and other religious traditions) and in a variety of non-religious political philosophies.[2] The term is sometimes used to describe the shared good of a relatively circumscribed group of persons—say, the common good of one's family or of a local community—or, at the other extreme, it can be used in a very expansive and inclusive sense—say, the global common good. For much of its history, at least with respect to created goods,[3] the term almost always referred to human communities of one sort or another. But, in recent times, conceptions of the common good have sought to include non-human inhabitants of earth and some have expanded the concept even further so that it includes the entire cosmos. Thus, Dan Scheid, for example, speaks of the "cosmic common good" (2016).

In this chapter, I will adopt a definition of the common good that has been relatively influential among contemporary moral theologians, the definition offered by the Second Vatican Council in *Gaudium et Spes* (1965). This text defined the common good as "the sum of those conditions of social life which allow social groups and their individual members relatively thorough and ready access to their own fulfillment" (n. 26). It will, of course, be necessary to elaborate upon this definition as we proceed, especially because the expansive scope of even this relatively circumscribed definition of the phrase implicates a range of other principles, values and commitments. These include a conception of the person (an anthropology) as a social and political being, a notion of justice related to this anthropology, a commitment to the dignity of *all* persons (with particular attention and care given to the poor and the marginalized), and the virtue of solidarity which helps to cultivate practices and critical modes of thinking that promote the common good. As we will see, these principles, values and commitments not only help to flesh out the meaning of the common good, but they will also help us to understand why the provision of health care generally, and of palliative care in particular, is (or should be) an integral part of the common good in contemporary societies.

[2]While it is true that "modern" political philosophies (those associated with the Enlightenment period forward) have tended to stress the rights of the individual over and against the demands of the common good, it is also true that some more recent political philosophers (e.g., the work of Michael Sandel, Martha Nussbaum, and Alasdair McIntyre) have begun to retrieve conceptions of the common good, conceptions that often stress that the good of individuals (and the protections of their rights) are not incompatible with a robust defense of the common good.

[3]Aquinas (1945) argued that the good to be sought by all is the uncreated Good (God): "The supreme good, God, is the common good, since the good of all things depends on God."

## 11.2 The Common Good

Whenever I speak about the common good, particularly in a US context, I can sense unease among many in the audience even, and sometimes especially, among Catholics. The phrase "common good" triggers, almost reflexively, associations to communism, socialism, or authoritarianism and however imprecisely those terms are deployed in these discussions their meaning is clear enough: the common good, whatever else it means, implies a diminishment of individual human freedom. Here, freedom means something like the freedom to do as one chooses without undue outside influence, especially from the government. As I write this, there are vigorous political discussions centered on the dismantling of large social initiatives undertaken by the US federal government, initiatives that include the Affordable Care Act (ACA), sometimes referred to as "Obamacare." For those who advocate the dismantling of the ACA, a potent justification for doing so is the claim that the ACA undermines individual freedom. Specifically, the freedom to choose what form of health care (if any) best suits individual or family needs. Consider, as one example, this statement in support of the repeal of the ACA by Paul Ryan, the Speaker of the US House of Representatives and an avowed Roman Catholic: "Freedom is the ability to buy what you want to fit what you need." For Ryan, our health care system works best when each of us, exercising our personal liberty, choose our own health care plan instead of, as he puts it, "Washington telling you what to buy regardless of your needs" (Covert 2017).

This pinched conception of the meaning of freedom which, in Ryan's formulation, sounds very much like consumer choice, implies that freedom is best realized, and justice is most effectively done, when people are left alone to freely exercise their personal choices. Here, the *common* good gives way almost entirely to the good of "free" *individuals*, each making discrete choices for their own *personal* good. Ryan's conception of freedom and implied notion of justice fits much more easily with the libertarianism of Robert Nozick (2013) than it does with Catholic social teaching and the common good tradition to which it is heir.

If we recall again the definition of the common good offered above—"the sum of those conditions of social life which allow social groups and their individual members relatively thorough and ready access to their own fulfillment"—we can begin to see why, in the Catholic tradition at least, Ryan's reduction of freedom to "the ability to buy what you want to fit what you need" is inadequate, not only with respect to Catholic understandings of the meaning of human freedom but also with respect to the just distribution of health care, something the Catholic tradition believes to be a human right. As Pope Francis puts it: "Health is not a consumer good but a universal right, so access to health services cannot be a privilege" (Wooden 2016). The definition of the common good offered by Vatican II makes it clear that *individual* well-being is connected to, and dependent upon, the establishment of certain *social* conditions. These conditions include social and political institutions, juridical frameworks, and social services necessary for human flourishing so as to provide, among other things, "the means necessary for the proper development of life, particularly food, clothing, shelter, [and] medical care…." (John XXIII 1963, n. 11). Because these goods are integral to human flourishing, modern Catholic social teaching asserts that persons have a *right* to them. For this reason, the Catholic tradition argues, "respect for freedom…involves more than not interfering with the activity of persons" (Hollenbach 1990). Rather, it involves affirmative steps be taken to insure the flourishing of all human beings. A person with inadequate nutrition or untreated sickness cannot be said to be truly free and it cannot be just to deny people access to the basic goods necessary for human flourishing. Accordingly, "obligations of justice include positive duties to aid persons in need, to participate in the maintaining of the public good, and to share in efforts to create the kinds of institutions that promote genuine mutuality and reciprocal respect" (Hollenbach 1990). Thus, for the Catholic tradition, the promotion of the common good, the establishment of justice, and the realization of human freedom, go hand in hand.

## 11.3 Persons as Social Beings

The idea that social and political institutions are necessary for human flourishing has long been a part of the Christian tradition and this belief is intimately linked to how the Church understands the human person, an understanding influenced by distinct but complementary theological and philosophical traditions. From a theological perspective, human beings are understood to be created in the image and likeness of God and thus possess a dignity that calls for reverence. This image of God in which persons are created is made up of three persons (the Trinity), and is, thus, *intrinsically* relational. Imaging a relational God, human beings are themselves meant to be in relationship with others: "It is not good that man should be alone," says the Book of Genesis (2:18).

The biblical vision of relationality is about more than individual relations, however. The biblical narrative makes clear that "human life is fulfilled in the knowledge and love of the living God in communion with others" (NCCB 1986, n. 30). Thus, God calls God's people out of Egypt and establishes a covenant with them, a covenant that specifies obligations expected of those whom God has rescued from political, economic, and social persecution. As the US Bishops make clear, "Far from being an arbitrary restriction on the life of the people" the demands of the covenant that God makes with God's people "made life in community possible. The specific laws of the covenant protect human life and property, demand respect for parents and the spouses and children of one's neighbor, and manifest a special concern for the vulnerable members of the community: widows, orphans, the poor, and strangers in the land" (NCCB, n. 36).

In Catholic social teaching, this theological understanding of the person as relational is reinforced and developed in conversation with ancient philosophical traditions. These latter traditions "spoke of the human person as a 'social animal' made for friendship, community, and public life" (NCCB 1986, n. 65). Both Aristotle and

Aquinas argued that human beings are essentially social and political. As Aristotle put it, "He who is unable to live in society, or who has no need because he is sufficient for himself, must be either a beast or a god" not a human being (1973). In his commentary on Aristotle's *Politics,* Thomas Aquinas endorses Aristotle's claim that human beings are social by nature and therefore they are dependent upon the human community, including the broader social and political community, for their flourishing:

> The fact that man is by nature a social [and political] animal—being compelled to live in society because of the many needs he cannot satisfy out of his own resources—has as a consequence the fact that man is destined by nature to form part of a community which makes a full and complete life possible for him. The help of such a communal life is necessary to him for two reasons. In the first place it is necessary to provide him with those things without which life itself would be impossible. But life in a community further enables man to achieve a plenitude of life; not merely to exist, but to live fully, with all that is necessary to well-being. (Aquinas 1948)

Thomas asserts that human beings need a life in community *both* to provide the things that we cannot provide by ourselves (e.g., among other things, health care) and because living in community allows them "to live fully," by which Aquinas has in mind, among other things, the experience of love and companionship, personal friendship, and the "civic friendship" that arises in the well-functioning political community. The cultivation and experience of these things by all persons in society is central to the realization of the common good.

## 11.4 Justice

The stress on the social nature of the human person also helps explain why recent Catholic social teaching often locates discussions of justice and injustice in relation to the importance of persons' *inclusion and participation* in society. Because human flourishing is secured in and through human society, it is crucial that all members of society be permitted and enabled to participate in it. Social *inclusion* can take many forms. It includes, for example, political participation, employment (and the right to participate in activities related to employment, e.g. joining unions), civic involvement, educational opportunities, participation in voluntary organizations, worship, and so on. Indeed a robust civil society, and the ability to fully participate in associations and groups within civil society, is an integral part of the common good and a crucial aspect of full human flourishing. This is why, in the definition of the common good offered in *Gaudium et Spes*, above, the Church speaks of the importance of creating the conditions necessary to give not just individuals "relatively thorough and ready access to their own fulfillment" but social groups as well.

*Exclusion* from participation in these many dimensions of social life—whether it be through active discrimination or a failure to provide the necessary institutions that enable persons to fully participate—is an injustice precisely because it impedes the flourishing of persons and social groups and in this way harms the common good. For this reason, the NCCB writes: "Basic justice demands the establishment of minimum levels of participation in the life of the human community for all persons. The ultimate injustice is for a person or group to be treated actively or abandoned passively as if they were nonmembers of the human race." (NCCB 1986, n. 77). It is this understanding of justice in terms of societal inclusion and injustice as social marginalization that leads the Bishops to define human rights as the "minimum conditions for life in community" (NCCB 1986, n. 77).

Working to insure that society is *structured* in a way to facilitate participation and access to those goods that contribute to human flourishing falls into the domain of two distinct aspects of justice—distributive justice and social justice—that concern the rights and duties of persons in our public life together, that is, in the state and civil society. *Distributive justice* "is the norm that states the obligation of society and the state to guarantee…participation by all in the common good" (Hollenbach 1990). In other words, distributive justice "establishes the equal right of all to share in all those goods and opportunities that are necessary for genuine participation in the human community" (Hollenbach 1990). Among the goods necessary for genuine participation in the human community is the good of health care.[4]

At the same time, modern health care is a public good, made possible by society itself. That is, society's ability to distribute the good of modern health care depends upon the prior commitment of society—both individuals within society and the range of social and political institutions that make up society—to developing the modern health care system. For this reason, public goods (like health care) should not be understood as

> the property of any individual or class in an exclusive sense, for all members of society are at least indirectly involved in their production through membership in public society. Even though participation in the creation of these public goods may be minimal or, in the case of children, infirm or aged persons, presently nonexistent, the tradition claims that membership in the human community creates a bond between persons sufficient to ground a right for all to share in the public good to the minimum degree compatible with human dignity. (Hollenbach 1990)

While distributive justice establishes the right of all persons to the goods and opportunities that allow for participation in society, *social justice* addresses itself to the "institutionalized patterns of mutual interaction and interdependence that are necessary to bring about the realization of distributive justice" (Hollenbach 1990). The goal of social justice is to protect human rights and to insure "mutuality and participation by all in social life" (Hollenbach 1990). When applied to the public good of health care, the task of social justice is to create institutions and processes that make it possible for all persons to receive adequate levels of health care and, in so doing help individual persons and the community to flourish. Thus Clark Cochran argues, "heal thcare is part of the proper functioning of a community life. People who are free from disease and whose injuries are repaired are better able to pursue their own goods, to contribute to the common good, and to share communal interactions with fellow citizens" (1999).

---

[4] Other goods understood to be social or public goods include "the fertility of hearth, the productivity of an industrialized economy," as well as social insurance programs (Hollenbach 1990).

## 11.5 Preferential Option for the Poor

The concern for justice—distributive, social, and more generally as enabling participation of all persons and groups in society—gives rise to another principle of Catholic social thought related to the common good. That principle is "preferential option for the poor," a principle that helps deepen our understanding of what a commitment to the common good requires. While the phrase "preferential option for the poor" was not used until relatively recently in the Church's social teaching,[5] the basic commitment expressed in the preferential option can be traced back to the Bible. In both the Hebrew and Christian scriptures we are told again and again that God has a special concern for the poor, downtrodden, marginalized, and oppressed. The defining narrative of the Jewish faith—the Exodus—depict a God that is moved by the plight of the Israelites who are portrayed as enslaved, impoverished, and socially and politically disempowered. Concern for the poor is also central to the Christian scriptures. In the Gospels, for example, Jesus is portrayed repeatedly as one who has a special concern for the poor and the marginalized of his society. His identification with the poor is perhaps best exemplified in the well-known biblical text "whatever you do to the least of these you do to me" (Matthew Ch. 25: 40–41) and the gospels make clear that the "least of these" includes those who are sick.

While there is within Christianity a long tradition of assisting the poor and the marginalized, the theological reflection that gave rise to the demand to make a "preferential option" for the poor has contributed to a greater awareness of the fact that those living in poverty, economic or otherwise, often do so because of unjust social structures. This consciousness of structural injustices marks an important change from the past wherein it was thought that "some people are born poor, other people are born rich, and both must accept this fate" (Gutierrez 2013). No longer is it enough to address the needs of those living in poverty (although this surely needs to be done); we must work to alleviate the social and political structures that help to make people poor and keep them poor. One sees this, for example, in *Economic Justice for All,* where the US Bishops insist that the decisions of the body politic "must be judged in light of what they do *for* the poor, what they do *to* the poor, and what they enable the poor to do *for themselves*. The fundamental moral criterion for all economic decisions, policies, and institutions is this: They must be at the service of *all people, especially the poor*" (NCCB 1966, emphasis in original). Similarly, no longer is it enough to attend to those who are sick, we must work to address the systemic injustices that deny people access to adequate health care, including palliative care.

---

[5]Within the Catholic tradition the first explicit development of this theme occurred in the Final Document of the Third General Conference of Latin American Bishops, which included a chapter titled "The Preferential Option for the Poor." There, the Latin American Bishops call attention to the fact that "the vast majority of our people lack the most elementary material goods. This is in contrast to the accumulation of wealth in the hands of a small minority, frequently the price being poverty for the majority. The poor do not lack simply material goods. They also miss, on the level of human dignity, full participation in sociopolitical life" (Consejo Episcopal Latinoamericano 1979).

## 11.6 Solidarity

Justice in its various forms, as well as the motivation and desire to make a preferential option for the poor are aided, and more likely to be realized, if persons and social groups manifest the virtue of *solidarity,* another aspect of Catholic social teaching that is integrally related to the common good. Thus Saint John Paul II (1987) described solidarity as "a firm and persevering determination…to commit oneself to the good of all and of each individual," that is, the common good, "because we are all really responsible for all" (n. 38). John Boswell argues that solidarity includes and encourages "relationships of community, sociability, conviviality, civility, fraternité, civic friendship, social consciousness, [and] public spirit" (2000). While justice "becomes more urgent in light of solidarity," so too does it move justice "beyond mere resource distribution" to include a concern to encourage "the fuller participation of those who have been marginalized" (Cahill 2004). Solidarity encourages us, as well, to recognize that "we can know a good in common that we cannot know alone" (Sandel 1982) and thus deepen our commitment to work toward greater inclusion of all in our social and political life together. In short, as David Hollenbach says, solidarity is "a commitment to being a community" (Hume 1999).

As this discussion makes clear, the common good is not a freestanding principle but is integrally related to other core commitments of the Catholic social tradition. Its full realization is connected to the promotion of human dignity, a recognition of the social nature of the person, a strong commitment to justice in its various forms, and a commitment that includes efforts to ensure the participation of all persons in society, especially those in poverty or marginalized in other ways. These commitments also depend upon the cultivation of the virtue of solidarity, a virtue that manifests in a deep commitment to seek the good of all persons. Because the fullest realization of our being comes through receiving from and giving to others, a commitment to the common good should not be seen as a threat to human freedom. To the contrary, authentic human freedom is more secure within societal structures and social interactions that support and enhance individual and communal life. The full realization of the common good necessitates the ability and willingness of citizens to see each other as friends and allies—not as strangers and aliens—working together to achieve a better world not just for ourselves, but for all.

## 11.7 The Common Good and Health Care

While we have already briefly noted connections between health care and the common good, in what follows I will explore in more depth both the need for, and impact of, integrating the principle of the common good into our discussions about health care. One way to begin to see the connection between the common good and health care is to start with a description of some of the challenges faced by the US health care system in recent years. When President Obama came into office US health care costs had been rapidly rising for some time and doing so at a rate that many analysts believed to be unsustainable.[6] Moreover, despite US health expenditures far exceeding most other OECD countries (Kaiser 2012b),[7] tens of millions of Americans had no health insurance coverage at all (DeNavas-Walt et al. 2010).[8] Given that so many lacked health care insurance it is unsurprising that the US lagged behind other countries in many health care outcomes, including rates of infant mortality and life expectancy where the US ranked 42nd (CIA 2016b) and 56th (CIA 2016a), respectively. Even these indicators, however, do not capture the degree to which the strengths and weaknesses of the current US health care system are unevenly distributed. For example, there are significant racial disparities in the distribution of health insurance, where "African Americans…[are] twice as likely, and Hispanics three times as likely, as [non-Hispanic] whites to be uninsured.… Almost one-third of all American Indians and Alaska Natives are uninsured, a rate almost as high as that for Hispanics" (Advisory Committee 2002). In addition, there are significant disparities in infant mortality and rates of life expectancy between African Americans and white Americans (Williams 2004). Similar inequalities are noted between rich and poor and between those living in the inner cities and certain rural areas of the United States. These inequalities in our health care system have been noticed for quite some time. For example, in 1993 the US Conference of Catholic Bishops described the situation of health care in the US in this way:

> The current health care system is so inequitable and the disparities between rich and poor and those with access and those without are so great that it is clearly unjust. The burdens of this system are not shared equally. One out of three Hispanics and one of five African Americans are uninsured. The health care in our inner cities and some rural communities leads to Third World rates of infant mortality. (USCCB 1993)

Just as troubling, these inequalities in health care distribution and concomitant inequalities in health outcomes exist within a health care system that is sustained and funded through vast public expenditures that make possible the very existence of the modern health care system in the United States. Indeed, the magnitude of public expenditures in the US health care system would likely stun many Americans. For example, in 2010, "Medicare and Medicaid accounted for 47% of total hospital spending, without which hospitals would be unable to pay their fixed costs" (Craig 2014). Likewise, in 2009, "public funding of medical school education outpaced by six to one tuition and fees that medical students paid for their education that year" (Craig 2014). Public funding of medical research is also substantial. In 2007,

[6]Between 2002 and 2012, the average cost for employer-sponsored family health coverage rose 97% to $15,700 per year, a rate of increase equal to three times the rate of inflation (The Henry J. Kaiser Family Foundation, "Family Health Premiums Rise," 2012a). Observing this rapid increase in the cost of health care insurance, the New America Foundation suggested (in 2008) that the average cost of an employee-sponsored health insurance plan would reach $24,000 in 2016, an amount they noted would mean that "at least half of American households will need to spend more than 45% of their income to buy health insurance" in that year (Axeen and Carpenter 2008).

[7]The US spent 48% more per capita than the next highest spending country, Switzerland.

[8]As President Bush left office, there were an estimated 46.3 million Americans lacking health insurance (DeNavas-Walt et al. 2009). And, as the Great Recession deepened, that number climbed to over 50 million.

federal, state, and local governments contributed $38 billion dollars to this effort, as compared to approximately $58 billion by private industry. But, as David Craig points out, "government is the principal funder of basic research into the causes of disease," something from which private industry derives huge benefits but has little interest in doing on their own since there is little direct profit to be made from such research (2014). Thus, it is clear that the modern health care system is the product of vast public expenditures and efforts without which the practice of modern health care would be impossible. Today's physicians are the product of medical schools that receive significant public support to facilitate the training of their students. They make use of discoveries of pharmaceuticals, techniques, and research that are the result of the efforts of many who have come before them, most of which were supported by publicly funded research. They depend upon a vast and intricate network of supporting industries to practice their art.

Reflecting on this situation, Michael Walzer (1983) points out that the increasing political unrest that accompanies the provision of health care is linked to the moral illegitimacy of a publicly funded health care system that restricts significant numbers of the public from access to the benefits of these expenditures. Walzer argues that as "long as communal funds are spent, as they currently are, to finance research, build hospitals, and pay the fees of doctors in private practice, the services that these expenditures underwrite must be equally available to all citizens."

If we look at this situation from the perspective of Catholic social teaching and the common good, several things deserve comment. The US health care system is far from an inclusive system. There are significant disparities in access to the system that are tied to race, class, and geography. This failure to include all persons in the provision of adequate health care effectively creates a "medical caste system," undermining not just the health and well-being of persons, but excluding persons from a life of basic dignity grounded in full participation in the human community (Cochran and Cochran 2003). The dignity of persons is protected and promoted when they are fully included in the life of society and able to participate in it. Those denied health care are not only prevented from accessing what they need to address the suffering associated with illness, they are also given the message that they are less valued than their fellow citizens. As Susanne DeCrane (2004) writes, adequate access to health care "strikes at the heart of being human—being alive, being healthy enough to participate in and enjoy life, being treated with dignity when one is vulnerable because of illness, being able to have the health of one's children adequately protected." The marginalization that comes with limited access to health care harms the most vulnerable but it also harms the common good. Society is diminished in a myriad of ways when all members cannot fully participate in it. Moreover, each of us is diminished. Part of being human—arguably the most important part—is to show care and concern for one's fellow human beings. Craig shows how Martin Luther King's vision of the Beloved Community conceptualized "all Americans as being on a moral journey together" (Craig 2014). This included attending to the health and well-being of one's fellow Americans. For King, there was no greater affront to equality than the denial of health care: "Of all the forms of inequality, injustice in health care is the most shocking and inhumane" (King 1966).

## 11.8 Palliative Care and the Common Good

Palliative care, embedded as it is in the general provision of health care, is not immune to the kinds of inequities and barriers that occur in our health care system overall. The pain and vulnerability that an individual requiring palliative care is likely to experience is an especially poignant instance of the need to remedy access to, and inequalities in, the provision of health care. Given that "poverty, among other social inequalities, is a major risk factor for suffering and premature death" (Krakauer 2008) the poor and socially marginalized suffer a double injustice. Burdened by the stress of poverty, they are more likely to fall ill and experience the suffering that accompanies sickness. And yet, they are the most likely to be excluded from all forms of health care, including palliative care.

As an awareness of both the need for, and the limited access to, palliative care has grown, access to palliative care has increasingly been characterized as integral to the practice of health care and, as such, it is due to persons as a basic right. As noted in the epigraph at the beginning of the chapter, Pope Benedict XVI speaks of palliative care as "a right belonging to every human being." So, too, do a number of scholars and research scientists. For example, Felicia Knaul (2015) and her co-authors assert that "Access to palliative care—the prevention and relief of physical, emotional, social, or spiritual suffering associated with any chronic or life-threatening illness, beginning from the time of diagnosis—is at the core of the right to health and is fundamental to health care." These individual assertions of a right to palliative care have been joined by a growing chorus of organizational declarations of this right. Among others, these include: The Barcelona Declaration on Palliative Care (1995), The Cape Town Palliative Care Declaration (Sebuyira et al. 2003), The Korean Declaration on Hospice and Palliative Care (2005), The Montréal Statement on the Human Right to Essential Medicines (2005), The International Children's Palliative Care Network Charter of Rights for Life Limited and Life Threatened Children (2008), Human Rights Watch (2009), and the European Association for Palliative Care (2013).

Although there has, in general, been an increase in the availability to palliative care, access to it is at best uneven, and non-existent at worst. While 63% of all US hospitals and over 85% of large hospitals (those with 300 beds or more) have palliative care teams, only "half of public safety net hospitals have a palliative care team" (Smith and Brawley 2014). Moreover, there is a growing body of literature that documents lower rates of access to pain treatment among those with a lower socio-economic status as well as lower rates of access by racial and ethnic minorities, regardless of socioeconomic status. This is attributed to a range of factors: limited availability of hospice services in low-income urban areas (O'Mahony et al. 2008); racial and ethnic minorities are less likely to have health insurance and therefore less likely to seek out palliative treatment (O'Mahony et al. 2008), a problem that also impacts low-income non-Hispanic whites (Hughes 2005); pharmacies located in a majority of non-white neighborhoods are less likely to stock opioids than pharmacies in predominately white neighborhoods (Morrison et al. 2000); limited access to transportation constitutes a practical barrier to palliative care services for low-income persons (Hughes 2005); and the fact that physicians "appear to deliver less information and communicate less support to African-American and Hispanic patients compared to white patients, even in the same care settings" (Smith and Brawley 2014). In addition, the obstacles that African-American and Hispanic patients and poorer Americans encounter when trying to obtain high quality health care in general appears to feed a suspicion that racial discrimination is at the heart of recommendations that their loved ones are being offered the "low-tech" option of palliative care rather than high-cost advanced technological interventions. This can contribute to a greater likelihood that African-American and Hispanic patients (and their families) will insist that aggressive medical interventions be undertaken when palliative treatment is the wisest choice (Gibson 2001).[9]

While much of this chapter has focused on the situation of health care and palliative care within the United States, it is important to acknowledge that a commitment to the common good extends to the well-being of persons outside our borders. And, with respect to palliative care, the situation outside the United States, particularly in developing countries, is considerably more dire. As Knaul and her colleagues put it, the "enormous global divide in access to pain control and palliative care constitutes an ongoing crisis that plays out almost entirely to the detriment of poor people" (2015). According to the World Health Organization (WHO), a staggering 80% of the world's population has either no or insufficient access to pain treatment. This includes 4 million cancer patients and 800,000 HIV/AIDs patients (World Health Organization 2014). As with so many other forms of consumption, North America and Europe lead the world in opioid use, accounting for some 89% of the world's morphine consumption. By contrast, low and middle-income countries contribute to just 6% of the worlds' morphine consumption, despite the fact that about half of all cancer patients and 95% of new HIV infections reside in these countries (Knaul et al. 2015). This underscores again the reality that the poor and marginalized lack access to adequate health care treatment, including access to palliative care. More than that, it illuminates for us the scale of immense human suffering that arises when society does not attend to the health care needs of *all* persons.

[9] Of course, these African-American and Hispanic patients are not alone in shunning palliative care even when it is the most appropriate intervention. As Ron Hamel has pointed out, if palliative treatments, when called for, are to be adopted with greater frequency, they need to be supported by a cultural shift in how Americans (and others) approach death. This will be a challenge since as "a society, we do not do well dealing with our mortality even in the abstract, let alone when confronted with a relatively imminent threat of our demise. We deny and fight against death with every fiber of our being and enlist medicine in the effort" (2011).

## 11.9 Conclusion

The principle of the common good does not lead *directly* to prescriptive remedies with respect to inadequacies in the provision of health care generally and to palliative care in particular. Nevertheless, it does real work on this and many other aspects of our social life together. In particular, it acts as an orienting principle, a measuring stick, and an aspirational goal, directing our attention to those who are excluded from full participation in human society, helping us to see how far we fall short in this respect, and inspiring us to do better. In this way, as Ron Hamel points out, the common good is both "formative and transformative." It "keeps before our eyes essential dimensions of who we are as individuals and communities, dimensions that are easily neglected in a society that prizes the individual. It also beckons us to be ever engaged in creating environments in which human beings can flourish" (Hamel 1999).

Similarly, Susanne DeCrane suggests that the principle of the common good "orients the community to a commitment to ongoing social-moral discernment, praxis, and evaluation, using the experiences of those who are most excluded from the benefits of the community as the evaluative measure of the degree to which the common good is authentically being sought" (DeCrane 2004). An ethic that focuses too narrowly on individual actions and freedom will likely cause us to overlook the ways in which some people are marginalized by current practices and it will lead us to overlook the degree to which the human community is diminished by such practices. Put more positively, the principle of the common good also helps to form and transform society's commitments towards a greater inclusion of all by reminding us of our dependence and interdependence on others and consequently of the responsibility we have for one another.

A renewed politics that incorporates the common good as its primary organizational principle is an important step to addressing the very serious and seemingly intractable social problems that we face today, including inequities in health care. This is so both because the principle of the common good is reflective of the reality of who persons are—social, interdependent beings, dependent upon social and political structures that support our flourishing—and because the common good tradition is capable of addressing the current circumstances of the world in which we live today, a world in which our "de facto technological, political, and economic interdependence grows stronger each year, indeed each day" (Hollenbach 1998). This is a world that will not best be navigated by political philosophies, such as libertarianism and other forms of liberalism, that tend to play the flourishing of individuals off against the community. We need, instead, a conception of social and political life that underscores our interdependence, enables and celebrates the full participation of persons in community, and acknowledges that freedom is most fully realized when society is characterized in this way.

## References

Advisory Committee on Social Witness Policy (Advisory Committee). 2002. *Resolution on advocacy on behalf of the uninsured.* Presbyterian Church USA. www.pcusa.org/resource/resolution-advocacy-behalf-uninsured. Accessed 1 Dec 2016.

Aquinas, Thomas. 1945. Summa contra gentiles, III.17. In *Basic writings of Saint Thomas Aquinas*, ed. Anton G. Pegis, vol. 2, 27–29. New York: Random House.

———. 1948. Book one: Commentary on the nicomachean ethics. In *Thomas Aquinas: Selected political writings*, ed. A.P. D'Entreves, 189–193. Oxford: Basil Blackwell.

Aristotle. 1973. Politics. In *Introduction to Aristotle*, ed. Richard McKeon, 584–659. Chicago: University of Chicago Press.

Axeen, Sarah, and Elizabeth Carpenter. 2008. *The cost of doing nothing: Why the cost of failing to fix our health system is greater than the cost of reform.* Washington, DC: New America Foundation. 2008.

Benedict XVI. 2006. *Message of his holiness benedict xvi for the fifteenth world day of the sick.* http://w2.vatican.va/content/benedict-xvi/en/messages/sick/documents/hf_ben-xvi_mes_20061208_world-day-of-the-sick-2007.html. Accessed 12 Dec 2016.

Boswell, John. 2000. Solidarity, justice and power sharing: Patterns and policies. In *Catholic social thought: Twilight or renaissance?* ed. J.S. Boswell and F.P. McHugh, 93–114. Leuven: Leuven University Press.

Cahill, Lisa Sowle. 2004. *Bioethics and the common good.* Milwaukee: Marquette University Press.

Central Intelligence Agency (CIA). 2016a. *World fact book: Infant mortality.* https://www.cia.gov/library/publications/the-world-factbook/rankorder/2091rank.html. Accessed 30 Aug 2017.

———. 2016b. *World fact book: Life expectancy at birth.* https://www.cia.gov/library/publications/the-world-factbook/rankorder/2102rank.html. Accessed 30 Aug 2017.

Cochran, Clarke E. 1999. The common good and heal thcare policy. *Health Progress* 80 (3): 41–44. 47.

Cochran Clark, E., and David Carroll Cochran. 2003. *Catholics, politics, and public policy: Beyond left and right.* Maryknoll: Orbis Books.

Consejo Episcopal Latinoamericano [CELAM]. 1979. Final document of the third general conference of latin american bishops. In *Puebla and beyond: Documentation and commentary*, ed. John Eagleson and Philip Scharper, 264–267. Maryknoll: Orbis Books.

Covert, Bryce. 2017. "Paul Ryan is wrong on freedom." *New York Times*, March 3, 2017, Page A27.

Craig, David. 2014. *Health care as a social good: Religious values and American democracy.* Washington, DC: Georgetown University Press.

DeCrane, Susanne. 2004. *Aquinas, feminism, and the common good.* Washington, DC: Georgetown University Press.

DeNavas-Walt, Carmen, Bernadette D. Proctor, and Jessica C. Smith. 2009. *U.S. census bureau, current population reports, P60–236, income, poverty, and health insurance coverage in the United States: 2008.* Washington, DC: U.S. Government Printing Office.

———. 2010. *U.S. census bureau, current population reports, P60–238, income, poverty, and health insurance coverage in the United States: 2009.* Washington, DC: U.S. Government Printing Office.

European Association for Palliative Care. 2013. *Prague charter palliative care: A human right.* http://www.eapcnet.eu/Themes/Policy/ PragueCharter.aspx. Accessed 30 Aug 2017.

Farley, Margaret A. 2011. Aging and dying, a time of grace. *Health Progress* 92 (1): 19–25.

Foley, Kathleen. 2002. Compassionate care, not assisted suicide. In *The case against assisted suicide*, ed. Kathleen Foley and Herbert Hendin, 293–309. Baltimore: Johns Hopkins University Press.

Gibson, Rosemary. 2001. Palliative care for the poor and disenfranchised: A view from the Robert Wood Johnson foundation. *Journal of the Royal Society of Medicine* 94: 486–489.
Gutierrez, Gustavo. 2013. Saying and showing to the poor: "God loves you.". In *In the company of the poor: Conversations with Dr. Paul Farmer and Fr. Gustavo Gutierrez*, ed. Michael Griffin and Jennie Weiss Block, 27–34. Maryknoll: Orbis Books.
Hamel, Ron. 1999. Of what good is the 'common good'? *Health Progress* 80 (3): 46–47.
———. 2011. Palliative care needs a culture that sustains it. *Health Progress* 92 (1): 70–72.
Hollenbach, David S.J. 1990. Modern catholic teachings concerning justice. In *Justice, peace, and human rights: American catholic social ethics in a pluralistic world*, 16–33. New York: Crossroad.
———. 1998. The catholic university and the common good. *Conversations on Jesuit Higher Education* 13: 5–15.
Hughes, Anne. 2005. Poverty and palliative care in the US: Issues facing the urban poor. *International Journal of Palliative Nursing* 11: 6–13.
Human Rights Watch. 2009. "Please, do not make us suffer any more..." *Access to pain treatment as a human right*. https://www.hrw.org/report/2009/03/03/please-do-not-make-us-suffer-any-more/access-pain-treatment-human-right. Accessed 30 Aug 2017.
Hume, Susan. 1999. Catholic theology informs thinking on health care reform (interview of David Hollenbach, SJ). *Health Progress* 80 (3): 43.
John Paul II, St. 1987. *Sollicitudo rei socialis*. http://w2.vatican.va/content/john-paul-ii/en/encyclicals/documents/hf_jp-ii_enc_30121987_sollicitudo-rei-socialis.html. Accessed 15 Nov 2016.
John XXIII, Pope. 1963. *Pacem in terris*. http://w2.vatican.va/content/john-xxiii/en/encyclicals/documents/hf_j-xxiii_enc_11041963_pacem.html. Accessed 30 Aug 2017.
King, Martin Luther Jr. 1966. *Presentation at the second national convention of the medical committee for human rights*. Chicago, March 25. http://www.pnhp.org/news/2014/october/dr-martin-luther-king-on-health-care-injustice. Accessed 28 Nov 2016.
Knaul, Felicia M., Paul E. Farmer, Afsan Bhadelia, Philippa Berman, and Richard Horton. 2015. Closing the divide: The Harvard global equity initiative—lancet commission on global access to pain control and palliative care. *The Lancet* 386: 722–724.
Korea Declaration on Hospice and Palliative Care. 2005. http://www.coe.int/t/dg3/health/Source/KoreaDeclaration2005_en.pdf. Accessed 30 Aug 2017.
Krakauer, Eric L. 2008. Just palliative care: Responding to the suffering poor. *Journal of Pain and Symptom Management* 36 (5): 505–512.
Montreal statement on the human right to essential medicines (2005). 2006. In *Health and human rights: Basic international documents*, ed. Stephen P. Marks. Cambridge: Harvard University Press.
Morrison, R. Sean, Sylvan Wallenstein, Dana K. Natale, Richard S. Senzel, and Lo-Li Huang. 2000. 'We don't carry that'—failure of pharmacies in predominantly nonwhite neighborhoods to stock opioid analgesics. *New England Journal of Medicine* 342: 1023–1026.
National Conference of Catholic Bishops (NCCB). 1986. *Economic justice for all: Pastoral letter on catholic social teaching and the U.S. economy*. Washington, DC: United States Catholic Conference.
Nozick, Robert. 2013. *Anarchy, state, and utopia*. New York: Basic Books.
O'Mahony, Sean, Janet McHenry, Daniel Snow, Carolyn Cassin, Donald Schumacher, and Peter Selwyn. 2008. A review of barriers to utilization of the medicare hospice benefits in urban populations and strategies for enhanced access. *Journal of Urban Health* 85 (2): 281–290.
Randall, Fiona, and R.S. Downie. 2006. *The philosophy of palliative care: Critique and reconstruction*. Oxford: Oxford University Press.
Sandel, Michael. 1982. *Liberalism and the limits of justice*. Cambridge: Cambridge University Press.
Scheid, Daniel P. 2016. *The Cosmic common good: Religious grounds for ecological ethics*. New York: Oxford University Press.

Sebuyira, Lydia Mpanga, Faith Mwangi-Powell, Jose Pereira, and Christopher Spence. 2003. The cape town palliative care declaration: Home-grown solutions for sub-saharan Africa. *Journal of Palliative Medicine* 6 (3): 341–343.

Smith, Cardinale and Otis Brawley. 2014. Disparities. In *Access to palliative care. Health affairs blog*. http://healthaffairs.org/blog/2014/07/30/disparities-in-access-to-palliative-care/ Accessed 5 Jan 2017.

The Henry J. Kaiser Family Foundation (Kaiser). 2012a. Family health premiums rise 4 percent to average of $15,745 in 2012, national benchmark employer survey finds." http://kff.org/private-insurance/press-release/family-health-premiums-rise-4-percent-to/. Accessed 28 November 2016.

———. 2012b. *Health care costs: A primer*. http://kff.org/report-section/health-care-costs-a-primer-2012-report/ Accessed 28 Nov 2016.

The International Children's Palliative Care Network. 2008. *Charter of rights for life limited and life threatened children*. http://www.icpcn.org/icpcn-charter. Accessed 30 Aug 2017.

United States Conference of Catholic Bishops (USCCB). 1993. *A framework for comprehensive health care reform: Promoting human life, promoting human dignity, pursuing the common good.* Washington, DC. http://www.usccb.org/issues-and-action/human-life-and-dignity/health-care/upload/health-care-comprehensive-care.pdf. Accessed 15 Jan 2017.

Vatican Council, Second. 1965. *Gaudium et spes*. http://www.vatican.va/archive/hist_councils/ii_vatican_council/documents/vat-ii_const_19651207_gaudium-et-spes_en.html. Accessed 30 Aug 2017.

Walzer, Michael. 1983. *Spheres of justice: A defense of pluralism and equality*. New York: Basic Books.

Wells, Francis. 2006. The doctor. *Tate, etc*.8 (Autumn): 108.

Williams, David R. 2004. Health and equality of life of African Americans. In *The state of black America*, ed. Lee A. Daniels, 115–138. New York: Urban League.

Wooden, Cynthia. 2016. *Health care is a right, not a privilege, pope says*. Washington DC: Catholic News Service/US Conference of Catholic Bishops. http://www.catholicnews.com/services/englishnews/2016/health-care-is-a-right-not-a-privilege-pope-says.cfm. Accessed 30 Aug 2017.

World Health Organization. 2014. *Global atlas of palliative care at the end of life*. http://www.who.int/nmh/Global_Atlas_of_Palliative_Care.pdf. Accessed 30 Aug 2017.

# Chapter 12
# Catholic Health Care and Models for Advancing Excellence in Palliative Care

**Tina Picchi**

## 12.1 Introduction

Palliative care is like a fine symphony orchestra. Each palliative care team member, like the musician, spends many hours seeking higher education in their distinctive field, fine-tuning their skills, and applying their learnings with delicacy and finesse. Each discipline brings to its particular modality a sense of mastery, and to the whole, a sense of personal responsibility. Palliative and end of life care is like a fine composition with common themes, variation, and, inevitably, a cadence and an ending. Palliative care professionals attend to each patient and family with sensitivity to the common themes of physical pain and emotional grief, and to the uniquely personal variables of each individual they care for. They listen to what truly matters to their patients and are attuned to their individual choices regarding care.

As palliative care teams expand their repertoire, from being primarily hospital-based to a variety of community settings, the challenges faced are like those of playing a new venue. Building confidence in working with new partners and adapting to the changing healthcare environment are crucial to performing well. Providing palliative care earlier in the trajectory of a serious disease will contribute to a higher quality of life. Ultimately, it is in listening to the experience of patients and families (who are the true conductor of this symphony) that palliative care excellence is achieved.

Palliative care seeks to relieve the pain and suffering associated with serious illness and enables patients and their loved ones to find sources of meaning and hope to sustain them through this difficult journey. When people are tempted to see their own lives as diminished in value or meaning because of profound illness or impending death, it is then that they most need the love and assistance of others to provide them the best care possible and to reassure them of their inherent worth. The United

T. Picchi (✉)
Independent Palliative Care Coach/Consultant, Camarillo, CA, USA



P. J. Cataldo, D. O'Brien (eds.), *Palliative Care and Catholic Health Care*, Philosophy and Medicine 130, https://doi.org/10.1007/978-3-030-05005-4_12

States Conference of Catholic Bishops' important pastoral letter, *To Live Each Day With Dignity*, begins with these words: "To live in a manner worthy of our human dignity, and to spend our final days on this earth in peace and comfort, surrounded by loved ones – that is the hope of each of us… A caring community devotes more attention, not less, to members facing the most vulnerable times in their lives" (USCCB 2011). It is imperative that Catholic health care model exceptional leadership in caring for the seriously ill and dying and collaborate with other like-minded organizations to address critical gaps in care.

Across Catholic health care, there are many fine palliative care programs that embody a commitment to provide compassionate, high quality and whole person care, anticipating, preventing and treating suffering. These programs assist the seriously ill to live fully in community and support survivors in their bereavement. Through such excellent care, God's healing love is revealed. Based on 2015 data from the American Hospital Association and the Catholic Health Association of the United States, 97.7% Catholic hospitals with 300 beds or greater report palliative care programs with 78.9% of all Catholic hospitals offering palliative care services (CHA 2017). Between 2001 and 2015 the percent of Catholic hospitals with palliative care programs more than doubled. In recent years, these programs have expanded their services beyond the hospital walls to many other community-based settings (SCC 2017).

As Executive Director of the Supportive Care Coalition, I was asked to contribute this chapter highlighting models of palliative care excellence in Catholic health care (SCC 1994). There are a number of exemplars within SCC member organizations as well as other Catholic systems that have raised the bar for palliative care; many have received national recognition for their innovative leadership. These programs exist in many settings, including hospitals, hospices, long-term care, outpatient clinics and home care. Each of these exemplars offer sustainable, replicable models that are making an appreciable difference in the lives of individuals and in the communities they serve and are contributing to a national orchestra comprised of high performers who are passionate about transforming care for the seriously ill and dying.[1]

## 12.2 National Organizations That Define Palliative Care Excellence

Palliative care excellence is built upon a strong foundation of consensus standards that have evolved through rigorous study and analysis. Many national professional groups contribute to a broad understanding of what constitutes excellence in palliative care. Through in-depth research and analysis these professional organizations have created consensus standards, endorsed quality measures, established

[1] I am grateful to my colleagues within the Supportive Care Coalition who shared descriptions of their palliative care programs for inclusion in this chapter.

certification and recognition criteria, designed leadership prototypes, examined health care policy and published significant and influential reports regarding palliative and end of life care (NCP 2018; TJC 2016; AHA 2017a, b, c; The Hastings Center 2017; CAPC 2003; NAS 2014). Collectively, they present national models to emulate and identify critical areas for improvement. Catholic health care has contributed significantly to the creation of this substantive foundational body of work and continues to expand upon it through innovation and excellence.

### *12.2.1 National Consensus Project (NCP) Guidelines corrections to content required due to new 2018 edition of these guidelines*

The National Consensus Project for Quality Palliative Care (NCP) has made a major contribution to advancing the field through the widely disseminated Clinical Practice Guidelines for Quality Palliative Care (NCP 2018). These guidelines address the structure and process of care in an effort to provide guidance for newly developing programs as well as to strengthen existing programs. The NCP recognized that the rapid growth in number and type of palliative care programs in health care settings across the country mandated a shared set of norms or standards so that patients could come to expect consistently high quality services wherever they might live or seek care (Ferrell et al. 2007). The NCP developed a set of guidelines that identify eight domains of quality palliative care, and together with the National Quality Forum, synergistically advance palliative care by formalizing the concept of palliative care and differentiate it from other types of care (NCP 2018). The 4th edition, completed in 2018, promotes access to quality palliative care, fosters consistent standards and criteria, and encourages continuity of palliative care across settings with an emphasis on collaborative partnerships within and between all care providers.

### *12.2.2 The National Quality Forum (NQF)*

The field of palliative care continues to develop and refine quality indicators that include evidence-based measures and benchmarks of excellence (NQF 2017). The National Quality Forum, a voluntary organization that sets consensus standards, seeks and endorses measures addressing various aspects of palliative and end-of-life care. These measures include but are not limited to: assessment of physical, emotional, social, psychological and spiritual aspects of care; access to and timeliness of care, patient and family experience with care; patient and family engagement; care planning; and caregiver support. Measures also track avoidance of unnecessary hospital or emergency department admissions, appropriateness of care and cost of care. The NQF was established as a unique public-private collaborative venture whose mission is to improve the quality of health care by standardizing the measurement and reporting of quality-related information (NQF 2016). In recent years, three Supportive Care Coalition board members – representing the fields of medicine, ethics and spiritual care from Catholic health care – have been appointed to the NQF Steering Committee for Palliative Care and End-of-Life Care to recommend measures for endorsement that are used for public accountability and quality improvement (NQF 2016).

### *12.2.3 The Joint Commission's Advanced Palliative Care Certification*

Building upon the work of the NCP Guidelines and measures endorsed by the National Quality Forum, The Joint Commission's Advanced Certification Program for Palliative Care, launched in 2011, recognizes programs that demonstrate exceptional patient- and family-centered care and optimize the quality of life for adult and pediatric patients with serious illness (TJC 2016). The program emphasizes the following standards: the creation of a formal, organized palliative care program led by an interdisciplinary team whose members have advanced training in palliative care; leadership endorsement and support of the program's goals for providing care, treatment and services; a special focus on patient and family engagement; processes that support the coordination of care and communication among all care settings and providers; and the use of evidence-based national guidelines or expert consensus to support patient care processes (TJC 2016). Supportive Care Coalition member organizations were among the early adopters of this advanced certification and served as mentors and role models for others to strive for this mark of distinction. St. Joseph Mercy Oakland, a 443-bed comprehensive, community teaching hospital and member of Saint Joseph Mercy Health System located in Pontiac, Michigan was among the very first hospitals in the country to achieve this advanced certification (St. Joseph Mercy Oakland 2017).

### *12.2.4 Center to Advance Palliative Care Leadership Centers (PCLC)*

The Center to Advance Palliative Care launched Palliative Care Leadership Centers in 2003 to catalyze palliative care program development, growth and excellence (CAPC 2003). Palliative Care Leadership Centers conduct a 2-day, in-person training followed by a year of mentoring. The focus is on the operational aspects of palliative care development and sustainability rather than on clinical training. Attendee teams choose a PCLC that best meets their needs and aligns with their organizational profile. The PCLCs represent diverse settings including health systems, community-based settings, hospices, academic medical centers, cancer centers, VA and safety net hospitals. Mount Carmel Palliative Care Service in Columbus, Ohio, a member of Trinity Health is one of eight exemplary palliative care programs in the United States designated as a Palliative Care Leadership Center (CAPC 2017). PCLC faculty members are leaders in the field and experts in applying the most effective palliative care program models. More than 1100 hospital and hospice teams have already attended a PCLC since 2003, and over 80% of PCLC graduates successfully established a palliative care program by their 2-year mark. Many palliative care teams in Catholic health care have successfully developed and sustained excellent programs after participating in this expert training and mentoring experience (CAPC 2003).

## 12.3 National Academy of Medicine and the *Dying in America* Report

The National Academy of Medicine (previously known as the Institute of Medicine) has played an important role in conversations and policies surrounding end-of-life care. The Academy convenes eminent professionals to examine policy matters pertaining to the health of the public and in recent years has published significant and influential reports regarding care at the end of life. In 1997, the Institute of Medicine produced the report *Approaching Death: Improving Care at the End of Life* (IOM 1997) and in 2003, it extended the conversation to pediatrics in its report *When Children Die: Improving Palliative and End-of-Life Care for Children and Their Families* (IOM 2003). Each of these reports had a major impact on end of life care and a number of new programs, policies, providers and systems of care have developed as a consequence.

In the National Academy of Medicine's more recent 2014 report, *Dying in America: Improving Quality and Honoring Individual Preferences Near the End of Life*, a committee of experts found that "improving the quality and availability of medical and social services for patients and their families could not only enhance quality of life through the end of life, but may also contribute to a more sustainable care system" (NAS 2014). This seminal report captured the experience of multiple stakeholders and identified significant gaps in care, providing key recommendations in five areas. These included the delivery of person-centered, family-oriented care; clinician-patient communication and advance care planning; professional education and development; policies and payment systems; and public education and engagement.

*Dying in America* received significant national attention and was credited with influencing The Centers of Medicare and Medicaid Services in July 2015 to pay doctors to counsel patients about end-of-life-care and engage in advance care planning. The ruling cited the report which discussed "vulnerabilities in the current health care system particularly to those who are approaching the end of life." The report states "that end-of-life care should be individualized based on patient values, goals, needs and informed preferences with a recognition that individual service needs and intensity will change over time" (NAS 2014).

To ensure implementation of the recommendations of *Dying in America*, the National Academy of Medicine sought public pledges of support from organizations as a call to action (NAS 2016b). Among the approximately 40 national organizations who stepped forward with an action plan to address critical gaps in palliative and end of life care were individual Catholic health systems as well as the Supportive Care Coalition. In May 2016 these same organizations were invited to Washington, DC to meet with members of the National Academies and the Committee on Approaching Death to discuss progress made and to identify remaining barriers to high quality end-of-life care (NAM 2016). This dialogue was instrumental in prioritizing the charter of the *Roundtable on Quality Care for People with Serious Illness* convened by the National Academies of Sciences, Engineering and Medicine. Several Catholic health care organizations, including the Supportive Care Coalition became inaugural sponsors of this roundtable (NAS 2016a).

## 12.4 Circle of Life Award: A National Palliative Care Award That Recognizes Excellence

Since 2000, the American Hospital Association and other major sponsors including the Catholic Health Association, the National Hospice and Palliative Care Organization (NHPCO), and the National Hospice Foundation, have recognized innovation in palliative and end-of-life care in hospices, hospitals, health care systems, long term care facilities and other organizations providing care, through its Circle of Life Award (AHA 2017a, b, c). This award shines a light on programs and organizations that demonstrate excellence in each of the eight domains of palliative care outlined in the NCP Guidelines. It celebrates and elevates a diversity of hospice and palliative care providers including large teaching hospitals as well as small rural programs that demonstrate creativity and innovation in addressing the unique cultural needs of their communities.

The Circle of Life Committee, which includes a representative from Catholic health care, has reviewed hundreds of innovative palliative and end-of-life care program applications throughout the years and conducted site visits on a select few. The program honors palliative care programs that:

- Serve people with life-limiting illness, their families and their communities;
- Demonstrate effective, patient/family-centered, timely, safe, efficient and equitable palliative and end-of-life care;
- Use innovative approaches to meet critical needs and serve as sustainable, replicable models for a segment of the field;
- Pursue quality improvement consistent with the National Consensus Project Clinical Practice Guidelines for Quality Palliative Care, NHPCO Standards of

Practice for Hospice Programs or other widely-accepted standards, within their resources and capabilities;
- Address medical, psychosocial, spiritual and cultural needs throughout the disease trajectory;
- Use innovative approaches to reach traditionally marginalized populations;
- Actively partner with other health care organizations, education and training programs, the community, providers of care, and/or insurers; and
- Use metrics that demonstrate significant impact and value for individuals, families and communities. (AHA 2017a, b, c)

Award recipients serve as teachers and mentors and inspire further development and advancement in the field of palliative care. Since the inception of this award, several Catholic hospitals and systems have been honored for their innovation and leadership, including the following selected programs (AHA 2017a, b, c).

## 12.5 Bon Secours Health System

The Mission of Bon Secours Health System is to bring compassion to health care and to be of good health to those in need, especially those who are poor and dying. "Unparalleled palliative care" is a key component of their quality plan. Bon Secours Palliative Medicine Program, based in Richmond, Virginia, was the recipient of the 2016 Circle of Life Award, and is a nationally recognized model for person-centered, family oriented palliative medicine (AHA 2016; Bon Secours 2017). Comprised of five hospitals ranging from a 100-bed urban facility to a 400-bed suburban facility, this system includes a multi-specialty medical group practice serving 300,000 distinct patients annually. A practice within the Bon Secours Medical Group, Palliative Medicine offers specialty consultation with a vision of providing palliative care "without walls," moving from facility-based consultation to ambulatory and home-based care delivery. In this way, the palliative medicine team cares for each seriously ill patient no matter where they are in their disease process and wherever they are receiving care. In many organizations, palliative care remains a silo, but this approach makes all the difference to the delivery of person-centered, family-oriented care. Bon Secours Palliative Medicine is a model for innovation in palliative care nationally. As part of the medical group, it uses primary and specialty palliative care as cornerstones of the program and is involved in innovative delivery and payment models. The program uses an interdisciplinary team to build, implement and refine care plans with patients and their loved ones regardless of the patient's care setting, including physicians, nursing professionals, social workers, chaplains and others who embrace a bio-psycho-social-spiritual model of care.

Patients are tracked during their care by 60 nurse navigators who are embedded throughout the health system. The nurse navigator is involved in the initial meeting with the patient and family and is responsible for the follow-up and ongoing care provided to them. Nurse navigators also remain vigilant for signs of potential crises that could prompt an unnecessary visit to the emergency department or a readmission to the hospital. Being a specialty practice in a larger medical group enhances relationships with primary care physicians and specialists, empowering primary care physicians and other providers to become more confident in approaching advance care planning and managing pain and other distressing symptoms.

## 12.6 Integrating Principles of Palliative Care and Whole Person Care: Three Providence St. Joseph Health Organizations

Three Providence Health and Services organizations (now Providence St. Joseph Health) in Southern California collaborate to provide innovative, inpatient, outpatient and home-based palliative care and hospice care for adult and pediatric patients. These programs are committed to excellence in hospice and palliative care as defined by national standards and consensus guidelines in the field. They are intent upon transforming the culture of healthcare by integrating principles of palliative care within a whole person and family-centered approach into practice at every level and setting. They have embraced the terminology and principles of Whole Person Care (WPC) as elaborated by Balfour Mount and colleagues at McGill University (McGill University 2017). This includes seeing each patient as a unique individual who deserves attention to bodily comfort and emotional, interpersonal, social, and spiritual dimensions of their lives.

Providence TrinityCare Hospice (TrinityCare) has two adult hospice teams and the only dedicated pediatric hospice team in Southern California with an average daily census of 178 adult and 158 pediatric patients (Providence TrinityCare Hospice 2017). TrinityCare also provides a 24/7 call center serving the hospice program and all of the region's palliative care programs. Providence Little Company of Mary Medical Center Torrance (PLCMT) is a 436-bed community medical center with a well-established, Joint Commission-certified inpatient palliative care program and outpatient/home-based palliative care program. These combined programs were honored with a 2017 Circle of Life Award (AHA 2017a, b, c).

Providence Institute for Human Caring was conceived by Ira Byock, MD, and founded within Providence Health & Services in 2014 (PIHC 2014a). The institute believes that raising expectations for health care that treats each of us as a whole person – eliciting and honoring our values, preferences and priorities – will inspire a growing demand for professionals who can make whole person care the new normal. The institute draws on the attitudes, knowledge and skills of palliative care to transform healthcare from a predominantly transactional model focusing on diagnoses and treatments, to a mode of practice that attends not just to medical problems, but also to the wellbeing of whole persons and their families throughout the continuum of care. This transition to person-centered, goal-aligned care advances goals of higher quality, improved health of the populations, responsible use of health care resources and health care professional wellbeing.

The Providence Institute for Human Caring's team of professionals conduct trainings in best practices and implement workflow refinements while measuring and monitoring quality. They have adopted the Ariadne Lab's Serious Illness Conversation Guide as their template for holding and charting goals of care conversations across primary care, acute care and home-based settings (Ariadne Labs 2017). They utilize ACP Decisions videos for introducing key issues such as advance care planning and CPR to patients and families to enhance communication and shared decision-making (ACP Decisions 2018; PIHC 2014b). In collaboration with StoryCorp™ they are developing *Hear Me Now*, a collection of story banks across Providence Health and Services to foster a culture that routinely gathers stories from patients, families and colleagues – and listens! (PIHC 2014c).

## 12.7 Educating Clinical Providers in Advance Care Planning

Most people nearing the end of life are not physically, mentally, or cognitively able to make their own decisions about care; consequently advance care planning is essential to ensure that patients receive care reflecting their values, goals, and preferences. *Dying in America* recommended quality standards for clinician-patient communication and advance care planning that are measurable, actionable and evidence based, with reimbursement tied to these standards (NAS 2014). The report indicated that clinicians should initiate high quality conversations about advance care planning and integrate the results of these conversations into the ongoing care plans of patients and communicate with other clinicians as requested by the patients.

OSF HealthCare (OSF), an integrated Catholic health care system headquartered in Peoria, Illinois and recognized with a 2016 Circle of Life Citation of Honor, has developed a model program for educating clinical providers and other members of the team in advance health care planning (OSF HealthCare 2017a, b; AHA 2016). Since 2009, OSF has provided advance care planning (ACP) facilitation to over 20,000 patients across their care delivery system using the OSF Care Decisions® advance care planning model (OSF HealthCare 2017a, b). This model stresses the importance of discussions surrounding patient and family specific values. This discussion then leads to identification of goals of care, which are used to determine health care preferences. Rather than a "to do" checklist of specific items, the discussion seeks to understand how patient values interact with particular situations, thereby enabling the designated agent to use good substituted judgment in making decisions for the patient when they are unable to do so. The output of the process includes both a valid legal designation of a Power of Attorney for health care, and a Discussion Record, both of which are accessible through the electronic health record across care settings and encounters.

In addition to offering this service to anyone who desires it, OSF has also developed an outreach program for patients who are at high risk of health care needs. By coordinating and integrating the patient's electronic health record with OSF's ACP database, they are able to identify patients with serious illness and/or changing status, and generate reports showing which of those patients have not yet had ACP. Care managers then educate these patients and encourage their participation.

## 12.8 Improving Access to Palliative Care Through Universal Screening

St. John Providence, based in Detroit, Michigan is part of Ascension Health. St. John Providence's (SJP) Palliative Care Program was recognized with a 2011 Circle of Life Award for bringing innovation into palliative and end-of-life care (AHA 2011; Ascension St. John Providence 2011). The organization was lauded for hardwiring palliative care into every service and educating staff so that no matter where a patient enters the health system, they will be evaluated for palliative care needs. This model program developed a unique and sustainable multi-pronged approach to improve access to quality palliative care. The key innovative elements of the program are the use of palliative care triggers for patients admitted to the hospital, full integration of spiritual care into the palliative care team, inclusion of all staff in palliative care education, and outreach/partnering with local faith communities. Each part of the strategy is led by multisystem, multidisciplinary workgroups.

The development and implementation of PC triggers are part of the admission assessment for all patients admitted to the hospital. These PC triggers were created using evidence-based data obtained during a pilot involving 1400 patients. During the pilot test for universal screening, palliative care consults increased threefold, the time from admission to palliative care referral shortened and consequently patients were able to benefit more fully from the palliative care services they received prior to being discharged from the hospital. This upstream approach to meeting palliative care needs ensures equitable access to timely palliative care for all patients entering a SJP hospital.

## 12.9 The Hastings Center Cunniff-Dixon Physician Awards: Caring for the Most Medically Fragile Patients at Home

The Hastings Center Cunniff-Dixon annual physician awards recognize doctors who give exemplary care to patients at the end of life (The Hastings Center 2017). The selection process is coordinated by The Hastings Center and Duke University's Institute on Care at the End of Life. The aim of The Hastings Center Cunniff-Dixon Physician Awards is to foster the skills and virtues of physicians, both young and old, who have shown their care of patients to be exemplary, a model of good medicine for other physicians, and a great benefit in advancing the centrality of end-of-life care as a basic part of the doctor-patient relationship.

There are many patients whose conditions or symptoms are not well served by the typical medical care model. Elderly patients with chronic illness and frailty often seek care repeatedly in the emergency department or hospital setting. Homebound patients are often no longer able to come in for routine primary care appointments. When these patients get urgent symptoms, families often call 911 so that ambulances can transport them for assessment and symptom management in the emergency department. Even during office hours, it may be too difficult to access immediate help during the day in the primary care office for urgent symptoms.

Providence Elder at Home (EAH), based in Portland, Oregon is part of the Providence St. Joseph Health. EAH is a primary care model designed to meet the needs of the most medically fragile elders by providing multidisciplinary support in the patient's home (Providence Health & Services 2017). The chief architect of this model, Marian Hodges, MD, was recipient of a 2016 Hastings Center Cunniff-Dixon Physician Award for her outstanding care for cognitively impaired seniors, and her leadership and mentorship in innovative elder care.

"Home" can be defined as a private home, an adult care home, an assisted living facility, or an intermediate-care nursing facility. Eligible patients are in the Providence Health Plan Medicare Advantage plan, meet certain requirements for burden of chronic illness and emergency department/hospital use from the previous 12 months (using a computer algorithm), and live within the Portland metro area. Providence is an integrated delivery system, which made it possible for the delivery system to take financial risk on the Medicare Advantage patients identified in the Providence insurance arm as "eligible." The EAH team works with the patient's primary care provider to monitor the patient's health, design a treatment plan, manage medications, address safety concerns and identify needed support services solutions. The patient's primary care physician remains engaged in the care plan, including decisions about referrals to specialists, transition to hospice, and providing helpful context about the patient's social and medical history. Communication is shared via the electronic medical record. Patients participating in EAH have access to 24/7 nursing triage support. If indicated, they receive an in-home assessment by RN and/or medical provider. There is no additional cost to the patient beyond their typical office visit "co-pay" if seen by the medical provider, and there is no financial risk to the primary care provider. The EAH core team has additional available support from dedicated clinical pharmacists, medical assistants, chaplains and field nurses.

When Providence Health and Services launched EAH in August 2015, there was no blueprint for this robustly designed 24/7 home-based care model within an integrated delivery system. In their first year of operation, Elder at Home engaged over 830 medically fragile homebound patients. The model achieved success beyond their first year projections; a preliminary evaluation revealed a 24% decrease in hospital admissions and a 26% decrease in inpatient hospital days. Almost 100 patients were referred to hospice. Most significantly, patients, families and primary care physicians reported positive experiences with this innovative care model. In fact, several families asked that EAH continue to follow the patient "socially" even after the patient's care had been assumed by the hospice team. The "goals of care" conversation held by the EAH team in the patient's home is often the pivotal point, early in their relationship, for forging relationship as well as clarifying goals.

## 12.10 Supportive Care Coalition: A Model of Excellence for Palliative Care in Catholic Health Care

Each of the palliative care exemplars described in this chapter are members of the Supportive Care Coalition (SCC), a national membership organization whose mission is to advance excellence in palliative care (SCC 2017). Formed in 1994, this Coalition is comprised of several Catholic organizations with health care ministries nationwide. SCC believes palliative care is a "hallmark of Catholic health care, intrinsic to our healing mission" (Picchi 2011). Palliative care leaders from each member organization serve on the Board of Directors and this provides an excellent forum for expert-to-expert engagement across Catholic health care. Member organizations share innovative palliative care practices that are often replicated by others in the Coalition. This testimonial from Ascension Health's director for palliative care affirms the value of this leadership model: "The Supportive Care Coalition is an invaluable resource in efforts to advance palliative care, share leading practices and benefit from lessons learned by other organizations. This interdisciplinary community is helping us to improve the spiritually-centered, holistic palliative care we provide to our patients and families" (SCC 2017).

Partnering with other like-minded organizations, the SCC leverages the power of its collective voice, expertise and resources to be an influential leader and advocate for needed improvements in palliative care education, research, clinical practice and financing mechanisms. SCC is a member of the Patient Quality of Life Coalition, comprised of over forty non-governmental organizations seeking high quality accessible palliative care (PQLC 2013). The SCC and the Catholic Health Association of the United States were early advisors and advocates for a federal bill entitled the Palliative Care & Hospice Education and Training Act (PCHETA) (Congress.Gov 2016). Through letter writing campaigns, Lobby Days on Capitol Hill and outreach to elected representatives, advocates from Catholic health care and the other PQLC member organizations have been able to achieve significant bipartisan congressional endorsement, and at the time of this writing, were hopeful the 115th Congress will pass this bill. PCHETA seeks to establish palliative care and hospice education centers to improve the training of interdisciplinary health professionals in palliative care, develop and disseminate curricula relating to palliative care; support the training and retraining of faculty; support continuing education; provide students with clinical training in appropriate sites of care; and provide traineeships for advanced practice nurses. This bill would also provide for the establishment of a national campaign to inform patients, families and health professionals about the benefits of palliative care and the services that are available to support patients with serious or life-threatening illness.

### *12.10.1 Integrating the Spiritual Aspects of Whole Person Care in Palliative Care Team Practice*

Spiritual care is an essential domain of quality palliative care that is person-centered and family-oriented. The National Consensus Project Guidelines recommend that the spiritual and existential dimensions of a person's care needs be assessed and responded to, based on the best available evidence, and skillfully and systematically applied by members of the interdisciplinary team. All palliative care disciplines should have education and appropriate training in pastoral care and be knowledgeable about the spiritual issues evoked by the presence of life-threatening illness (NCP 2018).

The Supportive Care Coalition created a model for chaplains to cross-train physicians and nurses in addressing spiritual suffering. Over a two and a half year period, this methodology was tested in 20 palliative care teams, all from SCC member organizations, representing geographically diverse programs nationally that included both inpatient and outpatient settings (English and Picchi 2014). Within the profession of medicine, there has been a growing emphasis on comprehensive treatment of patient suffering, including spiritual suffering. Chris Smyre and colleagues have reported from physicians surveyed that spiritual suffering tends to intensify physical pain and that physicians should seek to relieve such suffering. The authors suggested that a collaborative team approach inviting chaplain participation might be the best strategy to help physicians when they seek to relieve spiritual suffering (Smyre et al. 2014). The SCC model places chaplains in a prominent role not merely to be content experts but to actively teach other team members how to recognize spiritual suffering and to highlight how spiritual beliefs and values inform the patient's goals of care. The SCC model is an adaptation of the standard goals of care consultation model developed by David Weissman, which includes a 10-step process and incorporates cues to stimulate new behaviors which may lead to deeper spiritual grounding and engagement (Ambuel and Weissman 2015; SCC 2017).

Recognizing that the quality of caregiver presence is a critical therapeutic variable, the SCC model focuses on the development of a team culture that encourages palliative care providers to be more intentionally present to the spiritual dimension of patients and families (Hutchinson 2011). This model embraces the consensus conference definition of spirituality, which states that spirituality is that "aspect of humanity that refers to the way individuals seek and express meaning and purpose and the way they experience their connectedness to the moment, to self, to others, to nature, and to the significant or sacred" (Puchalski et al. 2009).

Spiritual inquiry is built on a foundation of trust. To that end, the SCC project borrowed from Parker Palmer's instructions for a Circle of Trust: "A circle of trust is a group of people who know how to sit quietly…with each other and wait for the shy soul to show up...The relationships in such a group…are not confrontational… they are filled with an abiding faith in the reality of the inner teacher and in each person's capacity to learn from it" (Palmer 2009). The SCC project participants were also introduced to Eric Cassell's work on suffering. "Suffering," Cassell writes, ". . . is the state of severe distress associated with events that threaten the intactness of the person…All aspects of personhood…are susceptible to damage and loss…[The] way to learn what damage is sufficient to cause suffering…is to ask the sufferer" (Cassell 1982). This project set out to establish a setting and scripting for an effective way to ask. As a result, palliative care clinicians became more attuned to spiritual suffering by intentionally focusing on the quality of their presence as caregivers and by learning specific skills designed to uncover spiritual concerns that inform patients' goals of care. This new way of approaching difficult conversations contributed to the team's sense of vocation as healers and a greater sense of resiliency.

In 2016, The National Association of Catholic Chaplains (NACC) recognized the Supportive Care Coalition with the Outstanding Colleague Award citing SCC's "partnership with the NACC to elevate the role of the chaplain as both mentor and role model for interprofessional team members, addressing spiritual suffering and drawing upon the spiritual strengths and resources of patients, families and colleagues to promote healing and to provide whole person care, thus contributing to the advancement of the profession of chaplaincy in a significant and lasting way" (NACC 2016).

## 12.11 Recognizing Five Domains of Pain: A Palliative Care Model for Long Term Care

The Carmelite Sisters for the Aged and Infirm in Germantown, New York, and their educational arm, the Avila Institute of Gerontology, have developed a guide for long term care facilities to establish holistic palliative care programs that has been applauded by other Supportive Care Coalition member organizations and the leadership of the End-of-Life Nursing Education Consortium (ELNEC) (Avila Institute of Gerontology 1988; AACN 1969). The project was brought to fruition through a series of pilot studies in six of the homes where the Carmelite Sisters either serve, sponsor or co-sponsor. The program formalizes the tradition of a philosophy of compassionate care begun by their Foundress, Venerable Mary Angeline McCrory, in 1929. Developed in collaboration with ELNEC, the curriculum trains caregivers to understand the importance of identifying and managing different types of pain, of understanding the resident's story, and of respecting each resident's culture and faith tradition (AACN 1969). It also educates families so that they can work as a team with nursing home caregivers to surround their loved one with the best care possible.

Palliative care begins on admission by interviewing each resident, using a series of questions that pertain to different types of pain that all people experience at one time or other in their lives. This process equips caregivers with knowledge to help them decide what type of interventions may be needed to help the resident become comfortable, depending on the type of pain identified. The success of this generalist model of palliative care lies in the interdisciplinary team approach to each resident's care. Implementing the program begins by establishing a steering committee made up of staff from all disciplines which, in turn, oversees and guides each unit's team. The role of every staff member in the long-term care facility is valued in both assessing residents needs and in contributing to their care plan.

The palliative care program also endows the residents with continuing autonomy, ensuring that they can have and do the things that are most meaningful to them throughout a debilitating disease and even at the end stages of life. With a focus on caring for the whole person, the training focuses on recognizing the five domains of pain for every resident admitted to the facility, by asking questions that go beyond a social visit conversation. The five domains of pain, identified by Dr. Michael Brescia, executive medical director of Calvary Hospital in New York, include physical, spiritual, psychiatric (mental), emotional and familial pain (Minda 2016). This unique approach recognizes that all nursing home residents are appropriate recipients of palliative care, and it seeks to provide them the highest quality of life possible. Sr. M. Peter Lillian Di Maria, O.Carm., director of the Avila Institute says, "We accompany residents by means of a program that respects their dignity and provides comfort. No matter what the pain or the suffering may bring, we love them, we support and journey with them through the many levels of pain which come from a life-changing illness or event."[2]

## 12.12 Conclusion

Catholic health care has a long and remarkable history of providing excellence in end-of-life care and championing whole person care. Imbued with the charism of their religious founders, Catholic health care ministries have historically modeled how our society should treat the sickest and most vulnerable among us, respecting the inherent dignity of each person. Pope Francis has captured this Catholic vision of palliative care:

> Palliative care is the properly human attitude, of taking care of one another, especially of those who suffer. It is a testimony that the human person is always precious, even if marked by illness and old age. Indeed, the person, under any circumstances, is an asset to him/herself and to others, and is loved by God. This is why, when their life becomes very fragile

[2] This quotation is from a personal communication with the author.

and the end of their earthly existences approaches, we feel the responsibility to assist and accompany them in the best way. (Pope Francis 2015)

In recent years, palliative care has become a recognized medical specialty that has adopted evidence-based quality standards for person-centered, family-oriented care. Catholic hospitals have experienced exponential growth in palliative care programs. Among these, many exemplar programs raise the bar and advance the field of palliative care. Through their innovation and leading practices, they are models for others to emulate. There are also many palliative care programs that are significantly under-supported, lacking administrative champions, adequate funding, and professionally trained and credentialed interdisciplinary professionals. Palliative care requires an interdisciplinary team with expertise and training, including board certified hospice and palliative care physicians, nurses, social workers, chaplains and others (NAS 2014). Like members of a fine symphony orchestra, each of these professionals performs a vital role in helping patients and families experience wholeness. These inter-professional teams serve as role models and mentors across disciplines and specialties, attending to physical, emotional and spiritual suffering. They focus on the goals of the patient – who is the true conductor of this symphony – to provide the best care for both patient and family.

For palliative care to thrive in Catholic health care (and in other-than-Catholic settings as well) there must be strong organizational commitment and a shared vision that palliative care is not only specialized care for the seriously ill and dying, but also a model of whole person care to be emulated across the entire health care delivery system. To sustain and build momentum for this vision, we need strong champions and advocates who recognize that quality palliative care is a hallmark of Catholic health care, intrinsic to its healing mission.

## References

ACP Advance Care Planning Decisions. 2018. https://acpdecisions.org. Accessed 08 Dec 2018.

Ambuel, Bruce, and David E. Weissman. 2015. Fast facts and concepts #16. Moderating an end of life family conference. http://www.mypcnow.org/blank-qy84d. Accessed 23 Jan 2017.

American Association of Colleges of Nursing (NACN). 1969. http://www.aacn.nche.edu/elnec. Accessed 23 Jan 2017.

American Hospital Association (AHA). 2016. 2016 circle of life winners. http://www.aha.org/about/awards/col/awardees.shtml. Accessed 23 Jan 2017.

———. 2017a. Circle of life award. http://www.aha.org/about/awards/col/index.shtml. Accessed 23 Jan 2017.

———. 2017b. Circle of life criteria. http://www.aha.org/about/awards/col/criteria.shtml. Accessed 23 Jan 2017.

———. 2017c. Circle of life award criteria and winners. http://www.aha.org/about/awards/col/index.shtml. Accessed 23 Jan 2017.

Ariadne Labs. 2017. Serious illness conversation guide. https://www.ariadnelabs.org/areas-of-work/serious-illness-care/resources/#Downloads& Tools. Accessed 23 Jan 2017.

Ascension St. John Providence. 2011. *Palliative care*. http://www.stjohnprovidence.org/palliative-care. Accessed 23 Jan 2017.

Avila Institute of Gerontology, Inc. 1988. http://www.avilainstitute.org/programs.htm. Accessed 23 Jan 2017.

Bon Secours. 2017. *Palliative medicine*. https://bonsecours.com/richmond/our-services/palliative-medicine. Accessed 23 Jan 2017.

Catholic Health Association of the United States (CHA). 2017. *Report: prevalence of palliative care in US Catholic hospitals*. Source: American Hospital Association Annual Survey of Hospitals 2001 and 2015.

Cassell, Eric J. 1982. The nature of suffering and the goals of medicine. *New England Journal of Medicine* 306: 639–645.

Center to Advance Palliative Care (CAPC). 2003. *Palliative care leadership centers overview*. https://www.capc.org/palliative-care-leadership-centers/. Accessed 23 Jan 2017.

———. 2017. *Palliative care leadership center (PCLC) at Mount Carmel Health System*. https://www.capc.org/palliative-care-leadership-centers/palliative-care-training-locations/pclc-mount-carmel/. Accessed 2 Feb 2017.

Congress.Gov. 2016. S.2748 *Palliative care and hospice training education act*. https://www.congress.gov/bill/114th-congress/senate-bill/2748. Accessed 23 Jan 2017.

English, Woodruff, and Tina Picchi. 2014. *Spiritual wisdom, a component of care*. *Health Progress*. https://www.chausa.org/publications/health-progress/article/january-february-2014/spiritual-wisdom-a-component-of-care. Accessed 23 Jan 2017.

Ferrell, Betty, Stephen R. Connor, Anne Cordes, Constance Dahlin, et al. 2007. The national agenda for quality palliative care: The national consensus project and the national quality forum. *Journal of Pain and Symptom Management* 33: 737–744.

Hutchinson, Ted. 2011. *Whole person care – A new paradigm for the 21st century*. New York: Springer Science & Business Media.

Institute of Medicine of the National Academies (IOM). 1997. In *Approaching death: Improving care at the end of life. Committee on care at the end of life*, ed. Marilyn J. Field and Christine K. Cassel. Washington, DC: Division of Health Care Services. National Academy Press. https://www.nap.edu/read/5801/chapter/1#ii. Accessed 30 Aug 2017.

———. 2003. *When children die: Improving palliative and end-of-life care for children and their families*. Washington, DC: The National Academies Press.

McGill University. 2017. *McGill programs in whole person care*. https://www.mcgill.ca/wholepersoncare. Accessed 23 Jan 2017.

Minda, Julie. 2016. Avila institute program help facilities address top sources of pain for residences. *Catholic Health World*. https://www.chausa.org/publications/catholic-health-world/archives/issues/december-1-2016/avila-institute-program-helps-facilities-address-top-sources-of-pain-for-residents. Accessed 23 Jan 2017.

National Association of Catholic Chaplains (NACC). 2016. 2016 *Outstanding colleague award*. http://www.nacc.org/conference/history/conference2016/awards/2016-outstanding-colleague-award/. Accessed 23 Jan 2017.

National Consensus Project (NCP). 2018. Clinical practice guidelines for quality palliative care 4th edition. https://www.nationalcoalitionhpc.org/wp-content/uploads/2018/10/NCHPC-NCPGuidelines_4thED_web_FINAL.pdf. Accessed 08 Dec 2018.

National Quality Forum (NQF). 2016. *Palliative and end of life care project 2015–2016*. http://www.qualityforum.org/ProjectDescription.aspx?projectID=80663. Accessed 23 Jan 2017.

———. 2017. http://www.qualityforum.org/Home.aspx. Accessed 23 Jan 2017.

OSF HealthCare. 2017a. *Supportive care*. https://www.osfhealthcare.org/supportive-care/. Accessed 23 Jan 2017.

———. 2017b. *Care decisions*. https://www.osfhealthcare.org/supportive-care/services/advance-care-planning/care-decisions/. Accessed 23 Jan 2017.

Palmer, Parker. 2009. *A hidden wholeness: The journey toward an undivided life*. San Francisco: Jossey-Bass.

Patient Quality of Life Coalition (PQLC). 2013. http://patientqualityoflife.org/members/. Accessed 23 Jan 2017.

Picchi, Tina. 2011. Palliative care: A hallmark of Catholic mission. *Health Progress*. https://www.chausa.org/publications/health-progress/article/january-february-2011/palliative-care-a-hallmark-of-catholic-mission. Accessed 23 Jan 2017.

Pope Francis. 2015. We must not abandon the elderly. http://www.news.va/en/news/pope-francis-we-must-not-abandon-the-elderly. Accessed 2 Feb 2017.

Providence Health & Services. 2017. *Providence elder at home program*. http://oregon.providence.org/our-services/p/providence-elder-at-home-program/. Accessed 23 Jan 2017.

Providence Institute for Human Caring (PIHC). 2014a. http://www.providence.org/institute-for-human-caring. Accessed 23 Jan 2017.

Providence Institute for Human Caring (PIHC). 2014b. Advance directive testimonial videos. http://www.providence.org/institute-for-human-caring/patient-and-family-resources/video-library. Accessed 23 Jan 2017.

———. 2014c. Hear me now. http://www.providence.org/hear-me-now. Accessed 23 Jan 2017.

Providence TrinityCare Hospice. 2017. *TrinityCare hospice*. http://california.providence.org/trinitycare/. Accessed 23 Jan 2017.

Puchalski, Christina, Betty Ferrell, Rose Virani, et al. 2009. Improving the quality of spiritual care as a dimension of palliative care: The report of the consensus conference. *Journal of Palliative Medicine*. 12: 885–904.

Smyre, Chris, John D. Yoon, Kenneth A. Rasinski, and Farr A. Curlin. 2014. Limits and responsibilities of physicians addressing spiritual suffering in terminally ill patients. *Journal of Pain and Symptom Management* 49: 562–569.

St. Joseph Mercy Oakland. Cancer care services. 2017. http://www.stjoesoakland.org/cancercare-services. Accessed 23 Jan 2017.

Supportive Care Coalition (SCC). 2017. http://supportivecarecoalition.org/. Accessed 23 Jan 2017.

The Hastings Center. 2017. *Hastings Center – Cunniff-Dixon Physician Awards*. http://www.thehastingscenter.org/who-we-are/service-to-bioethics/hastings-center-cunniff-dixon-physician-awards/. Accessed 23 Jan 2017.

The Joint Commission (TJC). 2016. Facts about the advanced certification program for palliative care. https://www.jointcommission.org/facts_about_palliative_care/. Accessed 23 Jan 2017.

The National Academies of Sciences, Engineering and Medicine (NAS). 2014. *Dying in America: Improving quality and honoring preferences near the end of life*. http://www.nationalacademies.org/hmd/Reports/2014/Dying-In-America-Improving-Quality-and-Honoring-Individual-Preferences-Nearthe-End-of-Life.aspx. Accessed 23 Jan 2017.

———. 2016a. *Integrating the patient and caregiver voice into serious illness care*: *Aworkshop*. http://www.nationalacademies.org/hmd/Activities/HealthServices/QualityCareforSeriousIllnessRoundtable/2016-DEC-15.aspx. Accessed 23 Jan 2017.

———. 2016b. *Organizational commitment statements*. http://www.nationalacademies.org/hmd/Reports/2014/Dying-In-America-Improving-Quality-and-Honoring-Individual-Preferences-Near-the-End-of-Life/Organizational-Commitment-Statements.aspx . Accessed 23 Jan 2017.

The National Academy of Medicine (NAM). 2016. *Assessing progress in end-of-life and serious illness care*.http://nam.edu/event/assessing-progress-in-end-of-life-and-serious-illness-care/. Accessed 23 Jan 2017.

United States Conference of Catholic Bishops (USCCB). 2011. *To live each day with dignity*. http://www.usccb.org/issues-and-action/human-life-and-dignity/assisted-suicide/to-live-each-day/index.cfm. Accessed 23 Jan 2017.

# Chapter 13
# Catholic Education on Palliative Care: Lessons Learned and Observations Made from the Field

M. C. Sullivan

## 13.1 Introduction

In the general election of 2012, voters in the Commonwealth of Massachusetts were faced with a ballot referendum which, if passed, would have legalized Physician-Assisted Suicide (PAS) by decriminalizing the writing by a physician of a lethal dosage of medication, the sole purpose of which was to end the life of the patient for whom the prescription was written and which was intended as well to be self-administered by that patient. Much has been written in this volume about PAS, and that is not the focus of this chapter.

Nor will this chapter focus on issues and questions such as Catholic teaching on pain management and the principle of double effect, risk and benefit in the withholding or withdrawing of treatment, the discussion of burdensomeness in cessation of treatment conversations, and so on. These issues are covered extensively in other chapters, although they will come up here by allusion.

Rather, this chapter will present the perspective of the fieldworker, charged with designing, constructing and implementing a follow up response to what was learned by public reactions, specifically those in the Catholic community, to the public debate about care of the seriously ill and dying precipitated by the PAS ballot referendum. For many, these reactions surfaced for the first time. The result was, for one Archdiocese, a comprehensive, strategic program to establish an office for palliative care and advance care planning, and to develop and implement tactics for pastoral education and ministry in those areas.

M. C. Sullivan (✉)
Archdiocese of Boston, Boston, MA, USA
e-mail: mc_sullivan@rcab.org



P. J. Cataldo, D. O'Brien (eds.), *Palliative Care and Catholic Health Care*, Philosophy and Medicine 130, https://doi.org/10.1007/978-3-030-05005-4_13

## 13.2 What Was Learned?

Among those involved in the work of defeating the ballot referendum in Massachusetts at that time, certain shared conclusions and observations were clearly and consistently made. As time has gone on, and other jurisdictions have faced their own referenda or legislative proposals, similar observations have been made by those who have worked to oppose PAS legalization measures. Some of the most-often cited reasons that people gave in 2012 for supporting PAS, "even just as an available option at the end of life," were the very concerns that led to the development of palliative care as a clinical subspecialty. These concerns about end of life care that gave rise to PAS may be roughly delineated into three categories:

- The perceived lack of control that most patients expressed in the presence of chronic, serious and life-limiting (both in terms of function and/or time) illness was all the more acutely felt in an era that came out of decades-long education about patient autonomy and shared participative decision-making;
- Un- or under-treated pain, in a time when developments in pharmaceuticals and complementary therapies aimed at pain management were enjoying heretofore unimaginable attention and availability; and
- The routinely perceived inattentiveness of providers to the other areas of life impacted by serious illness, such as strained relationships, financial pressure, employment concerns, and general family tension, that caused suffering in the lives of those patients.

Many of these issues were reflected in public opinion research that had been commissioned earlier by the Center to Advance Palliative Care (McInturff and Harrington 2011).[1]

For those involved in palliative care, the years leading up to its recognition as a Board-certified subspecialty for physicians, and the development of programs such as the End-of-Life Nursing Education Consortium (ELNEC)—which began in 2000 for nurses—have also led to specialty and certification programs in palliative care for social workers and chaplains.[2] This abundance of professional health care provider educational programs has been encouraging. Despite these strides, palliative care has confronted, and continues to face, challenges in being widely recognized and understood even among health care providers. The PAS debates have illuminated some of these problems (Center to Advance Palliative Care 2011).

There is a *general misunderstanding* of palliative care by many. This misunderstanding is shared by health care providers—who remain uninformed about and are perhaps threatened by losing control of the care of their patients to palliative care

---

[1] See the public opinion research by McInturff and Harrington (2011), which examines public perception of the concept, the language and the purpose of palliative care, as well of the perception among clinical providers, and underscores the need for better communication to those constituencies.

[2] Palliative care has been a specialty since 1996, and was officially recognized by Accreditation Council for Graduate Medical Education in 2006.

specialists and teams—and policy makers, church leaders across denominations, and the general public, which, of course, includes the practicing faithful. There is also a *general mistrust* of palliative care, which grows out of the misunderstanding of what exactly palliative care is. Nowhere was this more evident than in the public discourse that was generated by and follows today from the discussion of "death panels" during the debate about the Affordable Care Act when it was presented in 2010 and continues as that bill's successor is being debated and implemented.

All of this leads to a general resistance to palliative care—as it is generally (mis) understood—and in particular, to advance care planning, the bedrock and fundamental component of palliative care's Interdisciplinary Care Plan. This plan, created in ongoing discussions among patients, their providers from all the clinical disciplines involved in their care, their families and their health care proxies, also serves as an important guideline for providers and proxy decision-makers at a time when patients have lost their capacity to speak for themselves.

In the Archdiocese of Boston there was a recurring sense that if people better understood, supported, and knew how to access palliative care, there would be less sympathetic attention paid to the emotional arguments made by proponents of PAS. This led to the development of an ongoing strategic activity called the *Initiative for Palliative Care and Advance Care Planning*. The initiative comprises three parts: (1) pastoral outreach to and education of parish communities about palliative care and advance care planning; (2) ongoing construction and articulation of the moral arguments that can be used in discussing the ethical issues that are inherent in the ongoing public debate about assisted suicide; and (3) advocacy activities that involve working with the Catholic and secular media and other organizations, and educating policy makers to clarify Catholic positions on the issues and the moral teaching that informs them.

## 13.3 One "On-The-Ground" Model of Catholic Education About Palliative Care

Given that palliative care as a clinical subspecialty is not new and yet remains something not clearly understood within clinical circles or in the general public, two goals were established at the outset of the initiative in the Archdiocese of Boston.

First, it was very important that an easily understood and commonly adopted definition of palliative care be communicated broadly and simultaneously in a relatively short time to as many people in the Archdiocese as possible. Second, it was also clear that if palliative care and advance care planning were to be treated truly as ongoing strategic concerns of the pastoral life of the Archdiocese, they would need to be organically present in the ongoing activities of parish life.

In order for any of that to happen, Church leadership had to understand why palliative care was *per se* a 'good' and that it was consistent with and constitutive of Church teaching. Looking to the history of Catholic health care, one sees that the healing ministry of Jesus has long been carried out in the activities of religious orders of men and women who welcomed into their monasteries and convents the frail, the elderly, the seriously ill and the dying, caring for them until their passage into eternal life. That historical tradition has been carried on into contemporary times when even the dearth of vocations to religious life (in many areas of the world) has not caused the work to cease. Indeed, the healing ministries of the Church continue and thrive, and thus the care of the seriously ill and dying. This care is made possible, in part, through collaboration with lay colleagues, the establishment of new Public Juridic Persons that are entrusted with the canonical reserved powers and authority once exclusively held by leaders of religious congregations, and through religious congregations and various other canonically-approved relationships and joint ventures.

This extension of the healing ministry of Jesus is captured by the definition of palliative care from the Supportive Care Coalition, a national network of Catholic health systems who come together to promote the implementation of palliative care:

> Catholic health care describes itself as 'attending to the needs of the whole person'; palliative care implements a holistic, interdisciplinary care plan that identifies, assesses, and addresses the comprehensive needs of the seriously ill patient, including pain and other symptom management, psychosocial issues, emotional support and spiritual care. (Picchi and Sullivan 2013)

Thus, palliative care is both consistent with and reflective of the Christian anthropological perspective that caring for the vulnerable among us requires assessment of and attention to the integral physical, psychosocial, emotional and, above all, spiritual needs of those in our care.

In addition, palliative care provides substantial evidence of a model of care that allows and encourages patients and providers to base that care on a foundation that is supported by Catholic teaching. This support is found in the fact that both Catholic teaching and palliative care recognize that we are, above all, relational beings, who are shaped, influenced and informed by the relationships that support us. For example, palliative care emphasizes the integration of information from patients, caregivers, interdisciplinary provider teams, and health care proxies. Palliative care is also inclusive of advance care planning documents that capture the ongoing discussions about decisions made as a serious disease progresses, which are based on patients' religious beliefs and values and their informed prior decisions.

It is crucially important that Catholic leaders have this information about palliative care so that across the Church a consistent ethic of care for the seriously ill can be taught, but as importantly, so that palliative care can be endorsed and supported throughout the faith community. In a series of Archdiocesan-wide publications and media events as well as in a talk by the Archbishop of Boston to the Presbyteral Council and clergy convocation, Church leaders throughout the Archdiocese and the laity were told about the establishment of and the reasons for the *Initiative for Palliative Care and Advance Care Planning*.[3] Following its official public launch, and in follow up to communications and outreach activities that had preceded the official announcements, *Initiative* staff began to rollout the public campaign.

[3] Communications about the Initiative were published in the weekly Archdiocesan newspaper, a print magazine, which has since become digital, and broadcast through the Archdiocese of Boston television station,

In conjunction with colleagues from Catholic health, the first educational presentations were made on the campus of Mary Immaculate Health Center in Lawrence MA, a multi-level and multi-facility long term care organization. Organizers invited its residential population but also their families and their neighbors in this densely urban community. Soon thereafter, contact was made with as many parishes as possible in the same geographic region where this Catholic health facility was located. Coincidentally, this region contains the largest parish in the Archdiocese, and once invitations to present workshops on palliative care came from that parish, others soon followed.

The curriculum for the presentation was fairly simple, and the same information and materials were shared at each event. Definitions of palliative care that came from a variety of sources were carefully selected so that they both educated the audiences and corrected some of the more common misapprehensions that showed up in the public opinions about palliative care. These misconceptions included: that palliative care was a signal that you were dying; that palliative care meant you could no longer seek curative treatment; that palliative care was a sign that the doctor was giving up and sending you to hospice; that palliative care meant there was no hope; and that palliative care was prescribed when there was nothing more to do.

Rather, attendees are told that the Center to Advance Palliative Care defines palliative care this way: "Palliative care, also known as palliative medicine, is specialized medical care for people living with serious illness. It focuses on providing relief from symptoms and stress of a serious illness—whatever the diagnosis. The goal is to improve quality of life for both the patient and the family" (CAPC 2015). Similarly, the Mayo Clinic says, "Palliative care offers pain and symptom management and emotional and spiritual support when you face a chronic, debilitating or life-threatening illness" (Mayo Clinic 2015). Attendees are told that palliative care is a response to serious illness that has been found to be helpful around the world, and that the World Health Organization defines it in the same way we do: "Palliative care is an approach that improves the quality of life of patients and their families facing the problem associated with life-threatening illness, through the prevention and relief of suffering by means of early identification and impeccable assessment of pain and other problems, physical, psychosocial and spiritual" (WHO 2015).

It is explained to participants that definitions of what palliative care is are remarkably similar and consistent from one of these organizations to another, including in what they do not mention at all in their definitions of palliative care: that there is no mention of death, of dying, of hospice, or of 'no hope.'

Invariably, the people who attend the sessions—which are always held on Church premises or in the meeting rooms of Catholic health facilities—ask about which Catholic organizations deal with palliative care and how they talk about it. We offer them the description from the Supportive Care Coalition, which is similar to the definitions from other palliative care organizations that have been presented:

- Palliative care implements a holistic, interdisciplinary care plan that identifies, assesses and addresses the comprehensive needs of the seriously ill patient, including pain and other symptom management, psychosocial issues, emotional support and spiritual care.

- Palliative care begins at the moment of diagnosis with a life limiting illness; it is not restricted to end-of-life care, but it is the model used in hospice care of the dying.
- Palliative care can be delivered simultaneously with treatments aimed at curing, or delivered by itself when curing is no longer an option.
- Palliative care planning is initiated with the conversations between patients and providers, patient and family, patient and support network that results in the creation of advance care planning documents, particularly with the designation of a health care proxy for decision-making when the patient no longer has the capacity to do so (Picchi and Sullivan 2013).

The crossover and overlap among and between each of these statements are powerful in both driving the points of what palliative care is or isn't, but more importantly—especially in the context of this chapter—in reassuring Catholics who are hearing about this for the first time that there is nothing in palliative care that is inconsistent with Catholic teaching. More affirmatively, they learn that palliative care as a model of care for the seriously ill—and for those who love them, care for them and support them—is the embodiment of, and a living testimony to, Catholic teaching.

One part of the curricular presentation that is always instructive and at times surprising is when the fact is shared that all three Popes who have sat in the chair of Peter since palliative care has become a specialized kind of medical care for the seriously ill, have invoked it by name and exhorted the delivery of it to those who need it. For example, in a November 2004 address, Saint John Paul II said,

> Even when medical treatment is unable to defeat a serious pathology, all its possibilities are directed to the alleviation of suffering… The refusal of aggressive treatment is neither a rejection of the patient nor of his or her life. Indeed, the object of the decision on whether to begin or to continue a treatment has nothing to do with the value of the patient's life, but rather with whether such medical intervention is beneficial for the patient.
>
> Particularly in the stages of illness when proportionate and effective treatment is no longer possible, while it is necessary to avoid every kind of persistent or aggressive treatment, methods of "palliative care" are required. As the Encyclical *Evangelium Vitae* affirms, they must "seek to make suffering more bearable in the final stages of illness and to ensure that the patient is supported and accompanied in his or her ordeal". (n. 65)
>
> In fact, palliative care aims, especially in the case of patients with terminal diseases, at alleviating a vast gamut of symptoms of physical, psychological and mental suffering; hence, it requires the intervention of a team of specialists with medical, psychological and religious qualifications who will work together to support the patient in critical stages….
>
> *To provide this help in its different forms, it is necessary to encourage the training of specialists in palliative care at special teaching institutes where psychologists and health-care workers can also be involved.* (John Paul II 2004, nn. 2–5). (Emphasis added)

Two years later, his successor, Pope Benedict XVI spoke about palliative care during his address for the World Day of the Sick, when he urged all followers to adopt and support the practices of palliative care:

> The Church wishes to support the incurably and terminally ill by calling for just social policies which can help to eliminate the causes of many diseases and by urging improved care for the dying and those for whom no medical remedy is available. There is a need to promote policies which create conditions where human beings can bear even incurable illnesses and death in a dignified manner. Here it is necessary to stress once again the need for more palliative care centres which provide integral care, offering the sick the human assistance and spiritual accompaniment they need. This is a right belonging to every human being, one which we must all be committed to defend…Many such individuals—health care professionals, pastoral agents and volunteers—and institutions throughout the world are tirelessly serving the sick, in hospitals and palliative care units, on city streets, in housing projects and parishes. (Benedict XVI 2007)

Pope Francis has continued the unequivocal support for palliative care articulated by his two predecessors:

> Your work during these days has been to explore new fields of application for palliative care. It has until now been a precious accompaniment for cancer patients, but today there is a great variety of diseases characterized by chronic progressive deterioration, often linked to old age, which can benefit from this type of assistance …. (Francis 2015)

He continues his remarks, investing them with both a sense of urgency and mission:

> Thus I appreciate your scientific and cultural commitment to ensuring that palliative care may reach all those who need it. I encourage professionals and students to specialize in this type of assistance which is no less valuable for the fact that it "is not life-saving". Palliative care accomplishes something equally important: it values the person. I exhort all those who, in various ways, are involved in the field of palliative care, to practice this task keeping the spirit of service intact and remembering that all medical knowledge is truly science, in its noblest significance, only if used as aid in view of the good of man, a good which is never accomplished "against" the life and dignity of man. (Francis 2015)

Thus, the support of palliative care as being fully consistent with Catholic teaching on the human person and care for life is clearly evident in recent papal teaching.

A toolkit that has been developed for the parish and community presentations serves two purposes. It presents consistent information and language that is useful, accessible and accurate in a way that adult learners can assimilate, but it has also been used successfully with middle school and high school students, particularly with Confirmation classes. At the same time, the toolkit demonstrates, with the sources used and the overlapping definitions pointed out, that attendees are becoming part of a larger, national and even global, conversation.

The sustainability of this initiative and the impact hoped to come from it depends, as stated earlier in this chapter, on the focus of palliative care becoming organic to the ongoing pastoral life of the parishes and of the Archdiocese's public "face." The only way that can happen is to have the parishes emulate the central ministry organization of the Archdiocese and have dedicated time, attention and personnel for it. While this does not require additional parish staff, it does require attention from someone on the staff, perhaps a parish nurse, or a pastoral associate or a volunteer from among those who engage in home visits to the sick or pastoral visits and Eucharistic ministry to local hospitals or nursing homes.

The goal is to identify an individual or a small group of interested parties to become the Palliative Care Resource Group/Person. The parish is not going to offer clinical healthcare, and that is not the purpose of the Resource Group. Rather, these individuals offer to contact the patient or family when the pastoral staff or rectory gets a call to request prayers or to notify the staff that a member of the parish community has been diagnosed with serious illness that will impact her life going forward. These parish workers offer information and material that will help to inform them and any conversations they want to have with their health care provider about what palliative care is and its appropriateness for this situation, how to get a palliative care referral, what additional questions or need for information they may have, and so on. The Palliative Care Resource Group or individual is in place to provide information, not to interfere with the existing relationships between patients/families and their clinical caregivers that are so important.

Additionally, the parish palliative care resource people work with their parish staff to regularly organize educational sessions about issues in palliative care and Catholic teaching on those issues. These issues include pain management questions, decisions about ongoing treatment, medically administered nutrition and hydration, discussions about burdensomeness, risks and benefits of proposed care, and so on. The staff of the *Initiative for Palliative Care and Advance Care Planning* is available to work on program planning, deliver presentations and to make recommendations for other speakers, as needed.

In addition to the parish presentations, the *Initiative* sponsors and organizes two major Archdiocesan-wide events each year. Each fall, there is a Train-the-Trainer Workshop offered for those who may want to volunteer to speak at the parish educational programs or who may want to serve as part of a parish Palliative Resource group. The workshop is available to all interested parties, and they leave with copies of the PowerPoint presentation and handouts that would be used for the parish educational sessions. In the first few years of the Initiative, some attendees have come to multiple presentations of the same information, before feeling comfortable enough to volunteer in either of the capacities—speaker or resource person—for which the training is necessary.

The other major annual event is a Palliative Care Colloquium. This event is open to all, and is advertised beyond the Catholic community. It offers the opportunity for a broad audience of parish members, staff and volunteers, as well as Archdiocesan staff and invited guests, to hear from noted national speakers, some of whom are experts in their fields, others of whom have compelling personal experiences to share. All of them are people from whom we can learn because their experiences, even if not identical to our own, are educational, enlightening, and above all, inspiring.

The purpose of the Colloquium is twofold: to teach us about the technical or policy aspects of palliative care and to teach us about life and what it means to help those with serious illness to optimize the lives they're living, despite those serious illnesses. We often learn from the speakers that serious illnesses can lend a sense of urgency, that "there's not time to waste" and a sense of perspective and proportion, in the face of limited time or function or energy. We learn that patients, friends, caregivers, and providers ask, "What I want to do and how I want to be?"

### 13.3.1 A Final Observation

A fascinating by-product of the *Initiative* has been its ringing endorsement by the palliative care provider community. In the densely rich medical, and specifically palliative care environment in which the Archdiocese of Boston is located, the support of these colleagues is extremely gratifying. More importantly, and frankly, quite surprisingly, the growing number of providers—Catholic and non-Catholic—at programs offered by the *Initiative* has resulted in an ongoing conversation between them and *Initiative* staff. Their patients are Catholic, some of them or their colleagues are Catholic, and much of the ethical discussion of issues in palliative care—the ethical principles, analytical tools, and so on—have their roots in Catholic moral thought. Consequently, the dialogue that continues to thrive around the shared area of passion, commitment, hard work and challenges continues to encourage all involved in it.

## References

Benedict XVI, Pope. 2007. *Message of his holiness Benedict XVI for the fifteenth world day of the sick.* http://w2.vatican.va/content/benedict-xvi/en/messages/sick/documents/hf_ben-xvi_mes_20061208_world-day-of-the-sick-2007.html. Accessed 15 Mar 2017.

Center to Advance Palliative Care (CAPC). 2015. https://www.capc.org/about/palliative-care/. Accessed 15 Mar 2017.

Francis, Pope. 2015. *Address of his holiness Pope Francis to the plenary of the pontifical academy for life*. http://w2.vatican.va/content/francesco/en/speeches/2015/march/documents/papa-francesco_20150305_pontificia-accademia-vita.html. Accessed 15 Mar 2017.

John Paul II, Pope. 2004. *Address of John Paul II to the participants in the 19th international conference of the Pontifical Council for Health Pastoral Care*. http://w2.vatican.va/content/john-paul-ii/en/speeches/2004/november/documents/hf_jp-ii_spe_20041112_pc-hlthwork.html. Accessed 5 Mar 2017.

Mayo Clinic. 2015. *Palliative care overview*. http://www.mayoclinic.org/tests-procedures/palliative-care/home/ovc-20200491. Accessed 5 Mar 2015.

McInturff, B. and E. Harrington.2011. *2011 Public opinion research on palliative care: a report based on research by public opinion strategies*. Center to Advance Palliative Care. https://media.capc.org/filer_public/18/ab/18ab708c-f835-4380-921d-fbf729702e36/2011-public-opinion-research-on-palliative-care.pdf. Accessed 1 Feb 2017.

Picchi, T. and MC Sullivan. 2013. *Palliative care: a hallmark of catholic health care*. Supportive Care Coalition. http://supportivecarecoalition.org/index.php/resources/. Accessed 15 Mar 2017.

World Health Organization (WHO). 2015. *WHO definition of palliative care*. http://www.who.int/cancer/palliative/definition/en/. Accessed 15 Mar 2015.

# Chapter 14
# Catholic Moral Teaching and Tradition on Advance Care Planning

**Mark Repenshek and Leslie Schmidt**

Since the mid-1970s a series of court cases—Karen Ann Quinlin (1976), Brother Fox (1979), Claire Conroy (1985), Nancy Cruzan (1990), and Terri Schiavo (2005)—have attempted to address the complex question of making health care decisions for patients who are unable to make these decisions for themselves. In each case, prior to disease progression or injury, these individuals had the ability to articulate health care preferences. Notwithstanding specific processes for doing so, a general consensus has developed relative to decision making for patients unable to make their own health care decisions. First, priority should be given to the patient's wishes when stated in advance in some type of written document. Second, if such a document does not exist, where wishes of the patient are well established and well represented by a surrogate decision maker, these wishes are ordinarily to be followed. Thirdly, general preferences or direction offered by a previously decisional patient is to guide decision-making on the patient's behalf. Despite this general consensus, it is still unfortunately the case that non-decisional patients often have no written or verbal instructions about their health care preferences. In such situations, family and the health care team are left to make treatment decisions in the best interests of the patient. These situations are often incredibly complex and lack the critical perspective of the patient.

The Catholic theological and magisterial tradition is long and rich in guiding persons in making health care decisions. This tradition holds that life is a sacred gift from God, but persons' obligations to preserve that life in all circumstances is not absolute. The tradition relies on the principles of proportionate (ordinary) and disproportionate (extraordinary) means to guide health care decision making in the face of illness. Persons may reject life-prolonging procedures that do not meet the

M. Repenshek (✉)
Hospital Sisters Health System, Springfield, IL, USA
e-mail: Mark.Repenshek@hshs.org

L. Schmidt
Ascension, St. Louis, MO, USA



P. J. Cataldo, D. O'Brien (eds.), *Palliative Care and Catholic Health Care*, Philosophy and Medicine 130, https://doi.org/10.1007/978-3-030-05005-4_14

criteria of proportionate (ordinary) means. What constitutes proportionate or disproportionate means is a question dealt with elsewhere in this book. The question of who makes these decisions on behalf of a non-decisional patient in light of this rich tradition is the focus of this chapter. We will address this question first, through a brief background on the evolution of advance directives and advance care planning in order to set the stage for later exploration of how such directives are utilized in health care. Second, we will explore Catholic teaching on advance directives, specifically some of the neuralgic ethical debates that have continued since the creation of such documents. Finally, through an exploration of the ethical and operational challenges facing contemporary discussions on and utilization of advance directives, we explore the role palliative care can have in optimizing access to and increased utilization of advance care planning processes and documentation.

## 14.1 Historical Background

Today most hospitals, nursing homes, home health agencies, and insurers routinely provide information on advance directives at the time of admission. This current reality can be credited to the federal law called the Patient Self-Determination Act (PSDA) of 1990. The Act amends Medicare and Medicaid rules to:

> (1) inform patients of their rights under State law to make decisions concerning their medical care; (2) periodically inquire as to whether a patient executed an advanced directive and document the patient's wishes regarding their medical care; (3) not discriminate against persons who have executed an advance directive; (4) ensure that legally valid advance directives and documented medical care wishes are implemented to the extent permitted by State law; and (5) provide educational programs for staff, patients, and the community on ethical issues concerning patient self-determination and advance directives. (H.R. 4449 1990)

Under Section 2 of the Act, where it specifically prohibits the denial of "the initial provision of care or otherwise discriminate against an individual based on whether or not the individual has executed an advanced directive,", the Act effectively makes it against the law to require the document (H.R. 4449 1990).

In an effort to build upon the PSDA, the Centers for Medicare and Medicaid (CMS) sought to "better enable seniors and other Medicare beneficiaries" to have control over decisions regarding care before an illness progresses. On October 30, 2015, CMS issued the 2016 final Medicare payment rules for physicians, hospitals and other providers. The rules, among many other provisions, expanded reimbursement for the following Current Procedural Terminology (CPT) codes specific to advance care planning (ACP) services:

- 99497: ACP including the explanation and discussion of advance directives such as standard forms by the physician or other qualified health care professional; first 30 min, face-to-face with the patient, family member(s), and/or surrogate.

- 99498: ACP including the explanation and discussion of advance directives such as standard forms by the physician or other qualified health care professional; each additional 30 min (DHHS, CMS 2016).

These additional CPT codes provide specific reimbursement for discussions on advance directives that heretofore did not exist. Although reimbursement has been identified as a barrier to increased utilization of ACP services, the true impact has yet to be determined (DHHS 2008).

## 14.2 Advance Care Planning and Advance Directives

It is important to distinguish advance care planning from an advance directive. Advance care planning (ACP) refers to any planning by persons regarding future health care decisions prior to one's inability to do so. It is an iterative process of making decisions about future health care for patients, often in consultation with clinicians, family members and others important to the patient (Scott et al. 2013). Typically ACP is characterized by completing a document, regardless of the type, but ideally ACP involves an opportunity for persons to explore their values, beliefs, and attitudes regarding medical interventions at the end of life. An optimal use of this process captures more than a snapshot in time and can be profoundly beneficial for persons should there be a loss of one's health care decision making capacity. If a patient loses decision-making capacity, those involved in a patient's ACP are more likely to have a basis for carrying out medical decisions that are consistent with the patient's values and preferences (Hickman et al. 2005).

Advance care planning often culminates in a written document reflecting the person's health care wishes called an advance directive. These written documents typically take the form of either a proxy directive or an instructional directive. Proxy directives, also commonly called a power of attorney for health care, provide the opportunity for the person completing the document to assign an individual to serve as health care agent or proxy to carryout health care decision making should illness or injury result in an inability to make health care decisions for oneself. These documents may also allow the individual an opportunity to add specific direction on health care treatment interventions consistent with one's values and preferences.

Instructional directives, often referred to as living wills, typically identify situations wherein specific treatment interventions would or would not be desired. Instructional directives apply only to the specific clinical scenarios described in the document. In other words, without naming an individual to serve as a proxy for the patient, instructional directives do not identify anyone to whom the health care team can turn to guide decision making not necessarily addressed within the directive itself.

## 14.3 Catholic Teaching on Advance Directives

Although there is no official magisterial teaching on advance directives, there is official magisterial teaching regarding mechanisms for making appropriate decisions for incapacitated patients. This teaching is premised on the understanding that "included in the concept of self-determination is a concern for respecting human dignity" (Hamel 1988, p. 38). Attention to this principle extends to the individual who is incapacitated. In such instances, responsibility for the patient's self-determination consistent with her inherent dignity transfers to the surrogate, preferably through a proxy directive. The Catholic magisterial tradition speaks to the principle of self-determination in the context of clinical decision making as early as 1957 in Pope Pius XII's address to the International Congress of Anesthesiologists: "The rights and duties of the doctor are correlative to those of the patient. The doctor, in fact, has no separate or independent right where the patient is concerned. In general, he can take action only if the patient explicitly or implicitly, directly or indirectly, gives him permission" (Pius XII 1958, p. 397). For the patient who is incapacitated, Pope Pius XII affirms a process whereby the family speaks on the patient's behalf in order to represent the patient's previous stated or known preferences to the health care team. In the same address he notes, "The rights and duties of the family depend in general upon the presumed will of the unconscious patient if he is of age and '*sui juris*'" (Pius XII 1958, p. 397). Advance directives, more specifically proxy directives, provide the structure through which the "presumed will" of the patient can be known and communicated. So, although proxy directives were certainly not envisioned by Pope Pius XII in 1957, the tradition implicitly supports the structure.

A more contemporary representation of the Church's moral teaching on advance directives can be found in the *Ethical and Religious Directives for Catholic Health Care Services,* 6th *Edition* (ERD). The United States Conference of Catholic Bishops (USCCB) provides teaching on the topic through at least two directives:

> In compliance with federal law, a Catholic health care institution will make available to patients information about their rights, under the laws of their state, to make an advance directive for their medical treatment. The institution, however, will not honor an advance directive that is contrary to Catholic teaching. If the advance directive conflicts with Catholic teaching, an explanation should be provided as to why the directive cannot be honored. (USCCB 2018, n. 24)

> Each person may identify in advance a representative to make health care decisions as his or her surrogate in the event that the person loses the capacity to make health care decisions. Decisions by the designated surrogate should be faithful to Catholic moral principles and to the person's intentions and values, or if the person's intentions are unknown, to the person's best interests. In the event that an advance directive is not executed, those who are in a position to know best the patient's wishes—usually family members and loved ones—should participate in the treatment decisions for the person who has lost the capacity to make health care decisions. (USCCB 2018, n. 25).

Evident in these two excerpts from the ERDs one notes both the advocacy for proxy directives and a caution regarding their use. A more in-depth examination of the ethical cautions identified by the USCCB, as well as more recent concerns with medical decision making in advance of terminal illness or disease will be the focus of the next section of this chapter. Despite these cautions and concerns, an argument will be offered that proxy directives remain an instrument that can be consistent with the Catholic moral tradition.

## 14.4 Ethical Issues

It seems that the contemporary ethical issues associated with advance directives within Catholic moral teaching pertain to two areas: informed decision making on medical interventions and the idea of prospective decision making on medical interventions when the person is not yet faced with the circumstances of incapacity. The two are related insofar as both rest on the idea of providing the patient with all the information necessary to allow one to make a prudent treatment choice, but distinct in that the latter is concerned with application of that information in a variety of unforeseen circumstances.

### *14.4.1 Informed Decision Making on Medical Interventions*

Determination of what constitutes an appropriate medical intervention for a patient is, by its nature, a moral decision. It is a decision that is to be based on informed consent (President's Commission 1982). This fundamental principle in bioethics requires that there is adequate disclosure of all information necessary to allow a reasonable person to make a prudent treatment decision, and that the choice be voluntary. The requisite information would include "the essential nature of the proposed treatment and its benefits; its risks, side-effects, consequences, and cost; and any reasonable and morally legitimate alternatives, including no treatment at all" (USCCB 2018, n. 27). In instances where the patient no longer has the ability to make a health care decision for herself, the surrogate has a right to this same standard (USCCB 2018, ns. 26 and 28; Smith 2001).

Catholic moral teaching holds that life is a precious gift from God that must be cared for (John Paul II 1995, n. 2). This same teaching holds that recognition of this gift and the duty to care for it does not obligate persons to use every and all means available to sustain our lives (Repenshek and Slosar 2004). *Evangelium Vitae* notes, "it is precisely [our] supernatural calling which highlights the *relative character* of each individual's earthly life. After all, life on earth is not an 'ultimate' but a 'penultimate' reality" (John Paul II 1995, Introduction, Sect. 2).

Building on these same fundamental commitments, Catholic moral teaching centuries ago first began developing the principles of ordinary and extraordinary means, or proportionate-disproportionate means (Cronin 2011). Although the distinction itself is dealt with elsewhere in this volume, it is important to note that this principle's substance is reflected in the *Catechism of the Catholic Church* where it affirms that "discontinuing medical procedures that are burdensome, dangerous, *extraordinary, or disproportionate* to the expected outcome can be legitimate; it is the refusal of 'over-zealous' treatment. Here one does not will to cause death; one's inability to impede it is merely accepted" (CCC 1994, n. 2278). The determination of whether a medical intervention is "extraordinary" or "disproportionate" is a *prudential judgment* of the patient (USCCB 2018, ns. 56 and 57). The same is true for decisions made by a surrogate on the patient's behalf. Such prudential judgments, writes Ron Hamel, "must consider how the treatment will affect the patient not only medically, but also emotionally, psychologically, and spiritually" (Hamel 1988, p. 39).

Given the importance of the proportionate-disproportionate distinction relative to a particular patient, knowledge of the patient's life, beliefs, values, medical history and their evaluation, if any, of the hoped for benefits and risks of treatments, will be especially informative to those who must make prudential judgments on behalf of the incapacitated patient. Catholic moral teaching provides a robust account of how one goes about understanding the multidimensional aspects of decision-making on medical interventions. In its *Declaration on Euthanasia*, the Congregation for the Doctrine of the Faith states: "it will be possible to make a correct judgment as to the means by studying the type of treatment to be used, its degree of complexity or risk, its cost and the possibilities of using it, and comparing these elements with the result that can be expected, taking into account the state of the sick person and his or her physical and moral resources" (CDF 1980). It seems that, at least insofar as informed decision making on medical interventions is concerned, the final decision on implementation of the proportionate-disproportionate distinction for the incapacitated patient appropriately rests within the "consciences of the patient or those who act in the patient's behalf" (Myers 1991, p. 279).

### 14.4.2 Prospective Decision Making on Medical Interventions

In an article on "Advance Directives" in *Ethics & Medics* from 1991, Russell Smith writes that most Catholic theologians and bishops agree that there are significant problems with a Living Will. He goes on to say that "the most glaring problem is that it does not allow for adequate informed consent, because one must make a decision in the present moment about a future medical condition which cannot be known in advance…One is not deciding about an actual case" (Smith 1991, p. 3). The timing of this critique was significant in that it came on the heels of the passing of the Patient Self Determination Act of 1990, which, as noted earlier, requires hospitals (among other health care entities) to inquire of and help patients execute an advance directive (H.R. 4449 1990). It would seem that if this critique were to take a foothold in the Catholic moral tradition on prospective decision making on medical interventions, advance directives of all types would present equally glaring problems. Given that the ERDs clearly support the drafting of advance directives and appointing a representative to speak on one's behalf, a more thorough examination of this critique is critical to understanding the contemporary reemergence of this same critique (USCCB 2018, ns.24 and 25).

Although Smith's critique is of a specific form of an advance directive, his assertion about decision making on medical interventions in the future is striking in that later in that same article, he advocates for the Durable Power of Attorney (DPA), noting it is "*essentially* different from the type of 'Living Will' discussed above, because DPA establishes a *personal* proxy, not an *instrumental* one. DPA, therefore, allows for decisions based on an *actual* situation, not a *hypothetical one* [emphasis in the original]" (Smith 1991, p. 3). Yet, Smith rightly notes years later that a DPA, in designating an agent to speak on behalf of the patient were the patient to become incapacitated at some point requires a "long talk with the agent about his or her [the patient] values, expectations and desires about treatment during sickness" (Smith 2001, p. 18/2). Such conversation would seem to require a discussion on *hypothetical* situations that are not in the present moment in order to direct the agent about future medical care which cannot be known in advance. It seems therefore, in a contemporary context, the critique is both misplaced with regard to Catholic moral teaching and is internally self-contradictory. What is striking, however, is that this critique continues to the present day despite the fact that the same inconsistencies with Catholic moral teaching and internal contradictions persist (Ganon 2014; Picarello 2015). We will focus our discussion on the misplaced critique within Catholic moral teaching.

## 14.5 Prospective Decision Making on Medical Interventions: Catholic Moral Teaching

Peter Cataldo and Elliott Bedford responding to what they describe as use of the "reductionist in-the-moment" moral criterion for decision making, note that the Catholic moral tradition "strongly affirms" persons' ability to make decisions concerning medical interventions despite the inability to know the exact circumstances of their future use (Cataldo and Bedford 2015). They address the reductionist critique within the context of the Catholic moral tradition by fleshing out the tradition's understanding of circumstances and moral certitude. Turning to longstanding principles within the Catholic moral tradition—principle of double effect, principle of cooperation in evil, and the moral status of foreseen and unforeseen future consequences—the authors note the moral legitimacy of determining one's ethical obligations by what can be seen in the present moment *and* by what can be reasonably foreseen or expected in the future (p. 59). The principle of double effect serves as an excellent example of the use of reasonably foreseen circumstances as a relevant principle for decision making consistent with the Catholic moral tradition.

The principle of double effect (PDE) is used in the Catholic moral tradition to resolve dilemmas concerning human acts that bring about two effects: one good and permissible and the other evil and prohibited. It is generally accepted that there are four conditions, the fulfillment of which govern legitimate use of the PDE. Although variation exists concerning the precise content of these four conditions, Joseph Mangan constructed the modern formulation of the principle over 60 years ago (Mangan 1949). He wrote:

> A person may licitly perform an action that he foresees will produce a good and a bad effect provided that four conditions are verified at one and the same time: 1) that the action in itself from its very object be good or at least indifferent; 2) that the good effect and not the evil effect be intended; 3) that the good effect be not produced by means of the evil effect; 4) that there be a proportionately grave reason for permitting the evil effect. (Mangan 1949, p. 43)

The PDE turns, then, upon the distinction between intended and merely foreseen effects. Proper application of the PDE uses this distinction in relation to the other conditions to render permissible those acts that are good or indifferent in themselves, where an evil effect is foreseen, but where the agent does not intend this evil effect (Bole 1999; Norcross 1999; Keenan 1993).[1]

Analyses of human actions that require use of the PDE rest on the idea of reasonably foreseen circumstances as a necessary component for decision making to achieve a congruency among the fundamental components of the moral act. These analyses require incorporating reasonably foreseen circumstances to inform decision makers precisely at the moment when they are attempting to discern their ethical obligations on the basis of what can be reasonably foreseen beyond the present circumstances (Cataldo and Bedford 2015, p. 54). This construct is at the heart of decision making in the context of future medical interventions where the patient, or by extension the surrogate, is attempting to consider the foreseeable outcomes of the considered medical intervention. Advance directives, then, serve as a tool that attempts to capture the individual's medical decision making in the present moment while considering reasonably foreseen circumstances related to the progression of disease. The consideration of the latter when drafting advance directives does not render such documents de facto inconsistent with the Catholic moral tradition.

Clinical decision making must also make use of prognostication in order to properly inform a patient as to the likely outcome of an intervention as well as the likely outcome of no intervention. Certainly such decision making takes into consideration circumstances in the present moment, but must also make use of what the clinician can "reasonably portend for the future circumstances of the patient" (Cataldo and Bedford 2015, p. 56) By its very nature, prognostication is not a perfect science, but the lack of exactness does not render such information invalid. Quite the contrary, prognostication is an essential element of clinical decision making that is very much a part of the Catholic moral tradition regarding decisions

[1] The distinction between foreseen and intended effects is not without controversy. For an example of those who argue in favor of the distinction, see Bole (1999). For those who argue against the distinction in terms of its problematic conclusions, see Norcross (1999). Then there is Keenan (1993), who articulates a non-metaphysical account of the PDE and who argues that it functions neither as a principle that justifies certain human acts nor as a principle that grants exceptions to moral norms; rather the principle "has a heuristic and confirming function" which serves to confirm a case's congruency with paradigm cases in the Catholic moral tradition.

concerning medical interventions. The USCCB's Office of the General Counsel opens a letter dated September 4, 2015 to the Centers of Medicare and Medicaid on the topic of the Proposed Rule to amend various aspects of the Medicare program, specifically advance directives, noting:

> The Catholic Church has no objection to encouraging patients to consider treatment decisions that may have to be made in the future, in light of their personal values and medical condition, in case they become unable to communicate their wishes. On the contrary, the Church has a long and rich tradition on the parameters for such decision making, providing concepts and distinctions that have long played an important role in secular medical ethics as well. (Picarello 2015)

Cataldo and Bedford reference a number of instances within the Catholic moral tradition that recognize the moral relevance of information beyond that found in the present moment concerning decisions about medical interventions in their article's section on *Informed Decision Making on Medical Interventions.* In each of these excerpts, Cataldo and Bedford note that "present-moment circumstances represent a basis on which to project future circumstances, which in turn becomes a basis for evaluating one's moral obligations about treatment in the future" (Cataldo and Bedford 2015, p. 57) It is in this way that Cataldo and Bedford properly characterize the Catholic moral tradition as "strongly affirming" the ability of persons making moral judgments about future medical care despite limitations internal to the process.

It is at the intersection of informed decision making on medical interventions, the patient's present and reasonably foreseen clinical circumstances and the patient's preferences to these ends that advance directives encounter the clinical reality. When the process of completing an advance directive is imbued with thoughtful consideration of the person's intentions and values, the tool can greatly assist the otherwise vulnerable incapacitated patient and her family. Yet, the complexity of the current health system can add significant challenges to the effective utilization of even a well-crafted document. The remaining sections of this chapter will speak to this complex clinical reality as well as attempt to respond to the challenges posed through the lens of Palliative Care.

## 14.6 Advance Directives in the Clinical Setting

When a patient encounters the clinic or hospital setting she is routinely prompted for an advance directive. If one has been completed, the document is stored in the patient's medical record in order to be easily retrieved and routinely reviewed by members of the health care team where appropriate. Retrieving this document is critically important for the patient who may lose decision making capacity and therefore autonomous self-determination related to health care decision making. Yet, to utilize advance directives only at this point in the patient's illness progression or injury state merely reacts to the clinical condition and the patient's lack of decision making authority, as opposed to a process that appreciates treatment options and their congruency with the patient's values and preferences. In the most extreme of cases, such a reactionary approach may view the document as a mere tool to secure discharge disposition or "authorization" to proceed with treatment intervention.

A proper assessment of the patient's decision making capacity is a requisite step in the process of advance directive utilization. Where a patient is evaluated as non-decisional or incapacitated—that is, lacking the ability to understand, evaluate and communicate a health care decision—an advance directive then serves as a tool to guide surrogate decision making (Palmer and Harmell 2016). Once an advance directive document has been reviewed and the surrogate decision maker has been identified through a proxy directive, for example, the surrogate decision maker exercises their authority utilizing substituted judgment. This standard holds that the identified surrogate carries out decisions that reflect the values and preferences of the patient to the greatest extent possible. That said the process of doing so can be one that presents significant challenges for the family, proxy and health care team in the best of circumstances. Unfortunately, this challenge is one among many in today's complex health care setting.

## 14.7 Practical Challenges with Advance Directives

In 2014 the Institute of Medicine (IOM) released its consensus report titled, *Dying in America* (IOM 2014a). The report highlighted key findings in several areas including the delivery of person-centered care, professional education and development, policies and payment systems, public education and engagement and, most relevant to this chapter, clinician-patient communication and advance care planning. The findings specific to the latter include:

- Most people nearing the end of life are not physically, mentally, or cognitively able to make their own decisions about care. The majority of these patients will receive acute hospital care form physicians who do not know them. Therefore, advance care planning is essential to ensure that patients receive care reflecting their values, goals and preferences.
- Of people who indicate end-of-life care preferences, most choose care focused on alleviating pain and suffering. However, because the default mode of hospital treatment is acute care, advance planning and medical orders are need to ensure that these preferences are honored.
- Frequent clinician-patient conversations about end-of-life care values, goals and preferences are necessary to avoid unwanted treatment. However, most people, particularly younger, poorer, minority and less educated individuals do not have these conversations. Clinicians need to initiate conversations about end-of-life care choices and work to ensure that patient and family decision making is based on adequate information and understanding.

The Institute of Medicine followed these findings with recommendations for both professional societies and quality improvement organizations to establish standards for clinician-patient communication and advance care planning that is "measureable, actionable, and evidence based." The IOM emphasized that these standards will need to evolve over time and be responsive to "emerging evidence, methods and technologies." It not only prodded health care institutions and providers, but payers as well in an effort to tie such standards to reimbursement, licensing and credentialing to encourage:

- All individuals have the opportunity to participate actively in their health care decision making throughout their lives and as they approach death, and receive medical and relevant social services consistent with their values, goals, and informed preferences;
- Clinicians to initiate high quality conversation, integrate the results of these conversations into the ongoing care plans of patients, and communicate with other clinicians as requested by the patient; and
- Clinicians to continue to revisit advance care planning discussions with their patients because individuals' preferences and circumstances may change over time (IOM 2014b).

Despite these recommendations, time and training continue to impact the ability of providers to engage in conversations about advance directives and advance care planning (Lund et al. 2015). In part, the CMS expanded billing codes for advance care planning attempts to tie reimbursement to the IOM recommendations. Yet, the challenges identified in the IOM report persist. Particularly relevant to our final section, is the fact that the IOM report did not define a specific clinical context for the application of these recommendations. It is our contention that the continued evolution of palliative care provides an excellent clinical context within which these recommendations can be brought to fruition.

## 14.8 Palliative Care and Advance Care Planning

Contemporary palliative care models are beginning to incorporate services into the outpatient and community setting. This shift is significant in that more traditional models maintain palliative care within the hospital or inpatient setting. This contemporary shift creates an ideal opportunity to couple palliative care's move to the outpatient setting with the IOM recommendations on improving clinician-patient communication and advance care planning. Our preferred definition of palliative care is illustrative of this ideal opportunity: "Palliative care provides relief from pain and other symptoms, supports quality of life, and is focused on patients with serious advanced illness and their families. *Palliative care may begin early in the course of treatment for a serious illness and may be delivered in a number of ways across the continuum of health care settings, including in the home, nursing homes, long-term acute care facilities, acute care hospitals, and outpatient clinics* [emphasis added]" (IOM 2014a, p.2). When palliative care is able to operationalize early intervention, the opportune moment exists for integration of a substantive advance care planning process. In this way, discussions regarding patient's values, treatment preference and goals of care begin early in the course of illness, allowing patients, providers and loved ones the time necessary to adjust to the diagnosis and discuss its implications. Put more simply, the two become integrally related.

As the patient continues to move through the palliative care model, an integrated advance care planning program optimally leverages the clinical characteristics of the patient's disease trajectory. These clinical characteristics include "new diagnoses of life-limiting conditions; severe, irreversible deterioration in the patient's health status; loss of response to, or complications from, disease-specific treatments; unrealistic expectations or requests for care by the patient or their family; or an expressed desire of the patient or their family to discuss ACP" (Scott et al. 2013). Discussions among the health care team, the patient and family serve both the needs of palliative care and advance care planning. Ultimately, any advance directive documentation that may result from this integrated approach achieves the recommendations of the IOM report and, from the perspective of Catholic moral teaching, importantly focuses attention on the inherent dignity of the patient in health care decision making.

## 14.9 Conclusion

Advance care planning has clearly evolved since the Patient Self-Determination Act of 1990. In the midst of this evolution the Catholic moral tradition's teaching on advance directives and advance care planning remains instructive. In optimizing support for a process whereby persons can identify in advance a representative to carry out health care decision making in circumstances where they lose the capacity to do so, fidelity to Catholic teaching on care at the end-of-life is especially important for the Catholic health care setting. Yet, these teachings do not often provide readymade answers on specific medical interventions. Rather, it is a tradition that offers guidance on what constitutes proportionate or disproportionate means to aid decision making. It is a tradition that recognizes that the inherent dignity of the individual is not irreconcilable with refusing the very medical interventions that may preserve that life, but with grave burden. Nonetheless, this too at times is subject to misinterpretation.

In this chapter we highlighted these debates. Even in the midst of these debates, the richness of the Catholic moral tradition on end-of-life care and prospective decision making offers a vision for advance care planning that coincides well with the evolving model of palliative care. It is a model of care that emphasizes pain and symptom relief, supports quality of life, and attempts to do so as early as possible in the course of treatment for serious illness. By integrating advance care planning into this model the patient is offered an opportune moment to discuss values, treatment preferences and goals of care in collaboration with caregivers who offer expertise in precisely this arena. Through this integration the richness of the Catholic moral teaching and tradition on advance care planning offers an invaluable contribution for the patient, family and the entire the health care team facing complex health care decision making at the end of life.

## References


Bole, T.J., III. 1999. The theoretical tenability of the doctrine of double effect. *Journal of Medicine and Philosophy* 16: 467–473.

Cataldo, Peter, and Elliott Bedford. 2015. Prospective medical-moral decision making. *National Catholic Bioethics Quarterly* 15: 53–61.

CCC. Catechism of the Catholic Church. 1994. Washington, DC: United States Catholic Conference. No. 2278.

CDF. Congregation for the Doctrine of the Faith. 1980. Declaration on euthanasia. http://www.vatican.va/roman_curia/congregations/cfaith/documents/rc_con_cfaith_doc_19800505_euthanasia_en.html. Accessed 30 Mar 2017.

Cronin, Daniel Anthony. 2011. *Ordinary and extraordinary means of conserving life*. Philadelphia: National Catholic Bioethics Center.

DHHS. U.S. Department of Health and Human Services. 2008, August. *Advance directives and advance care planning: Report to congress*. https://aspe.hhs.gov/basic-report/advance-directives-and-advance-care-planning-report-congress. Accessed 7 Dec 2016.

DHHS. U.S. Department of Health and Human Services, Centers for Medicare & Medicaid (CMS). 2016. *Advance care planning*. https://www.cms.gov/Outreach-and-Education/Medicare-Learning-Network-MLN/MLNProducts/Downloads/AdvanceCarePlanning.pdf. Access 1 Apr 2017.

Ganon, Daniel. 2014. Favor dnr/dni orders over polst: A compromise without compromise. *Ethics & Medics* 39: 1 3.

H.R. 4449. 1990. *Patient self determination act of 1990*. https://www.congress.gov/bill/101st-congress/house-bill/4449. Accessed 30 Mar 2017.

Hamel, Ron. 1988. Advance directives compatible with Catholic moral principles. *Health Progress* 69: 36–40 88.

Hickman, S.E., B.J. Hammes, A.H. Moss, and S.W. Tolle. 2005. Hope for the future: Achieving the original intent of advance directives. *Hastings Center Report* Spec. No 35: S26–S30.

Institute of Medicine of the National Academies (IOM). 2014a. *Dying in America: Improving quality and honoring individual preferences near the end of life*. https://www.nationalacademies.org/hmd/~/media/Files/Report%20Files/2014/EOL/Report%20Brief.pdf. Accessed 6 Feb 2017.

———. 2014b. Recommendations and summary. In *Dying in America: Improving quality and honoring individual preferences near the end of life*. https://www.nap.edu/read/18748/chapter/2#12. Accessed 6 Feb 2017.

John Paul II, Pope. 1995. *Evangelium vitae*. *Origins* 24: no. 2.

Keenan, James F. 1993. The function of the principle of double effect. *Theological Studies* 54: 294–315.

Lund, Susi, Alison Richardson, and Carl May. 2015. Barriers to advance care planning at the end of life: An explanatory systematic review of implementation studies. *PLoS One* 10 (2): e0116629. https://doi.org/10.1371/journal.pone.0116629.

Mangan, Joseph T. 1949. An historical analysis of the principle of double effect. *Theological Studies* 10: 41–61 https://philpapers.org/archive/MANAHA-2.pdf. Accessed 1 Apr 2017.

Myers, Bishop John. 1991. Advance directives and the catholic health facility. *Origins* 21: 276–280.

Norcross, A. 1999. Intending and foreseeing death: Potholes on the road to hell. *Southwest Philosophy Review* 15 (1): 115–123.
Palmer, B.W., and A.L. Harmell. 2016. Assessment of health care decision-making capacity. *Archives of Clinical Neuropsychology* 31 (6): 530–540.
Picarello, Anthony R. 2015. *USCCB comments on Medicare's proposed rule on advance care planning*. Letter dated September 4, 2015. http://www.usccb.org/about/general-counsel/rule-making/upload/Comments-Advance-Directives-Medicare-9-15.pdf. Accessed 1 Apr 2017.
Pius XII, Pope. 1958. Address to the international congress of anesthesiologists: The prolongation of life. *The Pope Speaks* 4: 393–398.
President's Commission for the Study of Ethical Problems in Medicine and Biomedical and Behavioral Research. 1982. Making health care decisions. In *The ethical and legal implications of informed consent in the patient-physician relationship*. Washington, DC: US Government Printing Office.
Repenshek, M., and J.P. Slosar. 2004. Medically assisted nutrition and hydration: A contribution to the dialogue. *Hastings Center Report* 34: 13–16.
Scott, Ian, Geoffrey K. Mitchell, Elizabeth J. Reymond, and Michael P. Daly. 2013. Difficult but necessary conversations—the case for advance care planning. *The Medical Journal of Australia* 199 (10): 662–666 https://www.mja.com.au/journal/2013/199/10/difficult-necessary-conversations-case-advance-care-planning. Accessed 6 Feb 2017.
Smith, Russel E. 1991. Murder, she typed. *Ethics & Medics* 16: 1–4.
Smith, Russell E. 2001. Advance directives for health care decisions. In *Catholic health care ethics: A manual for practitioners*, ed. Edward J. Furton and Peter J. Cataldo, 2nd ed., 18/1–18/3. Philadelphia: National Catholic Bioethics Center.
USCCB. United States Conference of Catholic Bishops. 2018. *Ethical and religious directives for Catholic health care services*. 6th ed. Washington, DC: USCCB.

# Chapter 15
# Medical Mercy and Its Counterfeit

**Elliott Louis Bedford**

*Death be not proud, though some have called thee*
*Mighty and dreadfull, for, thou art not soe,*
*For, those, whom thou think'st, thou dost overthrow,*
*Die not, poore death, nor yet canst thou kill mee.*
*From rest and sleepe, which but thy pictures bee,*
*Much pleasure, then from thee, much more must flow,*
*And soonest our best men with thee doe goe,*
*Rest of their bones, and soules deliverie.*
*Thou art slave to Fate, Chance, kings, and desperate men,*
*And dost with poyson, warre, and sicknesse dwell,*
*And poppie, or charmes can make us sleepe as well,*
*And better then thy stroake; why swell'st thou then?*
*One short sleepe past, wee wake eternally,*
*And death shall be no more, death thou shalt die.* (Donne 1609)
John Donne

## 15.1 Introduction

In "Death be not proud," John Donne addresses death in a manner that echoes how Moses spoke to the Lord: "face to face" (Ex 33:11). Death, as Donne assumes, is a relatable reality. It is not hidden or unknown to human beings. We All exist in some relationship to it. But as his tone indicates, Death is not Donne's friend; Death is being dressed down and deflated. As Donne's disposition illustrates, the reality of death can and must be acknowledged soberly and confronted honestly. The deep, resonating power of the poem comes from Donne's strikingly confident tone and

E. L. Bedford (✉)
Ascension, St. Louis, MO, USA
e-mail: Elliott.Bedford@ascension.org



P. J. Cataldo, D. O'Brien (eds.), *Palliative Care and Catholic Health Care*, Philosophy and Medicine 130, https://doi.org/10.1007/978-3-030-05005-4_15

clear hopeful message: the power of death over human life is relative, not absolute, for death itself shall be conquered. Death is meaningful, and yet only a doorstep to something greater.

Donne's poem gives voice to an innate human need to find meaning in our finitude and physical death. Two contemporary manifestations of the same drive can be found in competing medical practices: palliative care and assisted suicide. Each practice, in its own distinctive way, seeks to find meaning and purpose amid human limitedness, pain, suffering and death. While they share a desire for meaning, when examined, these methods embody contradictory medical philosophies, right down to their anthropological roots. By gaining clarity about the points of fundamental divergence between them, the Catholic tradition can stake out an evangelization strategy in which medical practices that ensure human dignity, like palliative care, are promoted. Those practices, like assisted suicide, that offer falsely the pretense of mercy and absolute control over suffering must be abandoned as harmful and inadequately human.

In three parts, this chapter provides an analysis for palliative acts and acts of assisted suicide, sketching the framework for a precise moral evaluation of each. First, the respective terms and concepts are established. Second, the various concrete actions characterizing each method are analyzed according to traditional moral categories. Third, the concept of mercy—often invoked or attributed to each intervention—is examined along with its anthropological roots to illustrate that only one method authentically embodies mercy: palliative acts. As such, the Catholic tradition is correct to affirm the benefit and value of palliative medical practices, and to condemn assisted suicide as dehumanizing, false mercy that violates social justice by marginalizing those who suffer.

## 15.2 Palliative Care

In order to detail the fundamental difference between palliative care and assisted suicide, it is important to precisely define and distinguish these terms. The term palliate means: "to reduce the violence of (a disease); *also*: to ease (symptoms) without curing the underlying disease" (Merriam-Webster 2017e). Accordingly, the World Health Organization defines palliative care as:

> An *approach* that improves the quality of life of patients and their families facing the problem associated with life-threatening illness, through the prevention and relief of suffering by means of early identification and impeccable assessment and treatment of pain and other problems, physical, psychosocial and spiritual. Palliative care:
>
> - provides relief from pain and other distressing symptoms;
> - affirms life and regards dying as a normal process;
> - intends neither to hasten nor postpone death;
> - integrates the psychological and spiritual aspects of patient care;

- offers a support system to help patients live as actively as possible until death;
- offers a support system to help the family cope during the patient's illness and in their own bereavement;
- uses a team approach to address the needs of patients and their families, including bereavement counselling (sic), if indicated;
- will enhance quality of life, and may also positively influence the course of illness;
- is applicable early in the course of illness, in conjunction with other therapies that are intended to prolong life, such as chemotherapy or radiation therapy, and includes those investigations needed to better understand and manage distressing clinical complications. (World Health Organization 2017)

This expansive definition illustrates several key points. First, with regard to palliative care one can distinguish between palliative care as an "approach"—a philosophy of medicine—and a set of concrete actions that aim to manage symptoms in order to alleviate pain and suffering. Second, this approach and its consequent interventions acknowledge death as a factual reality and natural part of the life process. Also acknowledged are limitations of what medicine can offer and hope to accomplish. The focus is not on curing incurable diseases or conditions but helping patients and families live well despite illness. A distinction is drawn between prolonging death and shortening life, while asserting that palliative interventions are designed to avoid either extreme by "impeccable assessment" and symptom management. In other words, palliative care is distinctively prudent at stewarding human life. Third, the social dimension of human beings is taken as a given which is dispositive and essential to quality care. For instance, it is specified that palliative care is delivered by a *team* to patients and their families as a *unit*. This inherent team effort to ease the burdens of life-threatening illness and the dying process reinforces the social value of the patient by actively learning and addressing the patient's needs. Fourth, the palliative care approach is thoroughly holistic insofar as it expressly "integrates psychological and spiritual" aspects of patient care, including engagement and bereavement of the family. It is also noteworthy that palliative care is a broad ranging care model, encompassing a spectrum of patients of all ages with a wide variety of conditions.

To describe palliative care according to a traditional moral framework (i.e. object, intention, circumstance), focus must be given to the specific actions this care model employs. Lorenz and colleagues have, for instance, identified a sample of such best practice palliative treatments based on evidence of their effectiveness (Lorenz et al. 2008: 151). Based on this data, Qaseem and colleagues identify the following as effective and appropriate palliative care interventions for patients with serious illness at the end of life:

- Regularly assess patients for pain, dyspnea, and depression.
- Use therapies of proven effectiveness to manage pain. For patients with cancer, this includes nonsteroidal anti-inflammatory drugs, opioids, and bisphosphonates.

- Use therapies of proven effectiveness to manage dyspnea, which include opioids in patients with unrelieved dyspnea and oxygen for short-term relief of hypoxemia.
- Use therapies of proven effectiveness to manage depression. For patients with cancer, this includes tricyclic antidepressants, selective serotonin reuptake inhibitors, or psychosocial intervention.
- Ensure that advance care planning, including completion of advance directives, occurs for all patients with serious illness (Qaseem et al. 2008: 144–45).

Clearly, each intervention aims in some way to address the patient's pain and suffering, both concurrently and into the future. In light of the WHO definition, it is obvious this list of palliative care interventions is neither exhaustive nor comprehensive. Nevertheless, it illustrates that all palliative interventions share a common object of treating pain and managing symptoms—whether physical, psychological, or spiritual—for the purpose of relieving suffering and maximizing the quality of a person's functioning and experience.

## 15.3 Assisted Suicide

Suicide, by contrast, is defined as "the act or an instance of taking one's own life voluntarily and intentionally" (Merriam-Webster 2017f). Assisted suicide is "suicide committed by someone with assistance from another person" (Merriam-Webster 2017a). These definitions are somewhat straightforward and represent traditional categories on the matter. However, when considering assisted suicide in contemporary discussion, it becomes apparent that this phenomenon is now referred to under many names and euphemisms. Oftentimes, the terms are especially those associated with various political and social advocacy movements: "right to die," "aid-in-dying," or "death with dignity." For instance, the advocacy group *Death with Dignity*, attempts to separate itself from suicide:

> Suicide is the act of intentional and voluntary ending of one's life. Causes of suicide vary and are complex, though a majority of people who die by suicide have a mental illness, especially depression, at the time of their death. Sometimes described as a permanent solution to a temporary problem, suicide ends a life that would otherwise continue. (Death with Dignity 2017)

According to this reasoning, persons acting under the aegis of the Death with Dignity laws are not committing suicide:

> People using the Death with Dignity laws do not wish to be referred to as suicidal, mentally incompetent, or emotionally depressed. People with a terminal illness do not want to die but are, by definition, dying. They are facing an imminent death and want the option to avoid unbearable suffering and loss of autonomy in their final days. (Death with Dignity 2017)

As the "Death with Dignity" movement explains, this terminological change away from assisted suicide is very intentional and strategic; it is meant to portray such actions as virtuous and good. Here the patient is construed as an autonomous agent, working with health care faculty to manage and fulfill their care goals and life plan.

The assumption in this attempted redefinition is twofold: (a) suicide is associated with mental illness (i.e. diminished autonomy), not fully actualized autonomy; and (b) since death is causally inevitable, avoiding future pain and suffering by autonomously ending one's life more hastily does not constitute suicide simply because it is autonomously chosen. But this argument for terminological change amounts to a red herring and a *non sequitur*, or, as C.S. Lewis wrote, "the abuse of language" (Lewis 1943). The traditional definition of suicide as an intentional act of self-killing is an objective definition. It describes an act regardless of motivation or mental state. This strain of argument for terminological change does nothing to deny that the act in question is one of intentional self-killing; rather it presumes it is such an act. It simply attempts to distract from the issue by claiming that suicide is negatively associated with mental illness. The second prong of the redefinition, however, attempts to deny this is an act of intentional self-killing by making an argument from inevitability. Since the patient is dying due to their disease, so the argument goes, then taking action (whether by commission or omission) to deliberately end their life earlier so as to avoid future pain is believed to not qualify as intentional self-killing. But this simply does not follow: just because all humans are mortal – i.e. they will die sometime—does not mean that ending their lives prior to natural death is not killing them. According to Merriam-Webster, to kill is "to deprive of life" (Merriam-Webster 2017c). This definition is not circumscribed to only those that are not otherwise dying. It applies equally to all mortal beings. Thus, ending a person's life is, in fact, precisely the definition of killing them. Likewise, deliberately ending one's own life is, in fact, precisely the definition of "self-killing" (or *suicide*).

Yet, even with regard to the term "killing," there is debate about its precision or appropriateness. In their discussion of the principle of nonmaleficence, bioethicists Tom Beauchamps and James Childress note that when discussing the morality of interventions at the end of life, a conceptual difference is drawn between actions that are "killing" and those that are "letting die." "In ordinary language," they write, "killing is a causal action that brings about death, whereas letting die is an intentional avoidance of causal intervention so that disease, system failure, or injury causes death" (Beauchamp and Childress 2009: 172–3). For Beauchamps and Childress, however, this distinction is vague and unhelpful because it confuses or conflates moral responsibility with causality (e.g. killing is active, letting die is passive). However, because within a culturally pluralistic society there is widespread disagreement about underlying conceptions of causality, there is no general consensus understanding about what types and degrees of 'causing' a patient's death is considered morally justifiable or not.[1] The confusion that Beauchamps and Childress identify remains pertinent but not insurmountable.

Given the definition of the terms, a more helpful distinction is between killing and palliating. According to their respective definitions, these are categorically distinct and mutually exclusive phenomena. Palliating acts are specified insofar as

[1] For more on the underlying systematic concepts of causality see Bishop (2012). Bishop argues that modern medicine recognizes only two types of causes: material and efficient. This view would be contrasted, for instance, with the Catholic perspective that would also include formal and final causes.

they manage pain and symptoms. By definition, they do not eliminate the underlying source of the symptom—the disease or condition—much less the person suffering from said disease or condition. Thus, a bolus of morphine prudently titrated to alleviate a cancer patient's pain and help maintain their consciousness eliminates neither the cancer nor the patient while treating their pain. This is a palliative measure.

In contrast, administration of a bolus of morphine drawn in quantity to sufficiently suppress respirations and end the patient's life is something else entirely. This is eliminating the suffering by eliminating the person who suffers. Breathing is not a symptom of having cancer. Yet suppression of breathing to deprive the patient of life (and thereby avoid future pain) is precisely what the morphine would be administered to accomplish. This is a killing act. In other words, when the object of an act is pain and symptom management, it is a palliative *kind* of act; when its object is life depriving, it is killing. The two are mutually exclusive and categorically distinct. Consequently, the difference that Beauchamp and Childress identify between active/passive and omission/commission are not significant from a moral gravity standpoint.

Regardless of their sociologic and political efficacy, the attempts to redefine assisted suicide are philosophically untenable. The actions they promote remain, objectively, facilitated acts of intentional self-killing. The nature of such actions does not change due to the mental state of the patient or the inevitability of the patient's death from their underlying condition. Personal culpability may change because of diminished freedom, but not the objective nature (or *kind*) of act in question. Therefore, throughout this paper I will use the term assisted suicide for two primary reasons: it is historically prevalent and philosophically precise in its description of the act in question. In what follows, I will provide a comparative analysis of palliative acts as distinct from acts of intentional self-killing.

## 15.4 Analysis

This section analyzes, first, palliative acts and, second, acts of assisted suicide according to traditional moral categories: means and ends; moral object, intention, and circumstances (*Catechism of the Catholic Church* 2008, n. 1750). For present purposes, I will place palliative acts under two general categories of practices. First, palliative care (i.e. the board certified medical specialty caring for a patient at any time after having been diagnosed with a chronic, life limiting illness) is aimed at managing pain and symptoms of chronic life limiting conditions, which may be provided concurrently with curative measures. Second, hospice is an element of palliative care focused solely on maximizing patient functional and experiential quality near the end of life. In both cases, the ultimate end of palliative acts is for the patient (and their family) to live with the greatest quality of experience and function as possible in light of their limitations from disease and suffering. Pain and symptom management are the specific, proximate means employed to help achieve this objective. Thus, palliative acts and interventions are a distinct set of medical practices designed as a means for reducing suffering or symptoms without curing a patients' underlying condition.

These palliative acts can thus be described according to their circumstances, object, and intention.

- *Circumstances:* a patient experiencing chronic or life limiting illness or condition;
- *Object*: pain and symptom management;
- *Intent*: maximize the quality of the patient's remaining functioning and subjective experience.

Considered under this traditional rubric, palliative care and the interventions that fall under this paradigm cannot be said to be intrinsically defective—or evil—according to their object.

Acts of assisted suicide, defined as either formally or materially facilitating another's intentional self-killing, can also be characterized in terms of ends and means. For examples, some specific acts of assisted suicide include: Asphyxial Suicide,[2] Voluntary Stopping Eating and Drinking,[3] and ingesting prescribed medications such as secobarbital or pentobarbital (Oregon Office of Disease Prevention Epidemiology 2016). Research into the motivations that patients and their families perception of such motivations express when choosing assisted suicide illustrates that a common ultimate end of the action is avoidance of physical or existential suffering (e.g. future pain, loss of control, autonomy or independence), whether concurrent or future.[4] A recent consensus statement regarding the definition of the "wish to hasten death" (WTHD) summarizes this phenomenon with great precision:

> The WTHD is a reaction to suffering, in the context of a life-threatening condition, from which the patient can see no way out other than to accelerate his or her death. This wish may be expressed spontaneously or after being asked about it, but it must be distinguished from the acceptance of impending death or from a wish to die naturally, although preferably soon.
>
> The WTHD may arise in response to one or more factors, including physical symptoms (either present or foreseen), psychological distress (e.g. depression, hopelessness, fears, etc.), existential suffering (e.g. loss of meaning in life), or social aspects (e.g. feeling that one is a burden). (Balaguer et al. 2016, p. 8)

[2] This is suicide by placing a plastic bag over the head, especially in combination with inhalation of gases.

[3] This is a form of assisted suicide, accomplished by omitting food and water, while managing the symptoms of dehydration and starvation.

[4] See, Ganzini (2009). For instance, in their study of family member's perception as to why patients chose assisted suicide, Ganzini et al. report, "the most important reasons that their loved ones requested PAD [Physician Assisted Death], all with a median score of 4.5 or greater, were wanting to control the circumstances of death and die at home, and worries about loss of dignity and future losses of independence, quality of life, and self-care ability. No physical symptoms at the time of the request were rated higher than a median of 2 in importance. Worries about symptoms and experiences in the future were, in general, more important reasons than symptoms or experiences at the time of the request. According to family members, the least important reasons their loved ones requested PAD included depression, financial concerns, and poor social support" (Ganzini et al. 2008: 154).

In assisted suicide, intentional self-killing facilitated by another is the act that completes and fulfills this WTHD and thereby serves then as the means of avoidance. In short, death is the means chosen by the primary agent (patient) to avoid suffering; the cooperating agent then facilitates this death either formally or materially.

Acts of assisted suicide can thus be distinguished according to their object, circumstances and intent:

- *Circumstances:* The specific circumstances vary but commonly include a patient experiencing pain and suffering from chronic or life limiting illness or fear of future pain or suffering (e.g. perceived loss of autonomy, imposing a burden on loved ones) associated with illness (Oregon Office of Disease Prevention Epidemiology 2016);
- *Object*: one's own death;
- *Intent* (remote): to avoid pain and suffering or future pain and suffering, including illness related experiences, diminished sense of self and fears about the future.[5]
- *Intent* (proximate): to directly cause death, either by commission (e.g. administering high-dose sedatives for the *purpose* of causing death as the chosen *means* of relieving pain and suffering) or omission (e.g. withholding food and water for the immediate *purpose* of causing death as the chosen *means* of relieving pain and suffering);

Such acts are appropriately called 'assisted' insofar as a medical provider stands as a cooperating agent, contributing in some causal way, either formally (intentionally) or materially, to the patient's intentional self-killing act.

The diagram below outlines the distinctions between palliative acts and acts of assisted suicide drawn in this analysis.

| Components of moral act | Palliative acts | Acts of assisted suicide |
|---|---|---|
| Circumstances | Patient experiencing chronic or life limiting illness; medical interventions part of coordinated effort of social workers, physicians, nurses, therapists, chaplains, family, patient, aides, and other care providers in communication with the patient or surrogate, to maintain the greatest benefits for the patient | Patient experiencing chronic or life limiting illness or fear of future pain or suffering associated with illness; communicative effort of the physician and patient to consider the option of easing patient suffering by actively, intentionally hastening death |
| Object | Pain and symptom management | Self-killing, either by commission or omission, with the assistance of a cooperating agent |
| Intention (remote) | Maximize the quality of the patient's functioning and experience; maximizing prudent medical judgment | Maximize control (e.g. autonomy) while avoiding pain and suffering or future pain and suffering |
| Intention (proximate) | Provide treatment or care that is in itself therapeutic (non-lethal) | Directly cause death by means that are lethal by design or are used for the purpose of causing death |

[5] Cf. Pearlman et al. (2005: 236).

This grid allows for a clear comparison between the two actions under investigation to illustrate their respective moral status according to each traditional font of morality.

### 15.4.1 Circumstances

First, the circumstances for the two options are not mutually exclusive; in fact, generally speaking, the circumstances in which both palliative acts and acts of assisted suicide are requested are often very similar. They might even be the exact similar scenario since patients often suffer from serious, life limiting conditions and diseases, many of which are incurable and progressive. Patients are often receiving medical care from a number of clinicians (though palliative care is distinct insofar as the clinicians are a circumscribed, cooperative team unit). Generally speaking, the most that can be said is this: on a general level, it is not the circumstances of illness that are morally determinative for either assisted suicide or palliative care.

### 15.4.2 Object

Second, consider the object of each action. The treatment of pain and the management of symptoms are considered a moral imperative because they are good or at least morally neutral in their object or purpose.[6] In contrast, depriving a person of life, whether as an end in itself or as a means to avoid suffering, is an intrinsic affront to human dignity (as I will explain below in my discussion on mercy). Consequently, palliative acts are morally licit according to their object and ought to be encouraged; acts of assisted suicide are intrinsically flawed according to their object—they can never be justified and ought be avoided.

### 15.4.3 Intention

Finally, a distinction can be drawn between the direct intention of each act, that is, the intention as to why a specific means (or object) would be chosen (proximate intention) in pursuit of the ultimate end (remote intention) of avoiding suffering. For palliative acts, the intention contains a positive and negative element. That is, palliative acts carry the intention to maximize the function and experience of a patient while also minimizing their pain, discomfort and symptoms. Palliative acts do not carry a curative intent, though they may be executed concurrently with curative

---

[6] This idea is clearly expressed in the *Ethical and Religious Directives for Catholic Health Care Services*: "The task of medicine is to care even when it cannot cure" (USCCB 2009).

interventions. Insofar as their intention is to maximize benefit while minimizing harm or suffering to the patient, the intention of palliative acts are at least morally neutral if not morally good. In contrast, acts of assisted suicide carry with them an explicit intention for self-killing, through any number of means or interventions, to avoid suffering. This intention to cause a person's death explicitly as a means to help them avoid suffering is intrinsically deficient since it entails intending the death of a human person, offending their dignity and violating the virtue of justice.

To summarize this analysis, there are sharp distinctions between the two actions in question. Palliative acts and acts of assisted suicide are contradictory opposites in terms of their object and their intention. Palliative acts manage pain and symptoms with the life-affirming intent of maximizing function and experience. Assisted suicide causes death with the intention of avoiding suffering or future suffering. Thus, authentically palliative acts are morally good and praiseworthy, while acts of assisted suicide are intrinsically flawed and damaging to persons, both those who are killed and the moral character of those who facilitate such dehumanizing acts. Such actions that harm people and their moral character should be avoided.

Nevertheless, assisted suicide is often portrayed in entertainment and political spheres as an act of authentic mercy, allowing the patient die in a manner of their choosing.[7] In the final section, I will examine the applicability of the term mercy to palliative acts compared to acts of assisted suicide—as well as the anthropologic roots of such applications—to demonstrate that only palliative acts truly actualize and manifest mercy.

## 15.5 Mercy: Authentic vs. Counterfeit

The concept of mercy stands at the core of the divide between palliative acts and those of assisted suicide, especially in regard to how these activities are understood and justified in practice. For instance, one of the claims with strong emotional appeal in support of assisted suicide runs like this: to disallow patients access to assisted suicide is to force them to endure intolerable pain and suffering. Such a situation, so the argument goes, is cruel, unmerciful and absurd; and it is intolerable for society to impose this fate upon the patient by making assisted suicide illegal. The purported merciful course entails removing barriers to the patient's autonomous actions (e.g. providers right of conscientious objection (Stahl and Emanuel 2017)) to address their diminishment, pain and suffering. Hence, as the argument goes, it is merciful and compassionate to support assisted suicide.

This vision of mercy springs from some distinctive underlying anthropological assumptions. For instance, advocates for assisted suicide assert its value in terms of aiding and enhancing the autonomous choice of individuals. Thus, the anthropology that serves as a basis for demonstrating the purported value of assisted suicide is

[7] Here a distinction must be drawn. Laws like the Oregon "Death with Dignity" law expressly forbid "mercy killing", i.e. active euthanasia (Death with Dignity 2017).

that of individualistic autonomous chooser. Historically, this individualistic anthropology has been given its clearest expression in the work of John Stuart Mill, a prominent thinker in the wide tradition of Liberalism.[8] In his essay, *On Liberty* Mill asserts "The only part of the conduct of any one, for which he is amenable to society, is that which concerns others. In the part which merely concerns himself, his independence is, of right, absolute. Over himself, over his own body and mind, the individual is sovereign (Mill 2003: 68–69)." Here, Mill sketches the basis of an anthropology wherein the individual is absolute sovereign over their bodily life and even its continuance.

One might counter that Mill himself did not approve of suicide. To this the simple rejoinder is that Mill's own position of opposition is inconsistent with his concept of sovereignty. To the contrary, as commentator Tauriq Moosa writes, "The consistency of individual autonomy, as Mill outlined, indicates that just as we can live as we wish (with certain constraints), we ought to be able to die as we wish, too" (Moosa 2017). Here is a radical autonomy sprung from a radical individualism. Under this vision, mercy, then consists of facilitating the radical autonomy of the individual through both negative and positive obligations, i.e. removing barriers to autonomous action and actively fostering autonomous choice, even if that involves providing someone with the resources to intentionally kill themselves.[9]

Pope John Paul II addressed this precise rationale in his 1995 encyclical *Evangelium Vitae*. In stark contrast, he taught that euthanasia and assisted suicide each stand as a *"false mercy*, and indeed a disturbing 'perversion' of mercy" (John Paul II 1995: no. 66). This perversion becomes evident in light of anthropological factors. In contrast to the individualistic anthropology previously mentioned, the Catholic tradition has long affirmed the inherently social nature of humankind.

> But God did not create man as a solitary being, for from the beginning "male and female he created them" (Gen. 1:27). This partnership of man and woman constitutes the first form of communion between persons. For by his innermost nature man is a social being, and unless he relates himself to others he can neither live nor develop his gifts. (Gaudium et Spes 1988, n. 12)

This intrinsic relationality is the source and field for the various responsibilities and rights that human beings enjoy (John XXIII 2010, ns. 34–35.). And it only in this context, in which the common good of individuals is built through solidarity, that authentic compassion can exist.[10] As Pope John Paul II explains, "True 'compassion' leads to sharing another's pain; it does not kill the person whose suffering we cannot bear" (John Paul II 1995, n. 66). Assisting in another's suicide, then, is not merciful; it is a failure to be sufficiently merciful and willingly stay in solidarity with the one who suffers.

---

[8] Benjamin Wiker, for instance, explores how thinkers in the tradition of Liberalism have constructed this individualistic anthropology and the implications it has for society (Wiker 2013).

[9] For discussion on the negative and positive obligations for facilitating and fostering autonomy see Beauchamp and Childress (2009, pp. 103–05).

[10] "In keeping with the social nature of man, the good of each individual is necessarily related to the common good, which in turn can be defined only in reference to the human person" (*Catechism of the Catholic Church* 2008: n. 1905).

This sharp divide regarding the phenomenon of mercy indicates that it is not merely a difference of perspectives or interpretations of the same phenomenon. Rather, each line of reasoning seems to manifest fundamentally distinct conceptions of mercy itself. This point warrants a further examination.

As defined by Merriam-Webster, the term "Mercy" contains the following meanings: "compassion or forbearance shown especially to an offender or to one [who is] subject to one's [own] power" and "compassionate treatment of those in distress" (Merriam-Webster 2017d). Both definitions specify mercy as a noun, a state of being, characterized by "compassion." Hence, compassion and mercy are closely associated. The same dictionary defines compassion similarly as a state of being (noun): "sympathetic consciousness of others' distress together with a desire to alleviate it; a feeling of wanting to help someone who is sick, hungry, in trouble etc." (Merriam-Webster 2017b). Thus, compassion is a state of being containing two parts: *awareness* of another's distress and a *desire* for its alleviation. Thus, compassion is primarily a subjective feeling (awareness and desire) that makes no claim to the objective moral status of the acts carried out from this motivation.

Such definitions, however, differ from how the terms 'mercy' and 'compassion' have been traditionally understood in the Catholic moral tradition. For instance, according to St. Thomas Aquinas, mercy is characterized by its regard for a "certain special aspect, namely the *misery* of the person pitied" (Aquinas 1948: II–II, q.30, a3, r.3.). The object of mercy therefore has some objective ground: the misery of another. Aquinas then distinguishes between affective and effective mercy. "Affective mercy," explains Robert Stackpole, "is an emotion: the pity we feel for the plight of another" (Stackpole 2005). Effective mercy, in contrast, is a positive action taken to relieve the misery of the one suffering. The affective mercy is then meant to motivate the effective mercy. Stackpole explains the significance of the distinction in Thomas' thought,

> St. Thomas argues that the human virtue of mercy necessarily will be both affective and effective. However, to be the authentic virtue of "mercy," it must manifest two additional characteristics. First, it must be rooted in "right reason"—that is, in the truth about the sufferings of others, and what is in fact the objective "good" for the other whom we seek to help. Secondly, the virtue of mercy is proven in effective action for the good of others, as circumstances permit. If we merely "sympathize" with the plight of another and "share their pain" without making the best of the opportunities we have to help them, then virtue of mercy does not abide in us in any significant degree. (Stackpole 2005)

With the defect or lack that is the source of such misery as its object, the virtue of mercy is ordered toward removing such defect based on what is objectively good for—and in solidarity with—the one who is suffering. The virtue of mercy is manifested through compassionate actions, which must adhere to right reason (e.g. the natural moral law and its attendant rights and responsibilities) and can be measured for their consistency against objective standards. Thus, even an action undertaken with a merciful intent might prove to be not actually merciful.

Because of its nature as a virtue that considers interaction with other persons, mercy is also intrinsically related to the cardinal virtue that moderates such interpersonal relationships: justice. For Thomas and the Catholic moral tradition, it is therefore necessary to understand the relation between mercy and justice. Justice regards what is due another; what is theirs by right. Take the example of a mortgage. For the lender to act justly, they ought not extort, defraud or penalize the borrower, especially if they are meeting the terms of the loan. However, mercy would involve forgiving the debt altogether. It would remove the defect of debt from the borrower even though there is no obligation on the part of the lender to do so. Thus, mercy presumes justice, builds upon it and fulfills it. In the words of commentator Wojciech Zyzak, "Mercy is the source and perfection of justice because it is its ultimate fulfillment" (Zyzak 2015: 144). 'Compassion' without justice is counterfeit mercy.

These preceding insights about mercy, compassion and justice apply directly to the question at hand: are both palliative acts and acts of suicide merciful compassionate and just; only one; or neither? First, consider palliative acts. Persons with serious, chronic and debilitating conditions are often experiencing some form of misery. Care providers are acutely aware of this misery and motivated to help the patient. This is especially true of professional caregivers who are publically committed to caring for and alleviating the suffering of the sick and incapacitated. Palliative acts are specified as those actions that seek to relieve the pain and symptoms which are the source of the patient's suffering. To ensure that such acts are effective, professionals take steps to ensure that palliative measures are prudently designed and implemented. For instance, pain medications can be carefully titrated to ensure that pain is controlled but do not kill the patient. Further, there is nothing intrinsic to a palliative act that violates the virtue of justice; rather, they manifest the respect and comfort that is due to a patient based on their inherent human dignity. Thus, palliative acts are measureable against objective standards and seek the authentic good of the patient without violating justice. Palliative acts, truly so-called, are *ipso facto* merciful acts.

Acts of assisted suicide are motivated ostensibly by an authentic sympathy and desire to ease the suffering of the patient. Yet, assisted suicide specifically violates objective standards for what is authentically beneficial to the patient, for instance, by administering excessive amounts of pain medications in order to cease respirations. Further, for an act to stand as authentically merciful and compassionate, it requires a subject receiving that action. In order for someone to have their suffering relieved, they must continue to exist. One cannot experience relief if one does not exist.

Yet, proponents for assisted suicide claim that this activity is the ultimate expression of autonomy, while mercy is removing those impediments to autonomous action that disease and its symptoms place upon human persons. It is (ostensibly) the highest expression of self-determination to determine whether one exists. But direct experience from Europe, demonstrates that despite well-intentioned legal safeguards, assisted suicide policy and practices gives way to euthanasia. For instance, in 2011, Pereira wrote an article reviewing the efficacy of the legal safeguards designed to prevent abuses in the practice of assisted suicide. Pereira offered this substantive summary:

> In 30 years, the Netherlands has moved from euthanasia of people who are terminally ill, to euthanasia of those who are chronically ill; from euthanasia for physical illness, to euthanasia for mental illness; from euthanasia for mental illness, to euthanasia for psychological

> distress or mental suffering—and now to euthanasia simply if a person is over the age of 70 and "tired of living." Dutch euthanasia protocols have also moved from conscious patients providing explicit consent, to unconscious patients unable to provide consent. Denying euthanasia or PAS in the Netherlands is now considered a form of discrimination against people with chronic illness, whether the illness be physical or psychological, because those people will be forced to "suffer" longer than those who are terminally ill. Non-voluntary euthanasia is now being justified by appealing to the social duty of citizens and the ethical pillar of beneficence. In the Netherlands, euthanasia has moved from being a measure of last resort to being one of early intervention. Belgium has followed suit, and troubling evidence is emerging from Oregon specifically with respect to the protection of people with depression and the objectivity of the process. (Pereira 2011: 38–45)

Sanctioned, facilitated self-killing to relieve suffering gives way, inevitably, to sanctioned other-killing in order to relieve suffering. The logic is straightforward: if it is good enough for people to choose for themselves when they are suffering, why would it not be good enough for others to choose for them? But for those that value autonomy above all, this experience presents a fundamental problem. In practice, the ultimate, final self-determination gives rise to its opposite: ultimate, final other-determination.

This is not a moral argument based on 'slippery slope' logic; it is an observation of sociologic fact. Nevertheless, this sociologic observation illustrates a moral point: assisting someone's suicide violates justice, particularly social justice, which entails that individuals have a right and responsibility to participate in society, contribute to and receive its benefits and not be marginalized from the community. Suicide is the ultimate marginalizing act—to isolate one's self from all relationships and all existence. Assisting in this marginalization thus violates the suffering patient's right to participate and not be abandoned by society, simply because they are suffering. As Pereira illustrates, allowing or affirming this marginalization to persist in society blossoms into the phenomenon that people feel obligated to eliminate themselves when they no longer want 'be a burden' on their family or society. Thus, because acts of assisted suicide violate social justice they cannot be considered authentically merciful. And here is why assisted suicide is a false mercy: it eliminates a problem (i.e. misery) unjustly, by eliminating a person. Assisted suicide eases suffering by eliminating the one suffering. Indeed the person's misery is eliminated precisely because the person suffering the misery is eliminated. Consequently, acts of assisted suicide cannot be properly called merciful or compassionate. They are dangerous counterfeits.

## 15.6 Conclusion

In 2014, The Institute of Medicine published a national study examining the trajectory of end-of-life care in the United States (IOM 2015). They argue that an aging American population deserves end-of-life treatment that is patient-centered, communicative, and respecting of diversity. Because of this, they assert that it is increasingly necessary that palliative medicine becomes fully integrated into the hospital setting. However, the momentum of end-of-life care in the United States finds itself on a contrary trajectory of increasingly legalizing assisted suicide. Five states—Oregon, Washington, Vermont, California, and Colorado–and the District of Columbia have passed "Death with Dignity" or "End of Life" legislation while Montana has decriminalized it via state Supreme Court ruling. In 2016 alone, bills decriminalizing PAS have been submitted in 15 states and D.C. Only 1 bill had passed (Patient Rights Council 2017). Thus, it appears that the tension and competition between palliative care and assisted suicide will only increase.

While both assisted suicide and palliative care are rooted in a desire to seek compassionate care for terminally ill patients, they are incongruent and incompatible in several key areas. At their respective anthropologic root, assisted suicide rests upon an excessively individualistic philosophy of man while palliative care relies upon a social vision of the human person. Additionally, while both practices seek to fulfill a physician's social role, there is a significant difference in what that role entails: palliative care casts providers as healers and relievers of pain, assisted suicide construes providers as dispensers of medications and mere facilitators of patient autonomy. These differences have important ramifications for the future of humane health care.

The differences between palliative acts and acts of assisted suicide are more than just abstract and academic. Viewing these two factors, anthropological and social, in light of the object of the act itself, it is apparent that these two categories of action are diametrically opposed. The more one practice flourishes, the more the other will wither. Therefore, there is a drastic need to ensure that authentic palliative care, which is true medical mercy, grows and develops.

For Christians, the impetus to promote authentic medical mercy flows from the beating Sacred Heart of the Christian religion. Aquinas illustrates how mercy is but the footstep—albeit an essential one—to the highest calling of the Gospel: "The sum total of the Christian religion consists in mercy, as regards external works: but the inward love of charity, whereby we are united to God preponderates over both love and mercy for our neighbor" (Aquinas 1948: II–II, q.30, a4, r.2). Thus, medical practices—such as Palliative care—that are authentic expressions of mercy are quintessential expressions of the Christian religion while the counterfeit mercy of assisted suicide should be avoided and abandoned at together. A better, more humane option exists.

## References

Aquinas, Thomas. 1948. *Summa theologica*. New York: Benziger Bros.

Balaguer, Albert, Cristina Monforte-Royo, Josep Porta-Sales, Alberto Alonso-Babarro, Rogelio Altisent, Amor Aradilla-Herrero, Mercedes Bellido-Pérez, William Breitbart, Carlos Centeno, Miguel Angel Cuervo, Luc Deliens, Gerrit Frerich, Chris Gastmans, Stephanie Lichtenfeld, Joaquín T. Limonero, Markus A. Maier, Lars Johan Materstvedt, María Nabal, Gary Rodin, Barry Rosenfeld, Tracy Schroepfer, Joaquín Tomás-Sábado, Jordi Trelis, Christian Villavicencio-Chávez, Raymond Voltz, and Antony Bayer. 2016. An international consensus definition of the wish to hasten death and its related factors'. *PLoS One* 11: e0146184.

Beauchamp, Tom L., and James F. Childress. 2009. *Principles of biomedical ethics*. New York: Oxford University Press.
Bishop, Jeffrey Paul. 2012. *The anticipatory corpse: Medicine, power, and the care of the dying*. Notre Dame: University of Notre Dame Press.
Catechism of the Catholic Church. 2008. Collegeville: The Liturgical Press.
Death with Dignity. 2017. *Terminology*. https://www.deathwithdignity.org/terminology/. Accessed 10 Jul 2017.
Donne, John. 1609. *Death, be not proud*. https://www.poetryfoundation.org/poems-and-poets/poems/detail/44107. Accessed 10 Jul 2017.
Ganzini, L. 2009. Oregonians' reasons for requesting physician aid in dying. *Archives of Internal Medicine* 169: 489–492.
Ganzini, L., E.R. Goy, and S.K. Dobscha. 2008. Why oregon patients request assisted death: Family members views. *Journal of General Internal Medicine* 23: 154–157.
Gaudium et Spes. 1988. In *Vatican council II: The conciliar and post conciliar documents*, ed. Austin Flannery. Northport: Costello Publications.
Institute of Medicine of the National Academies (IOM). 2015. *Dying in America: Improving quality and honoring individual preferences near the end of life*.
John Paul II, Pope. 1995. *Evangelium vitae: The gospel of life*. New York: Random House.
John XXIII, Pope. 2010. Pacem in terris. In *Catholic social thought: The documentary heritage*, ed. David O'Brien and Thomas A. Shannon. Maryknoll: Orbis.
Lewis, C.S. 1943. *The abolition of man*. Oxford: Oxford University Press.
Lorenz, Karl A., Joanne Lynn, Sydney M. Dy, Lisa R. Shugarman, Anne Wilkinson, Richard A. Mularski, Sally C. Morton, Ronda G. Hughes, Lara K. Hilton, Margaret Maglione, Shannon L. Rhodes, Cony Rolon, Virginia C. Sun, and Paul G. Shekelle. 2008. Evidence for improving palliative care at the end of life: A systematic review. *Annals of Internal Medicine* 148: 147.
Merriam-Webster. 2017a. *Assisted suicide*. https://www.merriam-webster.com/dictionary/assisted%20suicide. Accessed 10 Jul 2017.
———. 2017b. *Compassion*. https://www.merriam-webster.com/dictionary/compassion. Accessed 7/10.
———. 2017c. *Kill*. https://www.merriam-webster.com/dictionary/kill. Accessed 10 Jul 2017.
———. 2017d. *Mercy*. https://www.merriam-webster.com/dictionary/mercy. Accessed 10 Jul 2017.
———. 2017e. *Palliate*. https://www.merriam-webster.com/dictionary/palliate. Accessed 10 Jul 2017.
———. 2017f. *Suicidez*. https://www.merriam-webster.com/dictionary/suicide. Accessed 10 Jul 2017.
Mill, John Stuart. 2003. *On liberty*. London: Penguin.
Moosa, Tauriq. 2017. *John Stuart Mill and the right to die. Big Think*. http://bigthink.com/against-the-new-taboo/john-stuart-mill-and-the-right-to-die. Accessed 10 Jul 2017.
Oregon Office of Disease Prevention Epidemiology. 2016. *Annual report on Oregon's death with dignity act*.
Patient Rights Council. 2017. *Assisted suicide laws*. http://www.patientsrightscouncil.org/site/assisted-suicide-state-laws. Accessed 10 Jul 2017.
Pearlman, Robert A., Clarissa Hsu, Helene Starks, Anthony L. Back, Judith R. Gordon, Ashok J. Bharucha, Barbara A. Koenig, and Margaret P. Battin. 2005. Motivations for physician-assisted suicide: Patient and family voices. *Journal of General Internal Medicine* 20: 234–239.
Pereira, J. 2011. Legalizing euthanasia or assisted suicide: The illusion of safeguards and controls. *Current Oncology* 18: e38–e48.
Qaseem, Amir, Vincenza Snow, Paul Shekelle, Donald E. Casey, J. Thomas Cross, and Douglas K. Owens. 2008. Evidence-based interventions to improve the palliative care of pain, dyspnea, and depression at the end of life: A clinical practice guideline from the american college of physicians. *Annals of Internal Medicine* 148: 141.

Stackpole, Robert. 2005. *St. Thomas Aquinas on the virtue of mercy*. http://www.thedivinemercy.org/library/article.php?NID=2214. Accessed 10 Jul 2017.

Stahl, Ronit Y., and Ezekiel J. Emanuel. 2017. Physicians, not conscripts—conscientious objection in health care. *New England Journal of Medicine* 376: 1380–1385.

United States Conference of Catholic Bishops (USCCB). 2009. *Ethical and religious directives for catholic health care services*. Washington, DC: USCCB.

Wiker, Benjamin. 2013. *Worshipping the state: How liberalism became our state religion*. Washington, DC: Regnery Pub.

World Health Organization. 2017. *WHO definition of palliative care*. http://www.who.int/cancer/palliative/definition/en. Accessed 10 Jul 2017.

Zyzak, Wojciech. 2015. Mercy as a theological term, the person and the challenges. *The Journal of Theology, Education, Canon Law and Social Studies Inspired by Pope John Paul II* 5: 137–153.

# Appendix: Magisterial and Episcopal Texts on Palliative Care and End of Life of Life Issues

## Letter of His Holiness Pope Francis to Participants in the European Regional Meeting of the World Medical Association on End-Of-Life Issues[1]

November 7, 2017

To My Venerable Brother

Archbishop Vincenzo Paglia President of the Pontifical Academy for Life

I extend my cordial greetings to you and to all the participants in the European Regional Meeting of the World Medical Association on end-of-life issues, held in the Vatican in conjunction with the Pontifical Academy for Life.

Your meeting will address questions dealing with the end of earthly life. They are questions that have always challenged humanity, but that today take on new forms by reason of increased knowledge and the development of new technical tools. The growing therapeutic capabilities of medical science have made it possible to eliminate many diseases, to improve health and to prolong people's life span. While these developments have proved quite positive, it has also become possible nowadays to extend life by means that were inconceivable in the past. Surgery and other medical interventions have become ever more effective, but they are not always beneficial: they can sustain, or even replace, failing vital functions, but that is not the same as promoting health. Greater wisdom is called for today, because of the temptation to insist on treatments that have powerful effects on the body, yet at times do not serve the integral good of the person.

Some 60 years ago, Pope Pius XII, in a memorable address to anaesthesiologists and intensive care specialists, stated that there is no obligation to have recourse in all circumstances to every possible remedy and that, in some specific cases, it is

[1] http://www.news.va/en/news/pope-addresses-end-of-life-issues



P. J. Cataldo, D. O'Brien (eds.), *Palliative Care and Catholic Health Care*, Philosophy and Medicine 130, https://doi.org/10.1007/978-3-030-05005-4

permissible to refrain from their use (cf. AAS XLIX [1957], 1027–1033). Consequently, it is morally licit to decide not to adopt therapeutic measures, or to discontinue them, when their use does not meet that ethical and humanistic standard that would later be called "due proportion in the use of remedies" (cf. CONGREGATION FOR THE DOCTRINE OF THE FAITH, *Declaration on Euthanasia*, 5 May 1980, IV: AAS LXXII [1980], 542–552). The specific element of this criterion is that it considers "the result that can be expected, taking into account the state of the sick person and his or her physical and moral resources" (ibid.). It thus makes possible a decision that is morally qualified as withdrawal of "overzealous treatment".

Such a decision responsibly acknowledges the limitations of our mortality, once it becomes clear that opposition to it is futile. "Here one does not will to cause death; one's inability to impede it is merely accepted" (*Catechism of the Catholic Church*, No. 2278). This difference of perspective restores humanity to the accompaniment of the dying, while not attempting to justify the suppression of the living. It is clear that not adopting, or else suspending, disproportionate measures, means avoiding overzealous treatment; from an ethical standpoint, it is completely different from euthanasia, which is always wrong, in that the intent of euthanasia is to end life and cause death.

Needless to say, in the face of critical situations and in clinical practice, the factors that come into play are often difficult to evaluate. To determine whether a clinically appropriate medical intervention is actually proportionate, the mechanical application of a general rule is not sufficient. There needs to be a careful discernment of the moral object, the attending circumstances, and the intentions of those involved. In caring for and accompanying a given patient, the personal and relational elements in his or her life and death – which is after all the last moment in life – must be given a consideration befitting human dignity. In this process, the patient has the primary role. The *Catechism of the Catholic Church* makes this clear: "The decisions should be made by the patient if he is competent and able" (loc. cit.). The patient, first and foremost, has the right, obviously in dialogue with medical professionals, to evaluate a proposed treatment and to judge its actual proportionality in his or her concrete case, and necessarily refusing it if such proportionality is judged lacking. That evaluation is not easy to make in today's medical context, where the doctor-patient relationship has become increasingly fragmented and medical care involves any number of technological and organizational aspects.

It should also be noted that these processes of evaluation are conditioned by the growing gap in healthcare possibilities resulting from the combination of technical and scientific capability and economic interests. Increasingly sophisticated and costly treatments are available to ever more limited and privileged segments of the population, and this raises questions about the sustainability of healthcare delivery and about what might be called a systemic tendency toward growing inequality in health care. This tendency is clearly visible at a global level, particularly when different continents are compared. But it is also present within the more wealthy countries, where access to healthcare risks being more dependent on individuals' economic resources than on their actual need for treatment.

In the complexity resulting from the influence of these various factors on clinical practice, but also on medical culture in general, the supreme commandment of *responsible closeness*, must be kept uppermost in mind, as we see clearly from the Gospel story of the Good Samaritan (cf. *Lk* 10:25–37). It could be said that the categorical imperative is to never abandon the sick. The anguish associated with conditions that bring us to the threshold of human mortality, and the difficulty of the decision we have to make, may tempt us to step back from the patient. Yet this is where, more than anything else, we are called to show love and closeness, recognizing the limit that we all share and showing our solidarity. Let each of us give love in his or her own way—as a father, a mother, a son, a daughter, a brother or sister, a doctor or a nurse. But give it! And even if we know that we cannot always guarantee healing or a cure, we can and must always care for the living, without ourselves shortening their life, but also without futilely resisting their death. This approach is reflected in palliative care, which is proving most important in our culture, as it opposes what makes death most terrifying and unwelcome—pain and loneliness.

Within democratic societies, these sensitive issues must be addressed calmly, seriously and thoughtfully, in a way open to finding, to the extent possible, agreed solutions, also on the legal level. On the one hand, there is a need to take into account differing world views, ethical convictions and religious affiliations, in a climate of openness and dialogue. On the other hand, the state cannot renounce its duty to protect all those involved, defending the fundamental equality whereby everyone is recognized under law as a human being living with others in society. Particular attention must be paid to the most vulnerable, who need help in defending their own interests. If this core of values essential to coexistence is weakened, the possibility of agreeing on that recognition of the other which is the condition for all dialogue and the very life of society will also be lost. Legislation on health care also needs this broad vision and a comprehensive view of what most effectively promotes the common good in each concrete situation.

In the hope that these reflections may prove helpful, I offer you my cordial good wishes for a serene and constructive meeting. I also trust that you will find the most appropriate ways of addressing these delicate issues with a view to the good of all those whom you meet and those with whom you work in your demanding profession.

May the Lord bless you and the Virgin Mary protect you.

## Address of His Holiness Pope Francis to Participants in the Plenary of the Pontifical Academy for Life[2]

Thursday, March 5, 2015.

*Dear Brothers and Sisters,*

I warmly welcome you on the occasion of your General Assembly, called to reflect on the theme "Assistance to the elderly and palliative care", and I thank the President for his kind words. I am especially pleased to greet Cardinal Sgreccia, who is a pioneer.... Thank you.

Palliative care is an expression of the truly human attitude of taking care of one another, especially of those who suffer. It is a testimony that the human person is always precious, even if marked by illness and old age. Indeed, the person, under any circumstances, is an asset to him/herself and to others and is loved by God. This is why, when their life becomes very fragile and the end of their earthly existence approaches, we feel the responsibility to assist and accompany them in the best way.

The biblical Commandment that calls us to honour our parents, reminds us in a broader sense of the honour that we owe to all elderly people. God associates with this Commandment a twofold promise: "that your days may be long" (Ex 20:12) and the other—"that it may go well with you" (Dt 5:16). Faithfulness to the fourth Commandment ensures not only the gift of land but the opportunity to enjoy it. Indeed, the wisdom that enables us to recognize the value of our elders and leads us to honour them, is the same wisdom that allows us to happily appreciate the many gifts we receive every day from the provident hand of the Father.

This precept shows us the fundamental pedagogical relationship between parents and children, between old and young, regarding the safekeeping and passing on of wisdom and religious teaching to future generations. To honour this teaching and those who transmit it is a source of life and blessing. On the contrary, the Bible reserves a severe admonition for those who neglect or mistreat their parents (cf. Ex 21:17; Lev 20:9). The same judgement applies today when parents, becoming aged and less useful, are marginalized to the point of abandonment; and we have so many examples of this!

The word of God is ever living and we clearly see how the Commandment proves central for contemporary society, where the logic of usefulness takes precedence over that of solidarity and of gratuitousness, even within the family. Therefore, let us listen with docile hearts to the word of God that comes to us from the Commandments which, let us always remember, are not bonds that imprison, but are words of life.

Today "to honour" could also be translated as the duty to have the utmost respect and to take care of those who, due to their physical or social condition, may be left to die or "made to die". All of medicine has a special role within society as a witness to the honour that we owe to the elderly person and to each human being. Evidence

---

[2] https://w2.vatican.va/content/francesco/en/speeches/2015/march/documents/papa-francesco_20150305_pontificia-accademia-vita.html

and effectiveness cannot be the only criteria that govern physicians' actions, nor can health system regulations and economic profits. A state cannot think about earning with medicine. On the contrary, there is no duty more important for a society than that of safeguarding the human person.

Your work during these days has been to explore new fields of application for palliative care. It has until now been a precious accompaniment for cancer patients, but today there is a great variety of diseases characterized by chronic progressive deterioration, often linked to old age, which can benefit from this type of assistance. The elderly need in the first place the care of their family members—whose affection cannot be replaced by even the most efficient structures or the most skilled and charitable healthcare workers. When not self-sufficient or having advanced or terminal disease, the elderly can enjoy truly human assistance and have their needs adequately met thanks to palliative care offered in conjunction with the supportive care given by family members. The objective of palliative care is to alleviate suffering in the final stages of illness and at the same time to ensure the patient appropriate human accompaniment (cf. Encyclical *Evangelium Vitae*, n. 65). It is important support especially for the elderly, who, because of their age, receive increasingly less attention from curative medicine and are often abandoned. Abandonment is the most serious "illness" of the elderly, and also the greatest injustice they can be submitted to: those who have helped us grow must not be abandoned when they are in need of our help, our love and our tenderness.

Thus I appreciate your scientific and cultural commitment to ensuring that palliative care may reach all those who need it. I encourage professionals and students to specialize in this type of assistance which is no less valuable for the fact that it "is not life-saving". Palliative care accomplishes something equally important: it values the person. I exhort all those who, in various ways, are involved in the field of palliative care, to practice this task keeping the spirit of service intact and remembering that all medical knowledge is truly science, in its noblest significance, only if used as aid in view of the good of man, a good which is never accomplished "against" the life and dignity of man.

It is this ability to serve life and the dignity of the sick, also when they are old, that is the true measure of medicine and society as a whole. I repeat St John Paul II's appeal: "respect, protect, love and serve life, every human life! Only in this direction will you find justice, development, true freedom, peace and happiness! (*ibid.*, n. 5).

I hope you continue this study and research, so that the work of the advancement and defence of life may be ever more effective and fruitful. May the Virgin Mother, Mother of life, accompany my Blessing. Please, do not forget to pray for me. Thank you.

## Message of His Holiness Benedict XVI for the 15th World Day of the Sick[3]

December 8, 2006

*Dear Brothers and Sisters,*

On 11 February 2007, when the Church keeps the liturgical memorial of Our Lady of Lourdes, the 15th World Day of the Sick will be celebrated in Seoul, Korea. A number of meetings, conferences, pastoral gatherings and liturgical celebrations will take place with representatives of the Church in Korea, health care personnel, the sick and their families. Once again the Church turns her eyes to those who suffer and calls attention to the incurably ill, many of whom are dying from terminal diseases. They are found on every continent, particularly in places where poverty and hardship cause immense misery and grief. Conscious of these sufferings, I will be spiritually present at the World Day of the Sick, united with those meeting to discuss the plight of the incurably ill in our world and encouraging the efforts of Christian communities in their witness to the Lord's tenderness and mercy.

Sickness inevitably brings with it a moment of crisis and sober confrontation with one's own personal situation. Advances in the health sciences often provide the means necessary to meet this challenge, at least with regard to its physical aspects. Human life, however, has intrinsic limitations, and sooner or later it ends in death. This is an experience to which each human being is called, and one for which he or she must be prepared. Despite the advances of science, a cure cannot be found for every illness, and thus, in hospitals, hospices and homes throughout the world we encounter the sufferings of our many brothers and sisters who are incurably and often terminally ill. In addition, many millions of people in our world still experience insanitary living conditions and lack access to much-needed medical resources, often of the most basic kind, with the result that the number of human beings considered "incurable" is greatly increased.

The Church wishes to support the incurably and terminally ill by calling for just social policies which can help to eliminate the causes of many diseases and by urging improved care for the dying and those for whom no medical remedy is available. There is a need to promote policies which create conditions where human beings can bear even incurable illnesses and death in a dignified manner. Here it is necessary to stress once again the need for more palliative care centres which provide integral care, offering the sick the human assistance and spiritual accompaniment they need. This is a right belonging to every human being, one which we must all be committed to defend.

Here I would like to encourage the efforts of those who work daily to ensure that the incurably and terminally ill, together with their families, receive adequate and loving care. The Church, following the example of the Good Samaritan, has always shown particular concern for the infirm. Through her individual members

[3] http://w2.vatican.va/content/benedict-xvi/en/messages/sick/documents/hf_ben-xvi_mes_20061208_world-day-of-the-sick-2007.html

and institutions, she continues to stand alongside the suffering and to attend the dying, striving to preserve their dignity at these significant moments of human existence. Many such individuals – health care professionals, pastoral agents and volunteers – and institutions throughout the world are tirelessly serving the sick, in hospitals and in palliative care units, on city streets, in housing projects and parishes.

I now turn to you, my dear brothers and sisters suffering from incurable and terminal diseases. I encourage you to contemplate the sufferings of Christ crucified, and, in union with him, to turn to the Father with complete trust that all life, and your lives in particular, are in his hands. Trust that your sufferings, united to those of Christ, will prove fruitful for the needs of the Church and the world. I ask the Lord to strengthen your faith in his love, especially during these trials that you are experiencing. It is my hope that, wherever you are, you will always find the spiritual encouragement and strength needed to nourish your faith and bring you closer to the Father of Life. Through her priests and pastoral workers, the Church wishes to assist you and stand at your side, helping you in your hour of need, and thus making present Christ's own loving mercy towards those who suffer.

In conclusion, I ask ecclesial communities throughout the world, and particularly those dedicated to the service of the infirm, to continue, with the help of Mary, *Salus Infirmorum*, to bear effective witness to the loving concern of God our Father. May the Blessed Virgin, our Mother, comfort those who are ill and sustain all who have devoted their lives, as Good Samaritans, to healing the physical and spiritual wounds of those who suffer. United to each of you in thought and prayer, I cordially impart my Apostolic Blessing as a pledge of strength and peace in the Lord.

## *The Gospel of Life*, n. 65[4]

Pope John Paul II

March 25, 1995

For a correct moral judgment on euthanasia, in the first place a clear definition is required. Euthanasia in the strict sense is understood to be an action or omission which of itself and by intention causes death, with the purpose of eliminating all suffering. "Euthanasia's terms of reference, therefore, are to be found in the intention of the will and in the methods used".

Euthanasia must be distinguished from the decision to forego so-called "aggressive medical treatment", in other words, medical procedures which no longer correspond to the real situation of the patient, either because they are by now disproportionate to any expected results or because they impose an excessive burden on the patient and his family. In such situations, when death is clearly imminent and inevitable, one can in conscience "refuse forms of treatment that would only secure a precarious and burdensome prolongation of life, so long as the normal care due to the sick person in similar cases is not interrupted". Certainly there is a moral obligation to care for oneself and to allow oneself to be cared for, but this duty must take account of concrete circumstances. It needs to be determined whether the means of treatment available are objectively proportionate to the prospects for improvement. To forego extraordinary or disproportionate means is not the equivalent of suicide or euthanasia; it rather expresses acceptance of the human condition in the face of death.

In modern medicine, increased attention is being given to what are called "methods of palliative care", which seek to make suffering more bearable in the final stages of illness and to ensure that the patient is supported and accompanied in his or her ordeal. Among the questions which arise in this context is that of the licitness of using various types of painkillers and sedatives for relieving the patient's pain when this involves the risk of shortening life. While praise may be due to the person who voluntarily accepts suffering by forgoing treatment with pain-killers in order to remain fully lucid and, if a believer, to share consciously in the Lord's Passion, such "heroic" behaviour cannot be considered the duty of everyone. Pius XII affirmed that it is licit to relieve pain by narcotics, even when the result is decreased consciousness and a shortening of life, "if no other means exist, and if, in the given circumstances, this does not prevent the carrying out of other religious and moral duties". In such a case, death is not willed or sought, even though for reasonable motives one runs the risk of it: there is simply a desire to ease pain effectively by using the analgesics which medicine provides. All the same, "it is not right to deprive the dying person of consciousness without a serious reason": as they approach death people ought to be able to satisfy their moral and family duties, and

[4] http://w2.vatican.va/content/john-paul-ii/en/encyclicals/documents/hf_jp-ii_enc_25031995_evangelium-vitae.html

above all they ought to be able to prepare in a fully conscious way for their definitive meeting with God.

Taking into account these distinctions, in harmony with the Magisterium of my Predecessors and in communion with the Bishops of the Catholic Church, I confirm that euthanasia is a grave violation of the law of God, since it is the deliberate and morally unacceptable killing of a human person. This doctrine is based upon the natural law and upon the written word of God, is transmitted by the Church's Tradition and taught by the ordinary and universal Magisterium.

Depending on the circumstances, this practice involves the malice proper to suicide or murder.

## Message of Pope John Paul II to the Staff and Residents of the Rennweg Hospice, Vienna[5]

Sunday, 21 June 1998

*To my beloved brothers and sisters of the Rennweg Hospice of Caritas Socialis and to all who live and work in the world of pain and suffering*

1. In the name of our Lord Jesus Christ, who has "borne our griefs and carried our sorrows" (Is 53:4), I greet you with great affection. My Pastoral Visit to Austria would have missed an important stop if I did not have the opportunity of meeting you, the sick and the suffering. In addressing this Message to you, I take the opportunity to express to all who work full- or part-time in hospitals, clinics, homes for the elderly and hospices my deep appreciation of their devotion to this self-sacrificing service. May my presence and my words support them in their commitment and their witness. Today, when I have the opportunity to visit the Caritas Socialis Hospice, I would like to confirm that the meeting with human pain contains good news. In fact, the "Gospel of suffering" (Apostolic Letter *Salvifici doloris*, n. 25), is not only written in Sacred Scripture, but in places like this it is rewritten day after day.
2. We are living in a society which seeks to remove pain, suffering, illness and death from personal and public awareness. But at the same time, the subject is being increasingly discussed in the press, on television and at conferences. The avoidance of death is also evident in the fact that many sick people die in hospitals or other structures, that is, outside their customary surroundings. Actually, most people would like to close their eyes to this world in their own home, among their relatives and trusted friends, but a great many families feel neither psychologically nor physically able to satisfy this desire. In addition, there are many people living alone who have no one to be close to them at the end of their life. Even if they die in a home, their heart is "homeless".

   To meet this need in past years, various ecclesial, municipal and private initiatives were undertaken to improve home, hospital and medical care, as well as to provide better pastoral care for the dying and competent help for their relatives. One of these important initiatives is the hospice movement, which has done exemplary work at the *Caritas Socialis* home in Rennweg. In it the sisters are inspired by the concern of their foundress, Hildegard Burjan, who wanted to be present at the focal points of human suffering as the "charismatic messenger of social love".

   No one who visits this hospice goes home disappointed. On the contrary, the visit is more than a tour. It becomes an encounter. By their mere presence, the sick, suffering and terminally ill patients invite the visitor who meets them not to hide the reality of suffering and death from himself. He is encouraged to be

[5] http://w2.vatican.va/content/john-paul-ii/en/speeches/1998/june/documents/hf_jp-ii_spe_19980621_austria-infermi.html

aware of the limits of his own life and to face them openly. The hospice makes one understand that dying means living before death, because even the last phase of earthly life can be lived consciously and organized individually. Far from being a "home for the dying", this place becomes a threshold of hope which leads beyond suffering and death.

3. Most sick people, after learning the results of the medical tests and the diagnosis of a terminal illness, live in fear of the progress of their disease. In addition to the suffering of the moment comes the fear of further deterioration and the feeling that their lives are meaningless. They are afraid of facing a path possibly marked by suffering. An anguish-filled future casts a shadow over the still bearable present. Perhaps those who have had a long and fulfilled life can wait for death with a certain tranquillity and accept their dying "full of years" (Gn 25:8). But for the majority death comes too soon. Many of our contemporaries, even the very elderly, hope for a quick, painless death; others ask for a little more time to take their leave. But fears, questions, doubts and desires are always present in this last phase of life. Even Christians are not spared the fear of death, which is the last enemy, as Sacred Scripture says (cf. 1 Cor 15:26; Rv 20:14).
4. The end of life raises profound questions for man: What will death be like? Will I be alone or surrounded by my loved ones? What awaits me after death? Will I be welcomed by God's mercy?

   To face these questions with gentleness and sensitivity — this is the task of those who work in hospitals and hospices. It is important to speak of suffering and death in a way that dispels fear. Indeed, dying is also part of life. In our time there is an urgent need for people who can revive this awareness. While in the Middle Ages "the art of dying" was known, today even Christians hesitate to talk to each other about death and to prepare for it. They prefer to be immersed in the present, seeking to distract themselves with work, professional recognition and amusement. Despite or perhaps because of today's consumer-, achievement- and experience-oriented society, there is an increasing thirst for transcendence among our contemporaries. Even if concrete concepts of life after death seem very vague, fewer and fewer believe that everything ends with death.
5. Death conceals even from the Christian the direct vision of what is to come, but the believer can trust in the Lord's promise: "Because I live, you will live also" (Jn 14:19). Jesus' words and the testimony of the Apostles reflect the new world of the resurrection for us in evocative language that expresses the hope: "We shall always be with the Lord" (1 Thes 4:17). To make the acceptance of this message easier for the critically ill and dying, it is necessary that all who approach them show by their own conduct that they take the words of the Gospel seriously. Therefore care and concern for people close to death is one of the most important signs of ecclesial credibility. Those who in the last phase of life feel supported by sincere Christians can more easily trust that Christ truly awaits them in the new life after death. Thus the pain and suffering of the present can be illumined by the joyful message: "So faith, hope, love abide, these

three; but the greatest of these is love!" (1 Cor 13:13). And love is stronger than death (cf. Song 8:6).

6. Just as the knowledge of being loved lessens the fear of suffering, so respect for the sick person's dignity helps him in this critical and difficult phase of life to discover something that fosters his human and Christian maturation. In the past, man knew that suffering was part of life and accepted it. Today he strives instead to avoid suffering in every way, as is shown by the wide range of pain-killing medicines for sale. Without detracting from their usefulness in many cases, it must still be pointed out that the overhasty elimination of suffering can prevent a person from facing it and acquiring greater human maturity through it. However, in this growth process, he needs competent people who can really accompany him. Giving practical help to another requires respect for his particular suffering and recognition of the dignity he still has despite the decline that suffering brings with it.
7. Hospice work arose from this conviction. Its goal is to respect the dignity of the elderly, sick and dying by helping them understand their own suffering as a process of growth and fulfilment in their life. Thus what I expressed as the leitmotiv of the Encyclical *Redemptor hominis*, that man is the way of the Church (cf. n. 14), is put into practice in the hospice. Its focus is not sophisticated, high-technology medicine, but man in his inalienable dignity.

   Willingness to accept the limits imposed by birth and death, learning to say "yes" to the basic passivity of our life, does not lead to alienation. It is rather the acceptance of one's own humanity in its full truth with the riches that belong to every phase of earthly life. Even in the frailty of the last hour, human life is never "meaningless" or "useless". A fundamental lesson for our society, tempted by modern myths such as the zest for life, achievement and consumerism, can be learned precisely from patients who are seriously ill and dying. They remind us that no one can determine the value or the non-value of another person's life, not even his own. As a gift of God, life is a good for which he alone can make the decisions.
8. From this standpoint, the decision actively to kill a human being is always an arbitrary act, even when it is meant as an expression of solidarity and compassion. The sick person expects his neighbour to help him live his life to the very last and to end it, when God wills, with dignity. Both the artificial extension of human life and the hastening of death, although they stem from different principles, conceal the same assumption: the conviction that life and death are realities entrusted to human beings to be disposed of at will. This false vision must be overcome. It must be made clear again that life is a gift to be responsibly led in God's sight. Hence the commitment to the human and Christian support of the dying which the hospice attempts to put into practice. From their different standpoints, doctors, nurses, pastors, sisters, relatives and friends strive to enable the sick and the dying personally to organize the last phase of their life, as far as their physical and psychological strength allows. This commitment has great human and Christian value. It aims to reveal God as One who "loves the

living" (Wis 11:26) and to perceive, beyond pain and death, the glad tidings: "I came that they may have life, and have it abundantly" (Jn 10:10).

9. We discover the face of God, who is a friend of life and of man, above all in Jesus of Nazareth. One of the most vivid illustrations of this Gospel is the parable of the Good Samaritan. The injured man lying by the wayside arouses the compassion of the Samaritan, who "came to where he was ... and went to him and bound up his wounds, pouring on oil and wine; then he set him on his own beast and brought him to an inn, and took care of him" (Lk 10:33ff.). In the Good Samaritan's inn lies one of the roots of the Christian hospice idea. Precisely along the medieval pilgrim routes, hospices used to offer travellers refreshment and rest. For the weary and the exhausted, they offered first aid and relief, for the ill and the dying they became places of physical and spiritual assistance.

   Down to our day, hospice work has been committed to this legacy. Just as the Good Samaritan stopped beside the suffering man, so those who accompany the dying are advised to pause, to be sensitive to the patients' wishes, needs and concerns. Many spiritual actions can spring from this sensitivity, such as listening to the word of God and praying together, and human ones, such as conversation, a silent but affectionate presence, the countless services which make the warmth of love tangible. Just as the Good Samaritan poured oil and wine on the wounds of the suffering man, the Church must not withhold the sacrament of the Anointing of the Sick from those who wish it. Offering this enduring sign of God's love is one of the duties of true pastoral care. This palliative care needs a spiritual element that will give the dying person the feeling of a "pallium", that is, a "mantle" for shelter at the moment of death.

   Just as the wounded man's suffering aroused the compassion of the Samaritan, so encountering the world of suffering in the hospice can make a community of suffering out of all those who accompany a patient in the last phase of his life. Feelings of closeness and sympathy can grow from this, as an expression of true Christian love. Only those who weep themselves can dry the tears of this world. A special role is played in this house by the sisters of Caritas Socialis, to whom the foundress wrote: "In the sick we can always care for our suffering Saviour and thus unite ourselves to him" (Hildegard Burjan, *Letters*, 31). Here is an echo of the Good News: "As you did it to one of the least of these my brethren, you did it to me" (Mt 25:40).

10. My deepest appreciation goes to all who are tirelessly involved in the hospice movement, including all who serve in hospitals and clinics, as well as those who care for their seriously ill or dying relatives. I am particularly grateful to the sick and dying, who teach us how better to understand the Gospel of suffering. *Credo in vitam*. I believe in life. Sister life and brother death take us in their midst when our hearts feel anxious before the last challenge we must face on this earth: "Let not your hearts be troubled. In my Father's house are many rooms" (Jn 14:1f.).

    I bless you with all my heart.

## Address of Pope John Paul II to the World Organization of Gastro-Enterology[6]

Saturday, March 23, 2002

*Distinguished Ladies and Gentlemen,*

1. I warmly greet all of you who are taking part in this Congress that aims to sensitize public opinion about the prevention of cancer of the digestive tract and of the colon. I want to greet Prof. Alberto Montori, President of the European Federation for Diseases of the Digestive Tract, and those of you who come from many countries for this important international meeting.

   I also wish to convey my deep gratitude to the organizers of the Congress, to the members of the Scientific Committee, the delegates, moderators, presenters, specialists, and all who are committed to the fight against the disease on whose cure you focus your attention.

   One can only be glad to see the growing availability of technological and pharmacological resources that allow for the speedy diagnosis of the symptoms of cancer in a great many cases, as well as rapid and effective treatment. I urge you not to stop at the results you have achieved but to continue with confidence and tenacity, in both research and treatment, making use of the most advanced scientific resources. Young doctors should learn from your example and, with your help, learn how to continue in this direction that is so beneficial for everyone's health.
2. Certainly, we cannot forget that man is a limited and mortal being. It is necessary to approach the sick with a healthy realism that avoids giving to those who are suffering the illusion that medicine is omnipotent. There are limits that are not humanly possible to overcome; in these cases, the patient must know how to accept his human condition serenely, which the faithful know how to interpret in the light of the divine will. The divine will is manifested even in death, the natural end of human life on earth. Teaching people to accept death serenely belongs to your mission.

   The complexity of the human being requires that, in providing him with the necessary treatment, the spirit as well as the body be taken into account. It would therefore be foolhardy to count on technology alone. From this point of view, an exasperated and overzealous treatment, even if done with the best of intentions, would definitely be shown to be, not just useless, but lacking in respect for the sick person who is already in a terminal condition.

   The concept of health, that we find in Christian thought, is quite the opposite of the vision that reduces it to a purely psycho-physical balance. Such a vision of health disregards the spiritual dimensions of the human person and would end by harming his true good. For the believer, as I wrote in my *Message for the Eighth*

[6] http://w2.vatican.va/content/john-paul-ii/en/speeches/2002/march/documents/hf_jp-ii_spe_20020323_congr-gastroenterologia.html

*World Day of the Sick*, health "strives to achieve a fuller harmony and healthy balance on the physical, psychological, spiritual and social level" (*ORE*, 6 August 1999, n. 13). This is the teaching and witness of Jesus, who was so sensitive to human suffering. With his help, we too must endeavour to be close to people today, to treat them and, cure them, if possible, without forgetting the requirements of the spirit.

3. Distinguished Ladies and Gentlemen, thanks to the cooperation of so many collaborators and volunteers, you are making a considerable effort to inform public opinion of the possibilities of enjoying better health by the rational regulation of daily habits and by submitting to preventive routine check-ups. I am delighted with your service and hope that your profession, by following the ethical norms that govern it, will always be inspired by the perennial ethical values that give it a firm foundation.

   Informing citizens with respect and truth, especially when they suffer from a pathological condition, is a true mission for those who take care of public health. Your Congress intends to make its contribution in this area, and I wish it every success. I strongly hope that you will have a large-scale response to the message you plan to launch, and that you will involve the mass media in an effective and informative campaign.

   I willingly support you with my prayer, and, as I commend your work to God, I cordially impart my Apostolic Blessing to you and gladly extend it to your loved ones and to all who work with you in this noble humanitarian mission.

## Address of John Paul II to the Participants in the 19th International Conference of the Pontifical Council for Health Pastoral Care[7]

Friday, November 12, 2004

*Your Eminence,*
*Venerable Brothers in the Episcopate,*
*Dear Brothers and Sisters,*

1. I am pleased to welcome you on the occasion of the *International Conference of the Pontifical Council for Health Pastoral Care,* which is taking place at this time. With your visit, you have wished to reaffirm your scientific and human commitment to those who are suffering.

   I thank Cardinal Javier Lozano Barragán for his courteous words on behalf of you all. My grateful thoughts and appreciation go to everyone who has made a contribution to these sessions, as well as to the doctors and health-care workers throughout the world who dedicate their scientific and human skills and their spirituality to relieving pain and its consequences.
2. Medicine is always at the service of life. Even when medical treatment is unable to defeat a serious pathology, all its possibilities are directed to the alleviation of suffering. Working enthusiastically to help the patient in every situation means being aware of the inalienable dignity of every human being, even in the extreme conditions of terminal illness. Christians recognize this devotion as a fundamental dimension of their vocation: indeed, in carrying out this task they know that they are caring for Christ himself (cf. Mt. 25: 35–40).

   "It is therefore through Christ, and in Christ, that light is thrown on the riddle of suffering and death which, apart from his Gospel, overwhelms us", the Council recalls (*Gaudium et Spes,* n. 22).

   Those who open themselves to this light in faith find comfort in their own suffering and acquire the ability to alleviate that of others. Indeed, there is *a directly proportional relationship between the ability to suffer and the ability to help those who are suffering.* Daily experience teaches that the persons most sensitive to the suffering of others and who are the most dedicated to alleviating the suffering of others are also more disposed to accept, with God's help, their own suffering.
3. Love of neighbour, which Jesus vividly portrayed in the Parable of the Good Samaritan (cf. Lk 10: 2ff.), enables us to *recognize the dignity of every person,* even when illness has become a burden. Suffering, old age, a comatose state or the imminence of death in no way diminish the intrinsic dignity of the person created in God's image.

---

[7] http://w2.vatican.va/content/john-paul-ii/en/speeches/2004/november/documents/hf_jp-ii_spe_20041112_pc-hlthwork.html

*Euthanasia* is one of those tragedies caused by an ethic that claims to dictate who should live and who should die. Even if it is motivated by sentiments of a misconstrued compassion or of a misunderstood preservation of dignity, euthanasia actually eliminates the person instead of relieving the individual of suffering.

Unless compassion is combined with the desire to tackle suffering and support those who are afflicted, it leads to the cancellation of life in order to eliminate pain, thereby distorting the ethical status of medical science.

4. True compassion, on the contrary, encourages every reasonable effort for the patient's recovery. At the same time, it helps draw the line when it is clear that no further treatment will serve this purpose.

   The refusal of *aggressive treatment* is neither a rejection of the patient nor of his or her life. Indeed, the object of the decision on whether to begin or to continue a treatment has nothing to do with the value of the patient's life, but rather with whether such medical intervention is beneficial for the patient.

   The possible decision either not to start or to halt a treatment will be deemed ethically correct if the treatment is ineffective or obviously disproportionate to the aims of sustaining life or recovering health. Consequently, the decision to forego aggressive treatment is an expression of the respect that is due to the patient at every moment.

   It is precisely this sense of loving respect that will help support patients to the very end. Every possible act and attention should be brought into play to lessen their suffering in the last part of their earthly existence and to encourage a life as peaceful as possible, which will dispose them to prepare their souls for the encounter with the heavenly Father.

5. Particularly in the stages of illness when proportionate and effective treatment is no longer possible, while it is necessary to avoid every kind of persistent or aggressive treatment, methods of "palliative care" are required. As the Encyclical *Evangelium Vitae* affirms, they must "seek to make suffering more bearable in the final stages of illness and to ensure that the patient is supported and accompanied in his or her ordeal" (n. 65).

   In fact, palliative care aims, especially in the case of patients with terminal diseases, at alleviating a vast gamut of symptoms of physical, psychological and mental suffering; hence, it requires the intervention of a team of specialists with medical, psychological and religious qualifications who will work together to support the patient in critical stages.

   The Encyclical *Evangelium Vitae* in particular sums up the traditional teaching on the licit use of pain killers that are sometimes called for, with respect for the freedom of patients who should be able, as far as possible, "to satisfy their moral and family duties, and above all... to prepare in a fully conscious way for their definitive meeting with God" (n. 65).

   Moreover, while patients in need of pain killers should not be made to forego the relief that they can bring, the dose should be effectively proportionate to the intensity of their pain and its treatment. All forms of euthanasia that would result from the administration of massive doses of a sedative for the purpose of causing death must be avoided.

To provide this help in its different forms, it is necessary to encourage the training of specialists in palliative care at special teaching institutes where psychologists and health-care workers can also be involved.

6. Science and technology, however, will never be able to provide a satisfactory response to the essential questions of the human heart; these are questions that faith alone can answer. The Church intends to continue making her own specific contribution, offering human and spiritual support to sick people who want to open themselves to the message of the love of God, who is ever attentive to the tears of those who turn to him (cf. Ps 39: 13). Here, emphasis is placed on the importance of *health pastoral care* in which hospital chaplaincies have a special role and contribute so much to people's spiritual well-being during their hospital stay.

   Then how can we forget the precious contribution of volunteers, who through their service give life to that *creativity in charity* which imbues hope, even in the unpleasant experience of suffering? Moreover, it is through them that Jesus can continue today to exist among men and women, doing good and healing them (cf. Acts 10: 38).

7. Thus, the Church makes her own contribution to this moving mission for the benefit of the suffering. May the Lord deign to enlighten all who are close to the sick and encourage them to persevere in their different roles and various responsibilities.

   May Mary, Mother of Christ, accompany everyone in the difficult moments of pain and illness, so that human suffering may be raised to the saving mystery of the Cross of Christ.

   I accompany these hopes with my Blessing.

## Letter of John Paul II to the President of the Pontifical Academy for Life On The Occasion Of a Study Congress on "Quality Of Life and Ethics of Health"[8]

February 19, 2005

*To my Venerable Brother Bishop Elio Sgreccia*
*President of the Pontifical Academy for Life*

1. I am pleased to send my cordial greetings to those who are taking part in the Study Congress that the Pontifical Academy for Life has sponsored on the theme:" *Quality of life and ethics of health"*. I greet you in particular, venerable Brother, and offer you my congratulations and good wishes on your recent appointment as President of this Academy. I also extend my greetings to the Chancellor, Mons. Ignacio Carrasco, to whom I also wish success in his new office. I next address thoughts of deep gratitude to eminent Prof. Juan de Dios Vial Correa, who has retired from the presidency of the Academy after 10 years of generous and competent service.

   Finally, a word of special thanks goes to all the Members of the Pontifical Academy for their diligent work, especially valuable in these times, marked by the manifestation of many problems in society related to the defence of life and the dignity of the human person. As far as we can see, the Church in the future will be increasingly called into question on these topics that affect the fundamental good of every person and society. The Pontifical Academy for Life, after 10 years of existence, must therefore continue to carry out its role of sensitive and precious activity in support of the institutions of the Roman Curia and of the whole Church.
2. The theme addressed at this Congress is of the greatest ethical and cultural importance for both developed and developing societies. The phrases "quality of life" and "promotion of health" identify one of contemporary society's main goals, raising questions that are not devoid of ambiguity and, at times, tragic contradictions. Thus, they require attentive discernment and a thorough explanation.

   In the Encyclical *Evangelium Vitae,* I said regarding the ever more anxious quest for the "quality of life" typical of the developed societies: "The so-called 'quality of life' is interpreted primarily or exclusively as economic efficiency, inordinate consumerism, physical beauty and pleasure, to the neglect of the more profound dimensions – interpersonal, spiritual and religious - of existence" (n. 23). These *more profound dimensions* deserve further clarification and research.
3. It is necessary first of all to recognize the *essential quality* that distinguishes every human creature as that of being made *in the image and likeness* of the Creator himself. The human person, constituted of body and soul in the unity of

[8] https://w2.vatican.va/content/john-paul-ii/en/letters/2005/documents/hf_jp-ii_let_20050219_pont-acad-life.html

the person – *corpore et anima unus*, as the Constitution *Gaudium et Spes* says (n. 14) -, is called to enter into a personal dialogue with the Creator. Man therefore possesses a dignity essentially superior to other visible creatures, living and inanimate. As such he is called to collaborate with God in the task of subduing the earth (cf. Gn 1: 28), and is destined in the plan of redemption to be clothed in the dignity of a child of God.

This level of *dignity* and *quality* belongs to the ontological order and is a constitutive part of the human being; it endures through every moment of life, from the very moment of conception until natural death, and is brought to complete fulfilment in the dimension of eternal life. Consequently, the human person should be recognized and respected in any condition of health, infirmity or disability.

4. Consistent with this first, essential level of dignity, a *second, complementary level* of quality of life should be recognized and promoted: starting with the recognition of the right to life and the special dignity of every human person, society must promote, in collaboration with the family and other intermediate bodies, the practical conditions required for the development of each individual's personality, harmoniously and in accordance with his or her natural abilities.

   All the dimensions of the person, physical, psychological, spiritual and moral, should be promoted in harmony with one another. This implies the existence of suitable social and environmental conditions to encourage this harmonious development. The *social-environmental context,* therefore, characterizes this second level of the quality of human life which must be recognized by *all people,* including those who live in developing countries. Indeed, human beings are equal in dignity, whatever the society to which they may belong.

5. However, in our time the meaning which the expression "quality of life" is gradually acquiring is often far from this basic interpretation, founded on a correct philosophical and theological anthropology.

   Indeed, under the impetus of the society of well-being, preference is being given to a notion of quality of life that is both *reductive* and *selective:* it would consist in the ability to enjoy and experience pleasure or even in the capacity for self-awareness and participation in social life. As a result, human beings who *are not yet* or *are no longer* able to understand and desire or those who can no longer enjoy life as sensations and relations are denied every form of quality of life.

6. The *concept of health* has also suffered a similar distortion. It is certainly not easy to define in logical or precise terms a concept as complex and anthropologically rich as that of health. Yet it is certain that this word is intended to refer to all the dimensions of the person, in their harmony and reciprocal unity: the *physical, the psychological,* and *the spiritual and moral* dimensions.

   The latter, the moral dimension, cannot be ignored. Every person is responsible for his or her own health and for the health of those who have not yet reached adulthood or can no longer look after themselves. Indeed, the person is also duty bound to treat the environment responsibly, in such a way as to keep it "healthy".

How many diseases are individuals often responsible for, their own and those of others! Let us think of the spread of alcoholism, drug-addiction and AIDS. How much life energy and how many young lives could be saved and kept healthy if the moral responsibility of each person were better able to promote prevention and the preservation of that precious good: health!

7. *Health is not,* of course, *an absolute good.* It is not such especially when it is taken to be merely physical well-being, mythicized to the point of coercing or neglecting superior goods, claiming health reasons even for the rejection of unborn life: this is what happens with the so-called "reproductive health". How can people fail to recognize that this is a reductive and distorted vision of health?

   Properly understood, health nevertheless continues to be one of the most important goods for which we all have a precise responsibility, to the point that it can be sacrificed only in order to attain superior goods, as is sometimes demanded in the service of God, one's family, one's neighbour and the whole of society.

   Health should therefore be safeguarded and looked after as the *physical-psychological and spiritual balance* of the human being. The squandering of health as a result of various disorders is a serious ethical and social responsibility which, moreover, is linked to the person's moral degeneration.

8. The ethical relevance of the good of health is such as to motivate a strong commitment to its *protection* and *treatment* by society itself. It is a duty of solidarity that excludes no one, not even those responsible for the loss of their own health.

   The ontological dignity of the person is in fact superior: it transcends his or her erroneous or sinful forms of behaviour. Treating disease and doing one's best to prevent it are ongoing tasks for the individual and for society, precisely as a tribute to the dignity of the person and the importance of the good of health.

   Human beings today, in large areas of the world, are victims of the well-being that they themselves have created. In other, even larger parts of the world, they are victims of widespread and ravaging diseases, whose virulence stems from poverty and the degradation of the environment.

   All the forces of science and wisdom must be mobilized at the service of the true good of the person and of society in every part of the world, in the light of that basic criterion which is the *dignity of the person,* in whom is impressed the image of God himself.

   With these wishes, I entrust the work of the Congress to the intercession of the One who welcomed the Life of the Incarnate Word into her life, while as a sign of special affection, I impart my Blessing to you all.

## Declaration on Euthanasia[9]

Sacred Congregation for the Doctrine of the Faith
May 5, 1980

### Introduction

The rights and values pertaining to the human person occupy an important place among the questions discussed today. In this regard, the Second Vatican Ecumenical Council solemnly reaffirmed the lofty dignity of the human person, and in a special way his or her right to life. The Council therefore condemned crimes against life "such as any type of murder, genocide, abortion, euthanasia, or willful suicide" (Pastoral Constitution *Gaudium et Spes*, no. 27). More recently, the Sacred Congregation for the Doctrine of the Faith has reminded all the faithful of Catholic teaching on procured abortion.[10] The Congregation now considers it opportune to set forth the Church's teaching on euthanasia. It is indeed true that, in this sphere of teaching, the recent Popes have explained the principles, and these retain their full force[11]; but the progress of medical science in recent years has brought to the fore new aspects of the question of euthanasia, and these aspects call for further elucidation on the ethical level. In modern society, in which even the fundamental values of human life are often called into question, cultural change exercises an influence upon the way of looking at suffering and death; moreover, medicine has increased its capacity to cure and to prolong life in particular circumstances, which sometime give rise to moral problems. Thus people living in this situation experience no little anxiety about the meaning of advanced old age and death. They also begin to wonder whether they have the right to obtain for themselves or their fellowmen an "easy death," which would shorten suffering and which seems to them more in harmony with human dignity. A number of Episcopal Conferences have raised questions on this subject with the Sacred Congregation for the Doctrine of the Faith. The

---

[9] http://www.vatican.va/roman_curia/congregations/cfaith/documents/rc_con_cfaith_doc_19800505_euthanasia_en.html

[10] DECLARATION ON PROCURED ABORTION, November 18, 1974: AAS 66 (1974), pp. 730–747.

[11] Pius XII, ADDRESS TO THOSE ATTENDING THE CONGRESS OF THE INTERNATIONAL UNION OF CATHOLIC WOMEN'S LEAGUES, September 11, 1947: AAS 39 (1947), p. 483; ADDRESS TO THE ITALIAN CATHOLIC UNION OF MIDWIVES, October 29, 1951: AAS 43 (1951), pp. 835–854; SPEECH TO THE MEMBERS OF THE INTERNATIONAL OFFICE OF MILITARY MEDICINE DOCUMENTATION, October 19, 1953: AAS 45 (1953), pp. 744–754; ADDRESS TO THOSE TAKING PART IN THE IXth CONGRESS OF THE ITALIAN ANAESTHESIOLOGICAL SOCIETY, February 24, 1957: AAS 49 (1957), p. 146; cf. also ADDRESS ON "REANIMATION," November 24, 1957: AAS 49 (1957), pp. 1027–1033; Paul VI, ADDRESS TO THE MEMBERS OF THE UNITED NATIONAL SPECIAL COMMITTEE ON APARTHEID, May 22, 1974: AAS 66 (1974), p. 346; John Paul II: ADDRESS TO THE BISHOPS OF THE UNITED STATES OF AMERICA, October 5, 1979: AAS 71 (1979), p. 1225.

Congregation, having sought the opinion of experts on the various aspects of euthanasia, now wishes to respond to the Bishops' questions with the present Declaration, in order to help them to give correct teaching to the faithful entrusted to their care, and to offer them elements for reflection that they can present to the civil authorities with regard to this very serious matter. The considerations set forth in the present document concern in the first place all those who place their faith and hope in Christ, who, through His life, death and resurrection, has given a new meaning to existence and especially to the death of the Christian, as St. Paul says: "If we live, we live to the Lord, and if we die, we die to the Lord" (*Rom.* 14:8; cf. *Phil.* 1:20). As for those who profess other religions, many will agree with us that faith in God the Creator, Provider and Lord of life – if they share this belief – confers a lofty dignity upon every human person and guarantees respect for him or her. It is hoped that this Declaration will meet with the approval of many people of good will, who, philosophical or ideological differences notwithstanding, have nevertheless a lively awareness of the rights of the human person. These rights have often, in fact, been proclaimed in recent years through declarations issued by International Congresses[12]; and since it is a question here of fundamental rights inherent in every human person, it is obviously wrong to have recourse to arguments from political pluralism or religious freedom in order to deny the universal value of those rights.

## I. The Value of Human Life

Human life is the basis of all goods, and is the necessary source and condition of every human activity and of all society. Most people regard life as something sacred and hold that no one may dispose of it at will, but believers see in life something greater, namely, a gift of God's love, which they are called upon to preserve and make fruitful. And it is this latter consideration that gives rise to the following consequences:

1. No one can make an attempt on the life of an innocent person without opposing God's love for that person, without violating a fundamental right, and therefore without committing a crime of the utmost gravity.[13]
2. Everyone has the duty to lead his or her life in accordance with God's plan. That life is entrusted to the individual as a good that must bear fruit already here on earth, but that finds its full perfection only in eternal life.
3. Intentionally causing one's own death, or suicide, is therefore equally as wrong as murder; such an action on the part of a person is to be considered as a rejection of God's sovereignty and loving plan. Furthermore, suicide is also often a refusal of love for self, the denial of a natural instinct to live, a flight from the duties of

[12] One thinks especially of Recommendation 779 (1976) on the rights of the sick and dying, of the Parliamentary Assembly of the Council of Europe at its XXVIIth Ordinary Session; cf. Sipeca, no. 1, March 1977, pp. 14–15.

[13] We leave aside completely the problems of the death penalty and of war, which involve specific considerations that do not concern the present subject.

justice and charity owed to one's neighbor, to various communities or to the whole of society – although, as is generally recognized, at times there are psychological factors present that can diminish responsibility or even completely remove it. However, one must clearly distinguish suicide from that sacrifice of one's life whereby for a higher cause, such as God's glory, the salvation of souls or the service of one's brethren, a person offers his or her own life or puts it in danger (cf. *Jn*. 15:14).

## II. Euthanasia

In order that the question of euthanasia can be properly dealt with, it is first necessary to define the words used. Etymologically speaking, in ancient times *Euthanasia* meant an *easy death* without severe suffering. Today one no longer thinks of this original meaning of the word, but rather of some intervention of medicine whereby the suffering of sickness or of the final agony are reduced, sometimes also with the danger of suppressing life prematurely. Ultimately, the word *Euthanasia* is used in a more particular sense to mean "mercy killing," for the purpose of putting an end to extreme suffering, or having abnormal babies, the mentally ill or the incurably sick from the prolongation, perhaps for many years of a miserable life, which could impose too heavy a burden on their families or on society. It is, therefore, necessary to state clearly in what sense the word is used in the present document. By euthanasia is understood an action or an omission which of itself or by intention causes death, in order that all suffering may in this way be eliminated. Euthanasia's terms of reference, therefore, are to be found in the intention of the will and in the methods used. It is necessary to state firmly once more that nothing and no one can in any way permit the killing of an innocent human being, whether a fetus or an embryo, an infant or an adult, an old person, or one suffering from an incurable disease, or a person who is dying. Furthermore, no one is permitted to ask for this act of killing, either for himself or herself or for another person entrusted to his or her care, nor can he or she consent to it, either explicitly or implicitly. nor can any authority legitimately recommend or permit such an action. For it is a question of the violation of the divine law, an offense against the dignity of the human person, a crime against life, and an attack on humanity. It may happen that, by reason of prolonged and barely tolerable pain, for deeply personal or other reasons, people may be led to believe that they can legitimately ask for death or obtain it for others. Although in these cases the guilt of the individual may be reduced or completely absent, nevertheless the error of judgment into which the conscience falls, perhaps in good faith, does not change the nature of this act of killing, which will always be in itself something to be rejected. The pleas of gravely ill people who sometimes ask for death are not to be understood as implying a true desire for euthanasia; in fact, it is almost always a case of an anguished plea for help and love. What a sick person needs, besides medical care, is love, the human and supernatural warmth with which the sick person can and ought to be surrounded by all those close to him or her, parents and children, doctors and nurses.

## III. The Meaning of Suffering for Christians and the Use of Painkillers

Death does not always come in dramatic circumstances after barely tolerable sufferings. Nor do we have to think only of extreme cases. Numerous testimonies which confirm one another lead one to the conclusion that nature itself has made provision to render more bearable at the moment of death separations that would be terribly painful to a person in full health. Hence it is that a prolonged illness, advanced old age, or a state of loneliness or neglect can bring about psychological conditions that facilitate the acceptance of death. Nevertheless the fact remains that death, often preceded or accompanied by severe and prolonged suffering, is something which naturally causes people anguish. Physical suffering is certainly an unavoidable element of the human condition; on the biological level, it constitutes a warning of which no one denies the usefulness; but, since it affects the human psychological makeup, it often exceeds its own biological usefulness and so can become so severe as to cause the desire to remove it at any cost. According to Christian teaching, however, suffering, especially suffering during the last moments of life, has a special place in God's saving plan; it is in fact a sharing in Christ's passion and a union with the redeeming sacrifice which He offered in obedience to the Father's will. Therefore, one must not be surprised if some Christians prefer to moderate their use of painkillers, in order to accept voluntarily at least a part of their sufferings and thus associate themselves in a conscious way with the sufferings of Christ crucified (cf. *Mt.* 27:34). Nevertheless it would be imprudent to impose a heroic way of acting as a general rule. On the contrary, human and Christian prudence suggest for the majority of sick people the use of medicines capable of alleviating or suppressing pain, even though these may cause as a secondary effect semi-consciousness and reduced lucidity. As for those who are not in a state to express themselves, one can reasonably presume that they wish to take these painkillers, and have them administered according to the doctor's advice. But the intensive use of painkillers is not without difficulties, because the phenomenon of habituation generally makes it necessary to increase their dosage in order to maintain their efficacy. At this point it is fitting to recall a declaration by Pius XII, which retains its full force; in answer to a group of doctors who had put the question: "Is the suppression of pain and consciousness by the use of narcotics ... permitted by religion and morality to the doctor and the patient (even at the approach of death and if one foresees that the use of narcotics will shorten life)?" the Pope said: "If no other means exist, and if, in the given circumstances, this does not prevent the carrying out of other religious and moral duties: Yes."[14] In this case, of course, death is in no way intended or sought, even if the risk of it is reasonably taken; the intention is simply to relieve pain effectively, using for this purpose painkillers available to medicine. However, painkillers that cause unconsciousness need special consideration. For a person not only has to be able to satisfy his or her moral duties and family obligations; he or she also has to prepare himself or herself with full consciousness for meeting Christ. Thus Pius

[14] Pius XII, ADDRESS of February 24, 1957: AAS 49 (1957), p. 147.

XII warns: "It is not right to deprive the dying person of consciousness without a serious reason."[15]

## IV. Due Proportion in the Use of Remedies

Today it is very important to protect, at the moment of death, both the dignity of the human person and the Christian concept of life, against a technological attitude that threatens to become an abuse. Thus some people speak of a "right to die," which is an expression that does not mean the right to procure death either by one's own hand or by means of someone else, as one pleases, but rather the right to die peacefully with human and Christian dignity. From this point of view, the use of therapeutic means can sometimes pose problems. In numerous cases, the complexity of the situation can be such as to cause doubts about the way ethical principles should be applied. In the final analysis, it pertains to the conscience either of the sick person, or of those qualified to speak in the sick person's name, or of the doctors, to decide, in the light of moral obligations and of the various aspects of the case. Everyone has the duty to care for his or he own health or to seek such care from others. Those whose task it is to care for the sick must do so conscientiously and administer the remedies that seem necessary or useful. However, is it necessary in all circumstances to have recourse to all possible remedies? In the past, moralists replied that one is never obliged to use "extraordinary" means. This reply, which as a principle still holds good, is perhaps less clear today, by reason of the imprecision of the term and the rapid progress made in the treatment of sickness. Thus some people prefer to speak of "proportionate" and "disproportionate" means. In any case, it will be possible to make a correct judgment as to the means by studying the type of treatment to be used, its degree of complexity or risk, its cost and the possibilities of using it, and comparing these elements with the result that can be expected, taking into account the state of the sick person and his or her physical and moral resources. In order to facilitate the application of these general principles, the following clarifications can be added: – If there are no other sufficient remedies, it is permitted, with the patient's consent, to have recourse to the means provided by the most advanced medical techniques, even if these means are still at the experimental stage and are not without a certain risk. By accepting them, the patient can even show generosity in the service of humanity. – It is also permitted, with the patient's consent, to interrupt these means, where the results fall short of expectations. But for such a decision to be made, account will have to be taken of the reasonable wishes of the patient and the patient's family, as also of the advice of the doctors who are specially competent in the matter. The latter may in particular judge that the investment in instruments and personnel is disproportionate to the results foreseen; they may also judge that the techniques applied impose on the patient strain or suffering out of proportion with the benefits which he or she may gain from such techniques. – It is also permissible to make do with the normal means that medicine can offer.

[15] Pius XII, Ibid., p. 145; cf. ADDRESS of September 9, 1958: AAS 50 (1958), p. 694.

Therefore one cannot impose on anyone the obligation to have recourse to a technique which is already in use but which carries a risk or is burdensome. Such a refusal is not the equivalent of suicide; on the contrary, it should be considered as an acceptance of the human condition, or a wish to avoid the application of a medical procedure disproportionate to the results that can be expected, or a desire not to impose excessive expense on the family or the community. – When inevitable death is imminent in spite of the means used, it is permitted in conscience to take the decision to refuse forms of treatment that would only secure a precarious and burdensome prolongation of life, so long as the normal care due to the sick person in similar cases is not interrupted. In such circumstances the doctor has no reason to reproach himself with failing to help the person in danger.

## Conclusion

The norms contained in the present Declaration are inspired by a profound desire to service people in accordance with the plan of the Creator. Life is a gift of God, and on the other hand death is unavoidable; it is necessary, therefore, that we, without in any way hastening the hour of death, should be able to accept it with full responsibility and dignity. It is true that death marks the end of our earthly existence, but at the same time it opens the door to immortal life. Therefore, all must prepare themselves for this event in the light of human values, and Christians even more so in the light of faith. As for those who work in the medical profession, they ought to neglect no means of making all their skill available to the sick and dying; but they should also remember how much more necessary it is to provide them with the comfort of boundless kindness and heartfelt charity. Such service to people is also service to Christ the Lord, who said: "As you did it to one of the least of these my brethren, you did it to me" (*Mt.* 25:40).

*At the audience granted prefect, His Holiness Pope John Paul II approved this declaration, adopted at the ordinary meeting of the Sacred Congregation for the Doctrine of the Faith, and ordered its publication.*

*Rome, the Sacred Congregation for the Doctrine of the Faith, May 5, 1980.*

**Franjo Cardinal Seper**
*Prefect*
**Jerome Hamer, O.P.**
*Tit. Archbishop of Lorium*
*Secretary*

## Address of His Holiness Pope Pius XII to an International Congress of Anesthesiologists[16]

November 24, 1957

Dr. Bruno Haid, chief of the anesthesia section at the surgery clinic of the University of Innsbruck, has submitted to Us three questions on medical morals treating the subject known as "resuscitation" [la réanimation].

We are pleased, gentlemen, to grant this request, which shows your great awareness of professional duties, and your will to solve in the light of the principles of the Gospel the delicate problems that confront you.

### Problems of Anesthesiology

According to Dr. Haid's statement, modern anesthesiology deals not only with problems of analgesia and anesthesia properly so-called, but also with those of "resuscitation." This is the name given in medicine, and especially in anesthesiology, to the technique which makes possible the remedying of certain occurrences which seriously threaten human life, especially asphyxia, which formerly, when modern anesthetizing equipment was not yet available, would stop the heartbeat and bring about death in a few minutes. The task of the anesthesiologist has therefore extended to acute respiratory difficulties, provoked by strangulation or by open wounds of the chest. The anesthesiologist intervenes to prevent asphyxia resulting from the internal obstruction of breathing passages by the contents of the stomach or by drowning, to remedy total or partial respiratory paralysis in cases of serious tetanus, of poliomyelitis, of poisoning by gas, sedatives, or alcoholic intoxication, or even in cases of paralysis of the central respiratory apparatus caused by serious trauma of the brain.

### The Practice of "Resuscitation"

In the practice of resuscitation and in the treatment of persons who have suffered head wounds, and sometimes in the case of persons who have undergone brain surgery or of those who have suffered trauma of the brain through anoxia and remain in a state of deep unconsciousness, there arise a number of questions that concern medical morality and involve the principles of the philosophy of nature even more than those of analgesia.

It happens at times – as in the aforementioned cases of accidents and illnesses, the treatment of which offers reasonable hope of success – that the anesthesiologist can improve the general condition of patients who suffer from a serious lesion of the brain and whose situation at first might seem desperate. He restores breathing either

[16] http://www.lifeissues.net/writers/doc/doc_31resuscitation.html

through manual intervention or with the help of special instruments, clears the breathing passages, and provides for the artificial feeding of the patient.

Thanks to this treatment, and especially through the administration of oxygen by means of artificial respiration, a failing blood circulation picks up again and the appearance of the patient improves, sometimes very quickly, to such an extent that the anesthesiologist himself, or any other doctor who, trusting his experience, would have given up all hope, maintains a slight hope that spontaneous breathing will be restored. The family usually considers this improvement an astonishing result and is grateful to the doctor.

If the lesion of the brain is so serious that the patient will very probably, and even most certainly, not survive, the anesthesiologist is then led to ask himself the distressing question as to the value and meaning of the resuscitation processes. As an immediate measure he will apply artificial respiration by intubation and by aspiration of the respiratory tract; he is then in a safer position and has more time to decide what further must be done. But he can find himself in a delicate position if the family considers that the efforts he has taken are improper and opposes them. In most cases this situation arises, not at the beginning of resuscitation attempts, but when the patient's condition, after a slight improvement at first, remains stationary and it becomes clear that only automatic, artificial respiration is keeping him alive. The question then arises if one must, or if one can, continue the resuscitation process despite the fact that the soul may already have left the body.

The solution to this problem, already difficult in itself, becomes even more difficult when the family – themselves Catholic perhaps – insist that the doctor in charge, especially the anesthesiologist, remove the artificial respiration apparatus in order to allow the patient, who is already virtually dead, to pass away in peace.

## A Fundamental Problem

Out of this situation there arises a question that is fundamental from the point of view of religion and the philosophy of nature. When, according to Christian faith, has death occurred in patients on whom modern methods of resuscitation have been used? Is Extreme Unction valid, at least as long as one can perceive heartbeats, even if the vital functions properly so-called have already disappeared, and if life depends only on the functioning of the artificial respiration apparatus?

## Three Questions

The problems that arise in the modern practice of resuscitation can therefore be formulated in three questions:

First, does one have the right, or is one even under the obligation, to use modern artificial respiration equipment in all cases, even those which, in the doctor's judgment, are completely hopeless?

Second, does one have the right, or is one under obligation, to remove the artificial respiration apparatus when, after several days, the state of deep unconsciousness does not improve if, when it is removed, blood circulation will stop within a few minutes? What must be done in this case if the family of the patient, who has already received the last sacraments, urges the doctor to remove the apparatus? Is Extreme Unction still valid at this time?

Third, must a patient plunged into unconsciousness through central paralysis, but whose life – that is to say, blood circulation – is maintained through artificial respiration, and in whom there is no improvement after several days, be considered *de facto* or even *de jure* dead? Must one not wait for blood circulation to stop, in spite of the artificial respiration, before considering him dead?

## Basic Principles

We shall willingly answer these three questions. But before examining them we would like to set forth the principles that will allow formulation of the answer.

Natural reason and Christian morals say that man (and whoever is entrusted with the task of taking care of his fellowman) has the right and the duty in case of serious illness to take the necessary treatment for the preservation of life and health. This duty that one has toward himself, toward God, toward the human community, and in most cases toward certain determined persons, derives from well ordered charity, from submission to the Creator, from social justice and even from strict justice, as well as from devotion toward one's family.

But normally one is held to use only ordinary means – according to circumstances of persons, places, times, and culture – that is to say, means that do not involve any grave burden for oneself or another. A more strict obligation would be too burdensome for most men and would render the attainment of the higher, more important good too difficult. Life, health, all temporal activities, are in fact subordinated to spiritual ends. On the other hand, one is not forbidden to take more than the strictly necessary steps to preserve life and health, as long as he does not fail in some more serious duty.

## Administration of the Sacraments

Where the administration of sacraments to an unconscious man is concerned, the answer is drawn from the doctrine and practice of the Church which, for its part, follows the Lord's will as its rule of action. Sacraments are meant, by virtue of divine institution, for men of this world who are in the course of their earthly life, and, except for baptism itself, presupposed prior baptism of the recipient. He who is not a man, who is not yet a man, or is no longer a man, cannot receive the sacraments. Furthermore, if someone expresses his refusal, the sacraments cannot be administered to him against his will. God compels no one to accept sacramental grace.

When it is not known whether a person fulfills the necessary conditions for valid reception of the sacraments, an effort must be made to solve the doubt. If this effort fails, the sacrament will be conferred under at least a tacit condition (with the phrase "*Si capax est*," "If you are capable," – which is the broadest condition). Sacraments are instituted by Christ for men in order to save their souls. Therefore, in cases of extreme necessity, the Church tries extreme solutions in order to give man sacramental grace and assistance.

## The Fact of Death

The question of the fact of death and that of verifying the fact itself (*de facto*) or its legal authenticity (*de jure*) have, because of their consequences, even in the field of morals and of religion, an even greater importance. What we have just said about the presupposed essential elements for the valid reception of a sacrament has shown this. But the importance of the question extends also to effects in matters of inheritance, marriage and matrimonial processes, benefices (vacancy of a benefice), and to many other questions of private and social life.

It remains for the doctor, and especially the anesthesiologist, to give a clear and precise definition of "death" and the "moment of death" of a patient who passes away in a state of unconsciousness. Here one can accept the usual concept of complete and final separation of the soul from the body; but in practice one must take into account the lack of precision of the terms "body" and "separation." One can put aside the possibility of a person being buried alive, for removal of the artificial respiration apparatus must necessarily bring about stoppage of blood circulation and therefore death within a few minutes.

In case of insoluble doubt, one can resort to presumptions of law and of fact. In general, it will be necessary to presume that life remains, because there is involved here a fundamental right received from the Creator, and it is necessary to prove with certainty that it has been lost.

We shall now pass to the solution of the particular questions.

## Answers to the Questions

### A Doctor's Rights and Duties

1. Does the anesthesiologist have the right, or is he bound, in all cases of deep unconsciousness, even in those that are considered to be completely hopeless in the opinion of the competent doctor, to use modern artificial respiration apparatus, even against the will of the family?

   In ordinary cases one will grant that the anesthesiologist has the right to act in this manner, but he is not bound to do so, unless this becomes the only way of fulfilling another certain moral duty.

The rights and duties of the doctor are correlative to those of the patient. The doctor, in fact, has no separate or independent right where the patient is concerned. In general he can take action only if the patient explicitly or implicitly, directly or indirectly, gives him permission. The technique of resuscitation which concerns us here does not contain anything immoral in itself. Therefore the patient, if he were capable of making a personal decision, could lawfully use it and, consequently, give the doctor permission to use it. On the other hand, since these forms of treatment go beyond the ordinary means to which one is bound, it cannot be held that there is an obligation to use them nor, consequently, that one is bound to give the doctor permission to use them.

The rights and duties of the family depend in general upon the presumed will of the unconscious patient if he is of age and *sui jurist*. Where the proper and independent duty of the family is concerned, they are usually bound only to the use of ordinary means.

Consequently, if it appears that the attempt at resuscitation constitutes in reality such a burden for the family that one cannot in all conscience impose it upon them, they can lawfully insist that the doctor should discontinue these attempts, and the doctor can lawfully comply. There is not involved here a case of direct disposal of the life of the patient, nor of euthanasia in any way: this would never be licit. Even when it causes the arrest of circulation, the interruption of attempts at resuscitation is never more than an indirect cause of the cessation of life, and one must apply in this case the principle of double effect and of "*voluntarium in cause*."

## Extreme Unction

2. We have, therefore, already answered the second question in essence: "Can the doctor remove the artificial respiration apparatus before the blood circulation has come to a complete stop? Can he do this, at least, when the patient has already received Extreme Unction? Is this Extreme Unction valid when it is administered at the moment when circulation ceases, or even after?"

We must give an affirmative answer to the first part of this question, as we have already explained. If Extreme Unction has not yet been administered, one must seek to prolong respiration until this has been done. But as far as concerns the validity of Extreme Unction at the moment when blood circulation stops completely or even after this moment, it is impossible to answer "yes" or "no."

If, as in the opinion of doctors, this complete cessation of circulation means a sure separation of the soul from the body, even if particular organs go on functioning, Extreme Unction would certainly not be valid, for the recipient would certainly not be a man anymore. And this is an indispensable condition for the reception of the sacraments.

If, on the other hand, doctors are of the opinion that the separation of the soul from the body is doubtful, and that this doubt cannot be solved, the validity of Extreme Unction is also doubtful. But, applying her usual rules: "The sacraments are for men" and "In case of extreme measures" the Church allows the sacrament to be administered conditionally in respect to the sacramental sign.

## When Is One "Dead"?

3. "When the blood circulation and the life of a patient who is deeply unconscious because of a central paralysis are maintained only through artificial respiration, and no improvement is noted after a few days, at what time does the Catholic Church consider the patient 'dead' or when must he be declared dead according to natural law (questions *de facto* and *de jure*)?"

   (Has death already occurred after grave trauma of the brain, which has provoked deep unconsciousness and central breathing paralysis, the fatal consequences of which have nevertheless been retarded by artificial respiration? Or does it occur, according to the present opinion of doctors, only when there is complete arrest of circulation despite prolonged artificial respiration?)

   Where the verification of the fact in particular cases in concerned, the answer cannot be deduced from any religious and moral principle and, under this aspect, does not fall within the competence of the Church. Until an answer can be given, the question must remain open. But considerations of a general nature allow us to believe that human life continues for as long as its vital functions – distinguished from the simple life of organs – manifest themselves spontaneously or even with the help of artificial processes. A great number of these cases are the object of insoluble doubt, and must be dealt with according to the presumptions of law and of fact of which we have spoken.

May these explanations guide you and enlightened you when you must solve delicate questions arising in the practice of your profession. As a token of divine favors which We call upon you and all those who are dear to you, We heartily grant you Our Apostolic Blessing.

## Catechism of the Catholic Church, ns. 2276, 2277, 2278, and 2279[17]



2276 Those whose lives are diminished or weakened deserve special respect. Sick or handicapped persons should be helped to lead lives as normal as possible.

2277 Whatever its motives and means, direct euthanasia consists in putting an end to the lives of handicapped, sick, or dying persons. It is morally unacceptable.

> Thus an act or omission which, of itself or by intention, causes death in order to eliminate suffering constitutes a murder gravely contrary to the dignity of the human person and to the respect due to the living God, his Creator. The error of judgment into which one can fall in good faith does not change the nature of this murderous act, which must always be forbidden and excluded.

2278 Discontinuing medical procedures that are burdensome, dangerous, extraordinary, or disproportionate to the expected outcome can be legitimate; it is the refusal of "over-zealous" treatment. Here one does not will to cause death; one's inability to impede it is merely accepted.

The decisions should be made by the patient if he is competent and able or, if not, by those legally entitled to act for the patient, whose reasonable will and legitimate interests must always be respected.

2279 Even if death is thought imminent, the ordinary care owed to a sick person cannot be legitimately interrupted. The use of painkillers to alleviate the sufferings of the dying, even at the risk of shortening their days, can be morally in conformity with human dignity if death is not willed as either an end or a means, but only foreseen and tolerated as inevitable Palliative care is a special form of disinterested charity. As such it should be encouraged.

[17] http://www.vatican.va/archive/ENG0015/__P7Z.HTM

## Ethical and Religious Directives for Catholic Health Care Services, Sixth Edition[18]

United States Conference of Catholic Bishops, 2018

Part Five, Introduction

The truth that life is a precious gift from God has profound implications for the question of stewardship over human life. We are not the owners of our lives and, hence, do not have absolute power over life. We have a duty to preserve our life and to use it for the glory of God, but the duty to preserve life is not absolute, for we may reject life-prolonging procedures that are insufficiently beneficial or excessively burdensome. Suicide and euthanasia are never morally acceptable options.

The task of medicine is to care even when it cannot cure. Physicians and their patients must evaluate the use of the technology at their disposal. Reflection on the innate dignity of human life in all its dimensions and on the purpose of medical care is indispensable for formulating a true moral judgment about the use of technology to maintain life. The use of life-sustaining technology is judged in light of the Christian meaning of life, suffering, and death. In this way two extremes are avoided: on the one hand, an insistence on useless or burdensome technology even when a patient may legitimately wish to forgo it and, on the other hand, the withdrawal of technology with the intention of causing death.

Part Five, Directives

56. A person has a moral obligation to use ordinary or proportionate means of preserving his or her life. Proportionate means are those that in the judgment of the patient offer a reasonable hope of benefit and do not entail an excessive burden or impose excessive expense on the family or the community.
57. A person may forgo extraordinary or disproportionate means of preserving life. Disproportionate means are those that in the patient's judgment do not offer a reasonable hope of benefit or entail an excessive burden, or impose excessive expense on the family or the community.
60. Euthanasia is an action or omission that of itself or by intention causes death in order to alleviate suffering. Catholic health care institutions may never condone or participate in euthanasia or assisted suicide in any way. Dying patients who request euthanasia should receive loving care, psychological and spiritual support, and appropriate remedies for pain and other symptoms so that they can live with dignity until the time of natural death.
61. Patients should be kept as free of pain as possible so that they may die comfortably and with dignity, and in the place where they wish to die. Since a person has the right to prepare for his or her death while fully conscious, he or she should not be deprived of consciousness without a compelling reason. Medicines capable of alleviating or suppressing pain may be given to a dying person, even if this therapy may indirectly shorten the person's life so long as the intent is not to hasten death. Patients experiencing suffering that cannot be alleviated should be helped to appreciate the Christian understanding of redemptive suffering.

[18] www.usccb.org/about/doctrine/ethical-and-religious-directives/

# Index




P. J. Cataldo, D. O'Brien (eds.), *Palliative Care and Catholic Health Care*, Philosophy and Medicine 130, https://doi.org/10.1007/978-3-030-05005-4

9783030050047